Library of Congress Cataloging in Publication Data
Main entry under title:

Molecular specialization and symmetry in membrane
    function.

    (Harvard books in biophysics; no. 2)
    Bibliography: p.
    Includes index.
    1. Cell membranes. 2. Biological transport.
I. Solomon, Arthur K., 1912-   II. Karnovsky,
Manfred, 1918-   III. Series.
QH601.M65   574.8'75   77-13465
ISBN 0-674-58179-2

Molecular Specialization and Symmetry in Membrane Function

EDITED BY ARTHUR K. SOLOMON

AND MANFRED KARNOVSKY

Harvard University Press

Cambridge, Massachusetts, and London, England

1978

Molecular Specialization and Symmetry in Membrane Function

Harvard Books in Biophysics, Number 2

Foreword

Molecular Specialization and Symmetry in Membrane Function addresses
one of the important facets of modern biology, the functions of the delicate and complex membranes that separate the inside and outside of cells
and organelles. It reflects the continuing evolution of the field of biophysics and expresses directly and indirectly some of the accomplishments of
the remarkable group of scientists who have worked at the Biophysical
Laboratory of Harvard Medical School, particularly the founder and
only director of the laboratory, Arthur K. Solomon, and his colleague
and coeditor of this volume, Manfred Karnovsky.

The Biophysical Laboratory has been an unusually productive unit of
the Harvard Medical School. Researches carried out in the laboratory
have contributed to our understanding of the molecular mechanisms and
cellular functions involved in the transport of ions and nonelectrolytes
across membranes of red blood cells, of bacteria, and of epithelia such as
frog skin, intestinal and gastric mucosa, and renal tubules. Perhaps even
more important has been the role played by the laboratory in educating
young biophysicists who have gone on to be effective independent investigators.

The lifespan of the laboratory has been coextensive with the emergence
of the discipline of biophysics. Biophysics unites an extremely heterogeneous group of scientists who share the idea that the insights and techniques of the physical sciences can contribute to the solution of biological
problems. In particular, they acknowledge that physical—for example,
electrical and mechanical—properties can be as important as chemical
properties in determining the behavior of biological systems. They further agree that much can be learned about living things by studying their

interactions with physical forces such as electromagnetic radiation, nuclear particles, pressure, and temperature. A theme of great interest has been the mechanism by which energy is converted from one form to another in the biosphere. It may be argued that the relevance of physics to biology was recognized long ago by alert investigators such as Helmholtz, Pfeffer, and van't Hoff. However, it took the exceptional productivity of research into the chemical dimensions of biology to reawaken this awareness in the minds of the many scientists who have flocked to the banner of biophysics. The work of these scientists is now recognized to be an important part of the interface between biology and the physical sciences. Biophysicists are and will remain an essential cadre in any faculty devoted to fundamental research in the biomedical sciences.

During the past twenty-five years, understanding of the structure and function of biological membranes has grown enormously. Indeed, prior to and even after the development of the electron microscope, the dimensions and even the existence of such membranes were matters of lively debate. The main features of the molecular architecture of biological membranes are now generally accepted. The membranes are bilayers of phospholipids and other amphiphilic molecules, less than 10^{-6} cm thick with hydrophilic groups facing outward toward the two aqueous phases (the cytosol and extracellular fluids) and with hydrophobic groups (the aliphatic chains of fatty acids) in the membrane interior. Thus, a molecule moving across a bilayer is exposed first to water, then to an interface between water and oil, then to oil, then to a second interface before reaching the aqueous solution bathing the opposite side of the membrane. This arrangement renders bilayers extremely impermeable to water-soluble substances, and, particularly, to ions. In recent years, we have come to appreciate that transport of these materials across biological membranes is often accomplished by specialized proteins that span the bilayer. The isolation and characterization of those proteins is a subject of intense investigation. Several examples, such as the ATPase that accomplishes active transport of Na and K and the band 3 protein thought to be involved in facilitated anion transport in red cells, are discussed in this volume (see Chapters 2, 3, and 5 through 11).

Another useful insight into membrane structure and function is the recognition that the two faces of a membrane are not, in general, identical. This is true not only for the head groups of the lipids that comprise the bilayer portions of the membrane (see Chapter 4), but also for the two ends of membrane-spanning transport proteins (see Chapters 8 and 9). The chemical structure and kinetic behavior of these asymmetries are issues of great importance and interest to membrane biologists. Equally important from a physiological point of view are the different properties found in different regions of the plasma membranes of the cells that perform transport across epithelia (see Chapters 12-15). Both types of asymmetry, that involving the two faces of a single membrane and that involving two different regions of membrane in the same cell, raise interesting questions about the mechanisms of membrane biogenesis (see Chapter 1).

This book, then, is important for scientists interested in the function of biological membranes. It reflects the achievements of a distinguished investigator, Arthur K. Solomon, and his colleagues past and present in the Biophysical Laboratory at Harvard; it celebrates the emergence of the discipline of biophysics, which has played such an important role in membrane research; and it presents interesting new work on two growing edges of the field, membrane asymmetry and the specialization of membrane molecules for transport functions.

Daniel C. Tosteson, M.D.

Preface

The occasion for the Symposium on Molecular Specialization and Symmetry in Membrane Function was the celebration of the thirtieth anniversary of the Biophysical Laboratory, which began its existence at the Harvard Medical School on July 1, 1946, with a staff of three and a single graduate student. Professor A. Baird Hastings, in company with Professors Joseph Aub and Shields Warren, was responsible for the creation of the laboratory, which was supported in its formative years by a generous grant from the Office of Naval Research. The initial charge to the laboratory was to facilitate the use of radioactive and stable isotopic tracers in biological and medical studies. Such studies were then in their infancy and commercial laboratory equipment for detection of radioactivity was not available. Hence, we had to design and construct our own counters and scalers to measure tracer radioactivity, and we had to build the laboratory's own mass spectrometer. This led to pioneering studies in the use of ^{14}C in biochemical tracer experiments, the measurement of total body water with D_2O, and the detection of brain tumors by ^{32}P. The laboratory also contributed to the physics of tracer measurement, by determining the maximum energy of ^{14}C beta particles and the halflife of ^{24}Na.

After an initial few years during which the laboratory specialized in the design and construction of radioactivity detectors, a flourishing industry sprang up and commercial firms took over this aspect of the laboratory's function. The laboratory was then able to shift its interest to biological membranes, where it has remained for the ensuing quarter of a century. At the outset, the laboratory's expertise with radioactive tracers was particularly apposite to measurement of biological membrane permeability. Initial excursions into ion transport in red cells were followed by studies of ion and water movement across epithelia. Studies on intestine and kidney tubules were followed by studies on gall bladder, frog skin, and, finally, pancreatic ducts. These research projects and other related ones were carried out by an exceptionally able and talented group of young scientists from the United States and abroad who now occupy distinguished positions in universities throughout the world.

When the laboratory was initially conceived, a close link to the Department of Biological Chemistry was essential, since the application of radioactive tracers to biochemical problems was particularly promising. Three successive members of that department had their laboratories and offices in the Biophysical Laboratory. Their interests in intermediary metabolism, protein biosynthesis, and lipid biochemistry were complementary to the research on biological membranes, and this special relationship, which lasted until the early sixties, was mutually supportive and scientifically productive.

The laboratory has been characterized by a tradition of designing and constructing its own scientific equipment, both mechanical and electrical. This practice grew from the early necessities of isotope detection, and its execution has been the responsibility of the laboratory mechanical and electrical shop staff, now comprising Robert Dooley, Bernard Corrow, and William Kazolias. Their expertise, developed and sharpened over thirty years, has been an essential component of our contributions to scientific research. Dr. Alfred Pandiscio from the Division of Applied Science has designed the laboratory's electronic devices, and his remarkable facility has enabled the instruments produced in the laboratory to take advantage of the newest technological developments.

Many of those who participated in the laboratory research attended the symposium; others sent greetings. Two central members of the group could not: Peter Curran, whose untimely death at the age of forty-three cut short a career of exceptional promise, and Aharon Katchalsky, whose death in the Lod terrorist attack left a personal and scientific void in the lives of friends and colleagues all over the civilized world. Many of those who participated in the symposium had close and affectionate links with Professors Curran and Katchalsky. Their two names were linked as authors of the initial volume in the Harvard Books in Biophysics series, so it is fitting that this book, devoted to the field of scientific research to which both had made such signal contributions, will appear as the second volume in that same series.

In the past twenty-five years there have been times when we in the laboratory have questioned the wisdom of remaining in the same scientific field and have felt that this might reveal an unhealthy parochialism in the laboratory's attitude toward discoveries emerging in other fields of science. Despite these misgivings, we have kept to our course. In the last decade, scientific interest in membranes has grown at a tremendous rate. Entire journals are now devoted to aspects of the field that had earlier yielded only a handful of scientific papers each year. This explosive growth has been occasioned by a similar explosion in scientific techniques available to explore the subject. Membrane proteins have been solubilized and may now be visualized in the electron microscope. New physical techniques, optical and other, have yielded important information about the proteins and lipids that comprise biological membranes. Methods of a primarily physical nature, such as electron spin resonance and nuclear magnetic resonance, have been used to explore the environment of molecules in and around membranes. The laboratory now has a nuclear magnetic resonance spectrometer and uses ^{31}P nuclear magnetic resonance as a tracer technique to study the environment of magnetically labeled molecules. Indeed, in a physical sense, the evolution of the laboratory's application of physical methods to biological problems may be compassed in the diminution of a single nucleon, in the transition from ^{32}P in the forties to ^{31}P in the seventies.

The symposium was planned by a committee whose members were Gerhard Giebisch, Joseph Hoffman, Manfred Karnovsky, Aser Rothstein, and Stanley Schultz. Planning the symposium was a great pleasure, since it occasioned convivial gatherings during which we developed the symposium format and selected topics and speakers. This contribution was basic to the success of the event, and I want to express my appreciation to the committee for their willingness to devote the time and energy required. I am particularly indebted to Manfred Karnovsky, not only for almost three decades of friendship in and out of the laboratory, but for his ready response when the prospect was first discussed and his subsequent devotion to its effective prosecution, culminating in his devoted service as editor of this volume. I should also like to thank Dean Henry Meadow and the Harvard Medical School for devising ingenious means to finance the symposium. We are particularly indebted to The Commonwealth Fund, whose generous contribution made it possible for us to proceed. The meeting could not have taken place without the organizational and secretarial skills of Mrs. Michael Jennings and the hard work of the laboratory staff, to each of whom I would like to extend my personal thanks. But most of all, the symposium depended upon the scientist members of the laboratory, past and present. To them and to the speakers and chairmen of the sessions I owe a special debt, which I am glad to acknowledge here.

Arthur K. Solomon

Contents

Part One. Specialization at the Molecular Level

1. Membrane Biogenesis   3
   GEORGE E. PALADE
2. Lipid-Protein Associations   31
   O. HAYES GRIFFITH AND PATRICIA C. JOST
3. Molecular Architecture of Oligomeric Membrane Proteins   61
   FERNANDO CABRAL, WALTER BIRCHMEIER,
   CARMEN E. KOHLER, AND GOTTFRIED SCHATZ
4. Transmembrane Compositional Asymmetry of Lipids in Bilayers and Biomembranes   78
   THOMAS E. THOMPSON
5. Ligand-Binding Properties of Membrane-Bound Cholinergic Receptors of *Torpedo Marmorata*   99
   JONATHAN B. COHEN
6. The Functional Roles of Band 3 Protein of the Red Blood Cell   128
   ASER ROTHSTEIN
7. Energetics and Molecular Biology of Membrane Transport   160
   H. R. KABACK, S. RAMOS, D. E. ROBERTSON, P. STROOBANT, AND H. TOKUDA

Part Two. Asymmetry in Transport

8. Asymmetry and the Mechanism of the Red Cell Na-K Pump, Determined by Ouabain Binding   191
   JOSEPH F. HOFFMAN
9. Simultaneous or Consecutive Occupancy by Sodium and Potassium Ions of Their Membrane Pump   212
   ROBERT L. POST

10. Proton Translocation in Submitochondrial Particles and
Reconstituted Segments of the Respiratory Chain   222
  PETER C. HINKLE

11. Studies on the Molecular Mechanism of Anion Transport across the Red Blood Cell Membrane   229
  HERMANN PASSOW AND LAILA ZAKI

Part Three. Polar Faces in Epithelia

12. The Double-Membrane Model for Transepithelial Ion
Transport: Are Homocellular and Transcellular Ion Trans-
port Related?   253
  STANLEY G. SCHULTZ

13. Differentiation of Cell Faces in Epithelia   272
  R. KINNE AND E. KINNE-SAFFRAN

14. Solute-Coupled Water Transport in the Kidney   294
  EMILE L. BOULPAEP

15. Modes of Sodium Transport in the Kidney   316
  GUILLERMO WHITTEMBURY, MADALINA CONDRESCU-
  GUIDI, MARGARITA PEREZ-GONZALEZ DE LA MANNA, AND
  F. PROVERBIO

Index   333

Molecular Specialization and Symmetry in Membrane Function

PART ONE

Specialization
at the Molecular Level

1   Membrane Biogenesis

GEORGE E. PALADE

Importance of Membranes in Cellular Organization

Membranes are mandatory building elements in the construction of all cells. Every cell—irrespective of its basic type (prokaryotic or eukaryotic)—has a limiting membrane, the plasmalemma, that separates the cell's own substance from the environment; in addition, all eukaryotic cells—and quite a number of prokaryotes—use a variety of differentiated intracellular membranes to construct distinct compartments within their bodies. Often these intracellular membranes exceed the plasmalemma by one or two orders of magnitude in amount as well as in aggregate area.

Irrespective of their location, cellular membranes are essentially diffusion barriers for ions and water soluble molecules, modified to achieve selective permeability properties specific for each membrane type. As far as we know at present, all cellular membranes consist of a bimolecular layer of polar lipids, fluid at the temperature of the cell's environment, which functions as a hydrophobic diffusion barrier. This lipid bilayer is modified by the insertion of a relatively large number of a variety of specific proteins, some of which could be regarded as permeability modifiers of a passive ("channels") or active ("pumps") type. By their existence and by the activity of their components, these membranes create and maintain chemical and electrochemical gradients between the cell and its environment, as well as among different intracellular compartments. Practically all important cell activities depend on these gradients, so that lasting impairment of membrane permeability and the ensuing decay of gradients inevitably leads to cell death.

Continuity Principles

In connection with these basic premises, it is worthwhile mentioning that work done in cell biology over the last decades indicates that all cellular membranes display *spatial continuity:* they form closed compartments by being continuous with themselves. The size and geometry of these compartments are highly variable: from cells of 10-100 μm diameter to small intracellular vesicles that measure not more than 30-70nm across, and from regular spheres to elaborate and complex networks. Notwithstanding this impressive variability, the limiting membranes of these compartments appear consistently as continuous structures down to the limit of resolution practically achieved in biological electron microscopy (∿15A). Cellular membranes never show free margins or free edges, and when resolvable discontinuities are detected they can be usually traced to defects in specimen preparation.

In addition to spatial continuity, cellular membranes appear to have *temporal continuity.* The usual set of membranes organized as boundaries of specific compartments persists throughout the life cycle of the cell; moreover, when a cell prepares to divide, it begins by nearly doubling the area of all its specialized membranes, irrespective of their location, and it ends by distributing them in nearly equivalent amounts between its daughter cells. In principle at least, it follows that a cell is not obliged to start from ground zero in developing its membrane systems, since it inherits from its mother (and in turn transmits to its daughters) a qualitatively complete set of cellular membranes. It can be safely assumed that continuity in space and time are fundamental requirements for *functional continuity.* Irrespective of the operations to which a cell submits its membranes—for example, replacement of components (turnover), growth, or division—the function of these selective permeability barriers is generally maintained. These basic continuity principles can be used as guides in formulating working hypotheses for membrane biogenesis; from the very beginning they restrict the choice to procedures compatible with spatial, temporal, and functional continuity.

Since membranes are so important for the existence of cells, it becomes important to find out how the cells manage to produce them. Besides being scientifically important, the problem is intriguing and challenging since its solution seems to have no parallel in human experience: Homo faber has not yet figured out a procedure by which a house can be converted into two houses without interrupting partitions and without seriously disturbing, or bringing to a halt, the activities that regularly go on in the house to be divided.

Site of Synthesis versus Site of Assembly

The molecular components of cellular membranes, that is, their polar lipids and their proteins, must be synthesized somewhere in the cell before being assembled into a given membrane. In some cases, the site of synthesis is in the membrane itself, a salient example being the synthesis of phospholipids and cholesterol (terminal steps) in the endoplasmic reticulum membrane of mammalian hepatocytes. Another example is the synthesis of certain membrane proteins by polysomes attached to the membrane in question (a topic to be discussed in more detail later on). But in many other cases, the site of synthesis is located in the cytoplasm at some distance from the membrane in which the corresponding product will be assembled. In this situation, specific mechanisms must insure the transport of the components from their site of synthesis to their site of assembly; moreover, these mechanisms must have a high degree of specificity, at least for proteins, so as to deliver the component for assembly at the right address. The site of synthesis of various membrane components is still unknown, but they are thought to be synthesized in either the membrane itself or the cytoplasm.

Modes of Assembly

A priori, a cell could use a one-step or a multistep assembly procedure for the production of its membranes. In the one-step variant, the cell has all the necessary components already available and assembles them in a single operation in a specific membrane. The corollary of this mode of assembly is a constant biochemical composition and an invariant functional competence. In the multistep assembly, the various components are inserted into the membrane at different times in accordance with some random or periodic pattern. In this case, the biochemistry of the membrane is continuously variable and its functional competence is expected to fluctuate. Experimental data, to be discussed later on, permit a choice between these two modes of assembly for the few types of membranes already investigated.

Topography of Assembly

Again a priori, but considering this time the topography of assembly, a cell could fit together all the components needed for a new membrane (irrespective of their sites of synthesis) into a structure entirely independent of any preexisting membrane around it. In this case the assembly process would have to start from scratch, so to speak, and could be described as *de novo* assembly. It should be noticed that this formula appears to be in contradiction with the general principles of continuity already discussed. Alternatively, new molecules could be inserted into the framework of a preexisting membrane and allowed to fit into its local molecular architecture according to their affinities. In this case membrane growth would be accomplished by the expansion of preexisting membrane. If all new molecules are inserted into the same region of the old membrane, the latter acquires an area of growth and a direction to its expansion. In the case of regional insertion, a distinction is still possible between old and new membrane, or more precisely between old and new areas within the continuity of the same membrane. If the new molecules are inserted, however, more or less evenly throughout the preexisting membrane, the latter is expected to expand in all directions and a distinction between old and new membrane, or membrane areas, is no longer applicable: in any given area a membrane growing by dispersive insertion of new components would be a mixture of old and new molecules, their ratio reflecting the rate of membrane growth. In principle, the topography of assembly could be independent of the mode of assembly so that one-step or multistep assembly could apply to any of the topographies considered. The following scheme summarizes the considerations affecting synthesis and assembly.

Membrane Biogenesis
I. Synthesis of major components
   A. Components
      1. Polar lipids (including glycolipids)
      2. Proteins and glycoproteins
   B. Site of synthesis
      1. In situ
      2. At other intracellular sites
II. Assembly
   A. Mode of assembly
      1. One-step assembly—corollary: invariable biochemistry and functional competence
      2. Multistep assembly—corollary: continuously variable biochemistry and functional competence
   B. Topography of assembly
      1. De novo
      2. By expansion of preexisting membrane
         a. Regional insertion of new molecules
         b. Dispersive insertion of new molecules

Experimental work on membrane biogenesis has been carried out on a number of different systems, but the information obtained is still limited and fragmentary. The emerging conclusions consistently support some of the alternatives considered in the preceding sections, yet they should be considered tentative primarily because of the small number of observations so far recorded. They promise to be useful, however, as starting premises for future work and as bases of reference for new observations.

Choice of Material

In general, the experimental work so far published has been concentrated on the plasmalemma in prokaryotes and on specialized membranes in eukaryotes. In the case of the latter, the investigators have tried to take advantage of situations in which a specialized membrane is produced in large amounts at a rapid rate. Such situations are encountered in the differentiating hepatocytes of a newborn rat and during drug induction in the fully differentiated hepatocytes of adult mammals. In both cases, the membrane undergoing relatively rapid growth is that of the endoplasmic reticulum (Dallner et al., 1966; Leskes et al., 1971; Ernster and Orrenius, 1965; Staubli et al., 1969).

The Thylakoid Membranes of Chlamydomonas Reinhardtii

A more convenient system is represented by the green alga *Chlamydomonas reinhardtii*. It grows in culture in a well-defined medium, and the cultured cells can be satisfactorily synchronized by repeated exposure to light (12 hours) and dark (12 hours). Since the cells undergo multiple divisions (two to three) in rapid succession during each cycle (which takes 24 hours), cell density in the culture increases four to eight times (six times, on the average) every cell cycle (Schor et al., 1970) (Fig. 1.1). In addition, the organism is haploid during the vegetative phase of its existence; hence, mutants can be easily produced, and the genetics of the organism are reasonably well documented (Goodenough and Levine, 1974). In the case of *Chlamydomonas*, the membrane of choice is that of its chloroplast's thylakoids, because of its relatively large amount and ease of separation from disrupted cells. It also turns out to be a relatively simple membrane if the number of its components, primarily its proteins, is considered.

The chloroplast of the alga is a large cup-shaped body that contains (among other structures) a large number of thylakoids or chloroplast disks (Fig. 1.2).[1] It is bounded by a chloroplast envelope that consists of two membranes in series and contains in its matrix (or stroma) a population of chloroplastic (70S) ribosomes (Fig. 1.3). The thylakoids have the tendency to form piles or grana by fusion of their apposed membranes (Fig. 1.3).

Thylakoid membrane biosynthesis has been studied during the greening process of a yellow mutant, y-1, unable to synthesize chlorophyll in the dark, but able to produce chlorophyll (by a photochemical rather than enzymatic reaction) in the light (Ohad et al., 1967; Eytan and Ohad, 1970, 1972a, b; Hoober, 1970). It has also been studied in the wild type in synchronized cultures (Schor et al., 1970; Bourguignon and Palade, 1976), a situation that mimics rather closely the conditions encountered by the alga in nature.

These studies show that chlorophyll accumulation (net synthesis) occurs in the light phase of the cycle (called for convenience "day"); it

[1] For more information on the morphology of this alga, see Sager and Palade (1957) and Ohad et al. (1967).

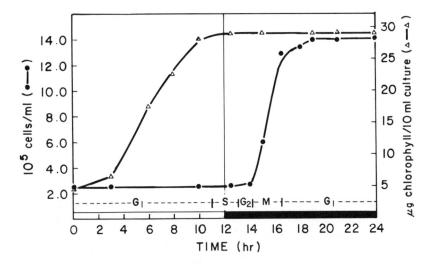

FIGURE 1.1 Variations in cell number (•————•) and chlorophyll content (Δ————Δ) in a synchronous culture of *Chlamydomonas reinhardtii* (WT 137[+]) during a full cell cycle of which the first half (12 hours) occurs in the light and the second half (12 hours) in the dark (white and black bands along the abscissa, respectively). The duration of the successive phases of the cell cycle (G_1, S, G_2, and M) are indicated above the abscissa. S refers to period of synthesis of nuclear DNA; chloroplast DNA is synthesized between 1L and 3L in G_1.

begins in the morning a few hours after exposure to light and continues for about 5 hours in the afternoon, but ceases in the evening and remains stationary (per ml culture) overnight. Preliminary observations indicate that the lipids of the thylakoid membranes are also synthesized in the morning (Beck and Levine, 1973).

Thylakoid Membrane Proteins in *Chlamydomonas*

Thylakoid membrane fractions have been prepared from algal cultures by a number of methods (De Petrocellis et al., 1970; Hoober, 1970; Bourguignon and Palade, 1976) (Fig. 1.4). The membranes have been solubilized and their polypeptides separated by gel electrophoresis using a variety of procedures (cf. Hoober, 1970; Eytan and Ohad, 1970), of which

FIGURE 1.2. (*Facing page*) *Chlamydomonas reinhardtii*. Longitudinal section through an algal cell. The nucleus is marked *n*, the contractile vacuole *cv*, Golgi complexes *go*, mitochondria *m*, and one of the two flagella, *f*. The large U-shaped profile marked *cl* represents the cup-shaped chloroplast of the cell, which contains a pyrenoid (*p*) surrounded by starch plates (*sp*), in addition to a large number of thylakoid stacks or grana (*gr*). X 19,500

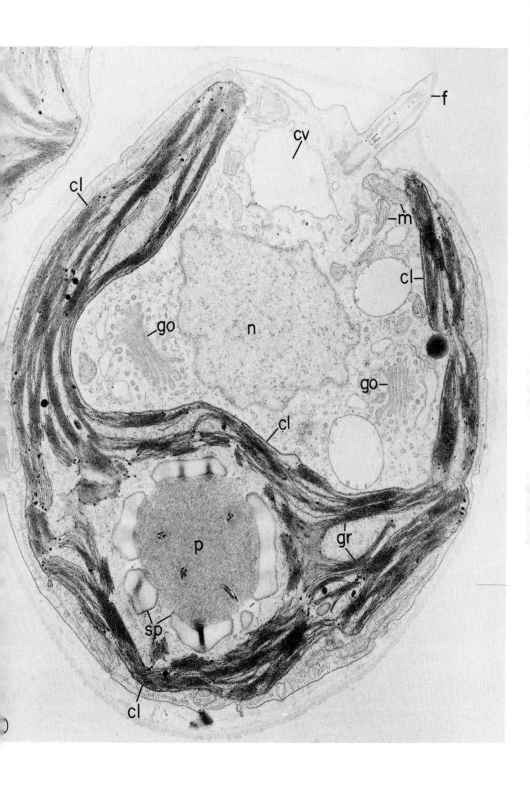

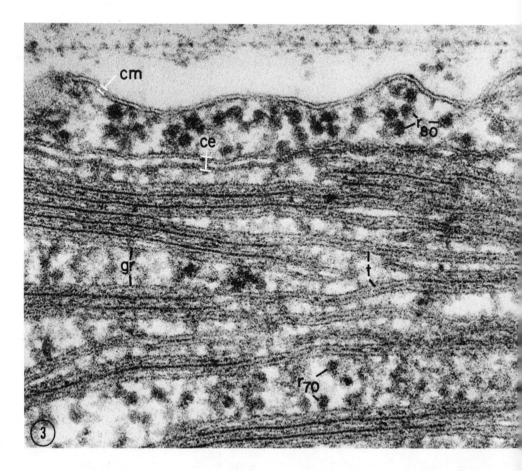

FIGURE 1.3. This small field at the periphery of a *C. reinhardtii* cell illustrates the chloroplast envelope (*ce*), thylakoids (*t*)—either free or stacked into grana (*gr*)—and chloroplast ribosomes (r_{70}). The cell membrane and cytoplasmic ribosomes appear at *cm* and r_{80}, respectively. X 190,000

FIGURE 1.4. (*Facing page*) Thylakoid membrane fraction isolated from *C. reinhardtii* cultures. Thylakoids represent the major component of the preparation. They appear as flat, relatively large sacks (the one marked by 5 *t*'s stretches from the upper right to the lower left corner of the figure), which occasionally branch (arrows) or bend over. Osmiophilic droplets (*os*) represent a minor component of the preparation. The thylakoids have been unstacked by treatment with 50mM Tricine-NaOH buffer, pH 7.3. Their stromal surface is marked by small particles (*p*). X 50,000

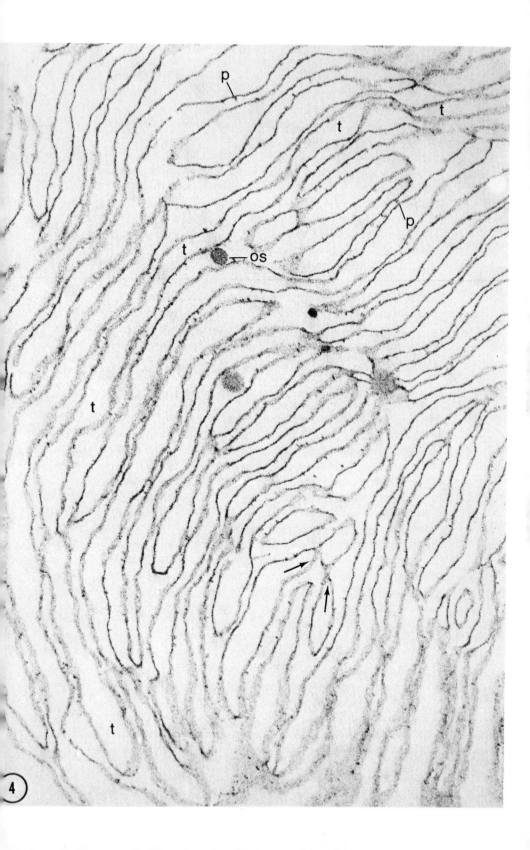

the most efficient to date involves solubilization in sodium dodecylsulfate (SDS) and electrophoresis in gradient polyacrylamide gels in the presence of the same detergent (Chua et al., 1975; Bourguignon and Palade, 1976). This procedure resolves more than 35 polypeptide bands, which fall naturally into three groups of large, medium, and low molecular weight (Fig. 1.5). Many of these bands probably contain more than one membrane polypeptide, and only two proteins (in this case each accounting for one band) have been identified so far as components of the two photosystems present in thylakoid membranes. By using appropriate mutants, Chua and Bennoun (1975) and Chua et al. (1975) have established that band L_5 (see Fig. 1.5) is the quenching factor associated with photosystem II, and band L_2 is the protein of the protein-chlorophyll complex of photosystem I. Band M_5 (which is in fact a multiple band) corresponds in part to bands previously designated as L protein by Eytan and Ohad (1970), and c peptide by Hoober (1970). The number of protein components so far identified in the electron transport chain of the membranes is small (about a dozen); hence, the membranes most probably contain additional proteins with other functions than photosynthesis, such as enzymes involved in the production of membrane lipids, pigments included.

Synthesis of Thylakoid Membrane Polypeptides

By using specific inhibitors for protein synthesis on chloroplastic versus cytoplasmic polysomes, it became apparent relatively early (Hoober et al., 1966; Hoober, 1970) that some thylakoid membrane proteins are synthesized in the chloroplast on chloroplastic polysomes, whereas some other polypeptides are synthesized in the cytoplasm and, hence, must be transported across the chloroplast envelope before being assembled into thylakoid membranes. Later on, Schor (1970) and Schor et al. (1970) showed that chloroplast ribosomes are inactive during the dark phase of the cell cycle (called, for convenience, "night"), and that certain components of the photosynthetic electron transport chain, namely the cytochromes 553, 559, and 563, accumulate in the membranes early in the morning before chlorophyll synthesis begins. These cytochromes may be part of the set of thylakoid membrane proteins synthesized in the chloroplast on chloroplastic polysomes. Recently, Chua and Gillham (1977) obtained evidence indicating that 9 out of the more than 35 thylakoid peptides are synthesized in the chloroplast on chloramphenicol-sensitive ribosomes, while the rest are imported from the cytoplasm. The two polypeptides identified by Chua et al. (1975) and Chua and Bennoun (1975) belong to the first group, that is, they are synthesized on 70S chloroplastic ribosomes.

With the relatively satisfactory (but far from perfect) resolution attained in electrophoretograms on SDS-gels, it became possible to carry

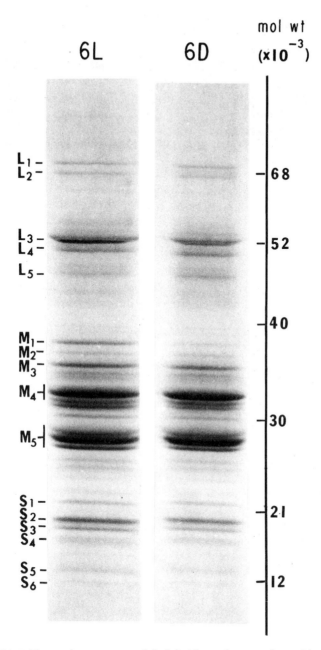

FIGURE 1.5. Electrophoretograms of thylakoid membrane polypeptides. Thylakoid membrane fractions were isolated from synchronized *C. reinhardtii* cells harvested at noon (*6L*) or midnight (*6D*). After solubilization, their peptides were separated by electrophoresis in gradient (3 to 20 percent) polyacrylamide gels (in the presence of sodium dodecylsulfate). With the exception of M_1, no differences are apparent between the two preparations.

out a comprehensive survey of the assembly of thylakoid proteins in the corresponding membranes as a function of time within the cell cycle. The object (Bourguignon and Palade, 1976) was to find out at what phase of the cycle new proteins of different molecular weight appear in the membranes. The inquiry relied on long-term (48 hour) labeling of the cultures with [^{14}C] arginine, followed by short-term (20 minute) labeling of the same culture with [^{3}H] arginine. Thylakoid membranes were isolated from these doubly labeled cells and, after solubilization, their polypeptides were separated by electrophoresis on SDS-polyacrylamide gels. The ratio of ^{3}H to ^{14}C radioactivity in each gel band was used to establish at what time in the cycle the polypeptide(s) of the band were synthesized and assembled at the highest rate (indicated by the highest ^{3}H to ^{14}C radioactivity ratio); and at what time these operations are carried out at the slowest rate (revealed by the lowest ^{3}H to ^{14}C radioactivity ratio). Within the limits of resolution attained in these gels, the results showed that only a few thylakoid membrane proteins are synthesized and assembled at a high rate during the day. For most of them the highest ratios were found shortly after midnight, which corresponds to the early part of G_1, and the lowest ratios were detected at the end of the night (Fig. 1.6), one hour before our experimental dawn (Bourguignon and Palade, 1976). Since exposure to [^{3}H] arginine is limited to 20 minutes in these experiments, the transport of protein from the site of synthesis (cytoplasmic polysomes in the majority of cases) to the chloroplast and the subsequent protein assembly in thylakoid membranes appear to be rapid and efficient processes.

Mode of Assembly of Thylakoid Membranes

Taken together with data on chlorophyll and lipid accumulation within these membranes, the findings described indicate clearly that the *Chlamydomonas* alga uses a multistep assembly procedure to produce more thylakoid membrane. Chlorophyll is added to these membranes during the middle part of the day, which corresponds to the last one-third of the G_1 phase; some proteins (among them probably the photosynthetic cytochromes) are inserted in the membranes in the morning, before chlorophyll is added, whereas other proteins are assembled at a high rate past midnight, shortly after cell division, during the first one-fifth of G_1. Better resolution of membrane polypeptides and more information on variations in the rate of synthesis and assembly of lipids into thylakoid membranes (as a function of time within the cell cycle) are needed to assess the true extent and fine details of this multistep operation, but it is already clear that some important and distinct steps can be discerned in the assembly process and that the biochemical composition of the thylakoid membrane as well as its function (Schor et al., 1970)

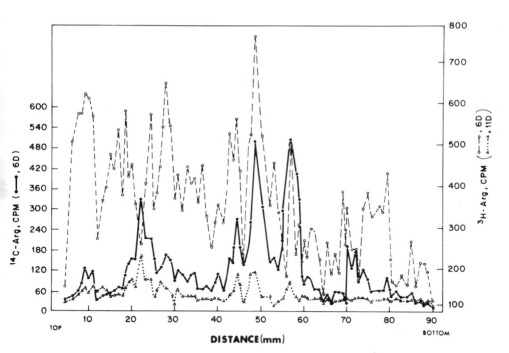

FIGURE 1.6. Radioactivity distribution, in counts per minute (cpm), in thylakoid membrane polypeptides after prolonged in vivo labeling (48 hours) with L-[^{14}C] arginine, followed by short (20 minutes) labeling with L-[^{3}H]arginine. The procedures used for the separation and solubilization of thylakoid membranes, the separation of membrane polypeptides, and the assay of their radioactivity are given in Bourguignon and Palade (1976). For labeling initiated at 6D the symbols used are •———• for L-[^{14}C]arginine (cpm) and o——— o for L-[^{3}H]arginine (cpm); for labeling initiated at 11D the symbol Δ· · · ·Δ represents L-[^{3}H] arginine (cpm). The L-[^{14}C]arginine profile (determined but not given) for 11D was nearly coincidental to that shown for 6D.

varies continuously during the cell cycle. These findings extend and, to some extent, modify earlier results obtained by Armstrong et al. (1971), Schor et al. (1970) and Beck and Levine (1974); in addition, they suggest that the assembly process might be resolved into a series of cyclic waves, coordinated with the cell cycle, each wave representing the insertion of a different component (or group of components) into the membrane.

A multistep assembly procedure is also used by differentiating hepatocytes (Dallner et al., 1966) and by drug-induced hepatocytes in adult animals (Kuriyama et al., 1969) for the production of additional membrane for the endoplasmic reticulum. In fact, it appears to be the preferred mode of assembly in all cases so far studied,[2] but their number is still too small to permit a generalization of the concept at this time.

Topography of Assembly of Thylakoid and Other Membranes

Thus far, there have been no direct studies on the topography of assembly of new molecules into the thylakoid membranes of wild strain *Chlamydomonas reinhardtii* grown in synchronized cultures. Some pertinent but indirect evidence will be discussed later on. In the y-1 mutant of the alga, however, this aspect has been studied by two different approaches at two different levels of resolution: first, by autoradiography, after exposing intact living cells to [³H] acetate, a nonspecific precursor of both lipids and proteins (Goldberg and Ohad, 1970), and, second, by following changes in the specific gravity of thylakoid membranes during the greening process (Eytan and Ohad, 1972). In the first approach, the label appeared distributed at random over all thylakoids with no evidence of concentration over some thylakoids or some regions of thylakoid membranes. Restricted localization would be expected for de novo assembly or for an expansion of preexisting membrane achieved through regional insertion of new molecules. The last conclusion assumes, of course, that there is no (or little) randomization of regionally inserted, labeled molecules by lateral diffusion in the plane of the membrane. The resolution attained in this approach is at best ∿200nm. The second approach gives information at a slightly finer level: ∿70nm. In this case, advantage was taken of the fact that the specific gravity of the thylakoid membranes increases during greening; hence, isolated membranes were fragmented, by sonication, into vesicles of 70nm average diameter, and their distribution was studied in a density gradient after centrifugation to equilibrium. The object was to find out whether or not distinct populations of light versus heavy vesicles could be detected. The first would represent residual old membrane, and the second, recently assembled new membrane. The results showed, however, that the specific gravity changed progressively for the entire population of vesicles as a function of greening time. Hence, the evidence favored expansion of preexisting membrane against de novo assembly, and dispersive insertion of new molecules—or efficient randomization—against conserved regional insertion.

The same topography of assembly was found in fact in all other systems investigated, that is, bacterial cell membranes (Mindich, 1970) and endoplasmic reticulum membranes during hepatocyte differentiation in

---

[2]One-step assembly was originally assumed to apply during the greening of the y-1 mutant of *Chlamydomonas* (Ohad et al., 1967), but more detailed studies (De Petrocellis et al., 1970) have shown that the greening thylakoid membrane of the mutant is actually assembled in a multistep process.

the newborn rat (Leskes et al., 1971). The last case is particularly informative since the approach used (localization of a new enzyme activity, glucose-6-phosphatase, by a cytochemical reaction carried out in parallel on intact liver tissue and on hepatic microsomal fractions) was definitely capable of detecting de novo assembly or conserved regional insertion, if extant. The results showed, in fact, that, in differentiating hepatocytes, glucose-6-phosphatase activity appeared randomly distributed down to domains of 4×10^4 nm^2 over the entire preexisting membrane of the endoplasmic reticulum.

It follows that in all cases so far investigated, and, again, they are not as many as desired, the cells appear to use the expansion of preexisting membranes to achieve membrane growth, irrespective of the type of membrane involved, irrespective of the process that requires an increase in membrane area (cell differentiation or growth in anticipation of cell division), and irrespective of the cell's position in evolution: bacterial, algal, or mammalian. The only well-documented exception is the formation of the protein-rich and lipid-poor inner membrane of pox viruses. These membranes appear to be assembled de novo, in complete isolation from the surrounding membranes of the infected eukaryotic cell. They have free edges during their assembly (Dales and Mosbach, 1968) and—by the nature of the infectious process—it is clear that they do not have (and do not need to have) temporal and functional continuity. In this respect, they are radically different from the cellular membranes of both prokaryotic and eukaryotic cells.

Polysomes Attached to Thylakoid Membranes

In the course of the work on synchronized cultures of *Chlamydomonas reinhardtii*, it became apparent that chloroplastic polysomes become attached to thylakoid membranes during the day period of the cell cycle. Their attachment, however, did not survive cell-disruption and cell-fractionation procedures unless special precautions were taken. The cells had to be rapidly chilled to near 0 °C before disruption and rapidly processed through cell fractionation, or they had to be pretreated with chloramphenicol or erythromycin—inhibitors of protein synthesis (more exactly of polypeptide elongation) on chloroplastic ribosomes—in order to retain polysomes on isolated membranes. These attached polysomes account for about one-third of the total 70S ribosome population of the alga; one-half of this one-third can be detached from the thylakoid membranes by a high salt (0.5 M KCl) wash, and the other half by a combined high salt-puromycin treatment (Chua et al., 1973). These are the means developed for the detachment of ribosomes from rough microsomes isolated from mammalian hepatocytes (Blobel and Sabatini, 1971), and the results they give are generally taken to indicate that the attachment involves, on the one hand, multiple ionic interactions between the large ribosomal subunit and the membrane and, on the other hand, still undefined interactions between the nascent polypeptide chain and the membrane. In the case of chloroplastic polysomes attached to thylakoid membranes, the nascent chain is most probably inserted into the membrane pari passu with its elongation, rather than being entirely transferred across the membrane, as in the case of secretory proteins.

By analogy with the situation found in mitochondria (Schatz and Mason, 1974), it can be assumed that the membrane polypeptides synthesized on chloroplastic ribosomes are hydrophobic; hence, synthesis directly coupled with insertion would appear as a logical solution to an otherwise difficult problem in transport from some distant site of synthesis to the site of assembly. Thylakoid membranes have an unusually high population of intramembranous particles (Branton and Park, 1967; Goodenough and Staehelin, 1971). In other membranes such particles are assumed to represent polymeric transmembrane proteins (Bretcher, 1971; Tillack et al., 1972; Segrest et al., 1973; Tomita and Marchesi, 1975); hence, many of the thylakoid polypeptides may have a transmembrane position. Vectorial insertion at the time of synthesis could position the nascent protein to various depths within the membrane or across the membrane, according to local interactions experienced by the corresponding polypeptides during insertion. It does not mean that all integral or transmembrane proteins of the thylakoid membranes have to be the products of attached chloroplastic polysomes; it means only that other means to achieve proper positioning must exist for integral proteins imported from the cytoplasm.

By electron microscopy, the vast majority of polysomes attached to thylakoid membranes (isolated from chloramphenicol-treated algae) appear as circular assemblies that vary in size from tetramers to octamers, hexamers being the predominant form (Figs. 1.7 and 1.8). They are found attached to the "exposed" thylakoid surfaces made available by a concomitant, wholesale dismantling of the chloroplast's grana (Evans-Castle, Bourguignon, and Palade, unpublished observations), which occurs early every morning. As expected, no polysomes and practically no ribosomes are left on such membranes after high salt-puromycin treatment, and no polysomes are found on the membranes if measures are not taken to prevent readout and detachment of ribosomes during cell disruption and cell fractionation. Among the factors that control attachment, light seems to be particularly important. Irrespective of precautions, no polysomes attached to the membranes can be isolated during the night-phase of the cycle, and transition from light to dark at any time during the day-part of the cycle results in detachment in about 10 minutes (Chua et al., 1976).

The fraction of chloroplastic ribosomes attached to thylakoid membranes during the day is probably greater than one-third of the total, since the frequency of attached polysomes is noticeably higher on thyla-

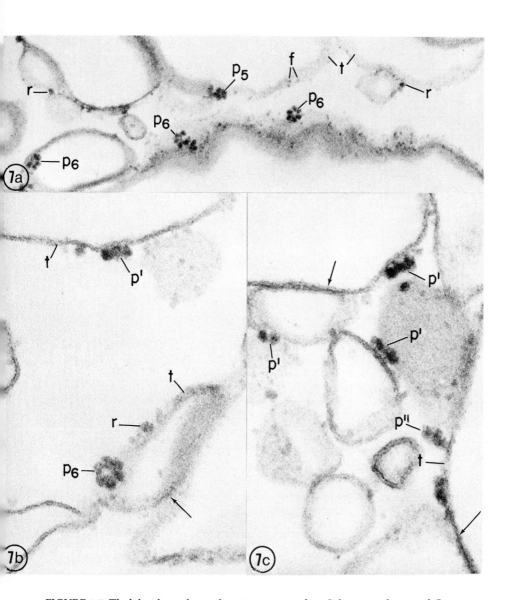

FIGURE 1.7. Thylakoid membrane fractions prepared at 4L from synchronized *C. reinhardtii* cells treated for 10 minutes with 100µg/ml chloramphenicol. The fraction treatment for *a* was 25mM KCl, no puromycin; for *b* and *c* it was 500mM KCl, no puromycin. In all three figures, chloroplastic polysomes are found only on the free, stromal faces of the thylakoids (*t*). They appear as circular, closed or open, pentamers (p_5) and hexamers when viewed in full face on grazing sections, and as linear series of two to four ribosomes, fully (*p'*) or partially (*p''*) attached to the thylakoid membranes, when viewed from the side on normal sections. Individual ribosomes are marked *r*, and coupling factor particles, *f*. The arrows point to areas of persisting thylakoid stacking. *a*: X 72,000; *b* and *c*: X 150,000. (From Chua et al., 1977.)

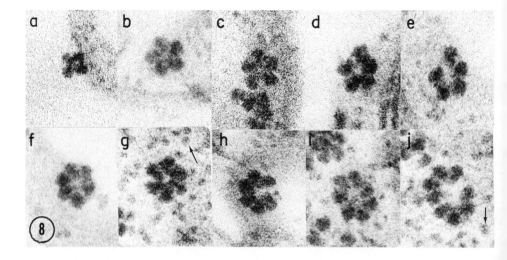

FIGURE 1.8. Gallery of chloroplastic polysomes seen in full face view in thylakoid membrane fractions prepared as for Figure 1.7 (4L, 100µg/ml chloramphenicol). A tetramer is seen in *a*, closed pentamers in *b* and *c*, open pentamers in *d* and *e*, closed hexamers in *f* and *g*, an open hexamer in *h*, a closed heptamer in *i*, and an open octamer in *j*. Coupling factor particles are indicated by arrows. *a*: X 150,000; *b* to *j*: X 200,000. (From Chua et al., 1977.)

koids found in partially disrupted chloroplasts (Fig. 1.9) than on isolated thylakoids.

It should be stressed that the proteins synthesized by attached chloroplastic polysomes have not yet been identified; hence, it is not known whether all thylakoid proteins produced in the chloroplast, or only part of them, are synthesized pari passu with their assembly; and it is not known either whether they are processed, or not, in any way at the time of their insertion into the membrane. Attached polysomes have been implicated in the synthesis of membrane proteins for other cell organs. Kuriyama and Luck (1973) have described mitochondrial polysomes attached to the inner mitochondrial membrane in *Neurospora* and have presented evidence involving them in the synthesis of proteins for that membrane. Moreover, evidence implicating attached cytoplasmic ribosomes (80S) in the synthesis of proteins for the membrane of the endoplasmic reticulum in mammalian hepatocytes has been available for some time (Dallner et al., 1966; Omura et al., 1967; Kuriyama et al., 1969; Omura and Kuriyama, 1971). In this case, however, the polysomes synthesizing membrane proteins are (most probably) a small fraction only of a large population of attached polysomes of which the large majority is engaged in the production of proteins for export to the extracellular medium (blood plasma, in the case of hepatocytes). It may prove difficult

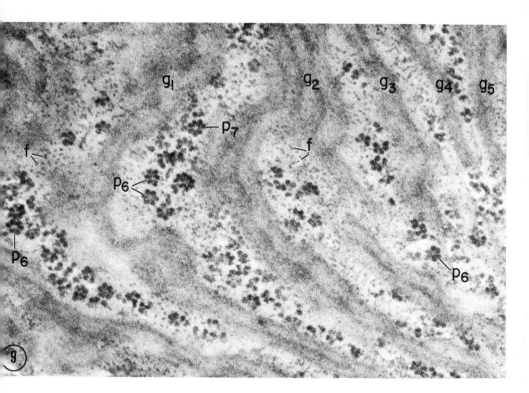

FIGURE 1.9. Chloroplast fragment obtained at 4L from synchronized *C. rein-hardtii* cells treated for 10 minutes with 100µg/ml chloramphenicol. Numerous polysomes (*p*) can be seen against the background of light irregular bands that correspond to oblique sections through terminal thylakoids (the darker bands, g_1 to g_5, are oblique sections through grana). Most polysomes are circular and vary in size from pentamers to hexamers (p_6), and heptamers (p_7). Coupling factor particles (*f*) appear scattered among the polysomes on the free surface of terminal thylakoids. X 72,000. (From Chua et al., 1977.)

to identify this minority component against the high background provided by polysomes working for secretion. The advantage of the thylakoid model is the apparent absence of this kind of background.

Since newly synthesized proteins for the membrane of the endoplasmic reticulum have been detected by purification or immune precipitation on both free and attached polysomes (Sargent and Vadlamudi, 1968; Ragnotti et al., 1969; Omura and Kuriyama, 1971), it may be assumed that in their case initiation occurs, and elongation begins, on polysomes free in the cytosol and that the attachment of these polysomes to the membrane of the endoplasmic reticulum takes place during elongation but before termination and discharge of the corresponding polypeptide chains. Evidence for such a sequence of events has been obtained recently for polysomes synthesizing secretory proteins (Blobel and Dobberstein, 1975; Devillers-Thiery et al., 1975). In the case of the attached chloroplastic ribosomes studied by Chua et al. (1973, 1976) there is also suggestive evidence that the attachment involves preformed polysomal assemblies; as such, it may represent a post-initiation event.

Finally, cytoplasmic ribosomes (80S) attached to the outer mitochondrial membrane have been described in yeast cells, and tentatively implicated in the synthesis (coupled with vectorial transport) of proteins for inner mitochondrial structures (Butow et al., 1975). So far, no such polysomes have been detected on chloroplast envelopes.

Dispersive Insertion

Attached polysomes are found distributed apparently at random on all exposed outer surfaces of thylakoid membranes. In this case, therefore, it is clear that newly synthesized proteins are dispersively, rather than regionally, inserted in the preexisting membrane. This is in fact the only case in which the insertion event has been visualized and reasonably well documented; hence, in this case at least, the random distribution of some membrane components can be ascribed to an originally dispersive insertion. Further randomization by lateral diffusion in the fluid bilayer of the membrane is evidently not excluded.[3]

Conclusions

The survey of the evidence so far secured allows an assessment of what is satisfactorily established as opposed to what is dimly understood in this field. It also gives some measure of what remains unknown at present and must be brought to light, if we are to arrive at a reasonable understanding of membrane biogenesis, a problem of obvious importance in biology because of the ubiquity and critical functional importance of membranes in all biological systems.

We know now that the majority of the component molecules of most membranes are synthesized somewhere else in the cell (rather than in situ). The only possible exception is the membrane of the endoplasmic reticulum in mammalian hepatocytes and perhaps in other eukaryotic cells.[4] It is clear, therefore, that we have to consider the existence of rather elaborate intracellular transport operations for both membrane lipids and membrane proteins. At present we have only a few disparate and, in some cases, puzzling pieces of evidence on this general process. We

[3]Lateral diffusion of thylakoid intramembranous particles has been demonstrated in *Chlamydomonas* by Ojakian and Satir (1974) and by Staehelin (1976).

[4]All membrane lipids but none of the proteins are synthesized in situ in bacterial cell membranes (cf. Cronan and Gelmann, 1975).

know, for instance, that mammalian eukaryotes have a number of phospholipid-exchange proteins (Wirtz, 1974), but we do not know how net transport is effected from the site of synthesis—the membrane of the endoplasmic reticulum—to the sites of assembly—probably all other cellular membranes, with the partial exception of the inner mitochondrial membrane.

Some membrane proteins appear to be synthesized by polysomes attached directly to the corresponding membrane, synthesis being apparently coupled with the insertion of the protein into the membrane matrix, most probably into the lipid bilayer. This procedure can be rationalized in the case of the membrane of the endoplasmic reticulum by the existence of distinct hydrophobic domains in some of its proteins (Ito and Sato, 1968, 1969), and in the case of mitochondria (and, by analogy, chloroplasts) by the general hydrophobic character of the proteins synthesized by polysomes attached to inner organellar membranes (Schatz and Mason, 1974). Other membrane proteins appear to be synthesized on free polysomes and transported through the cytosol to the site of assembly, where they may be modified (by partial proteolysis) before, or upon, being fitted into the membrane, but the number of well-documented cases of this kind is small (Lodish, 1973; Lodish and Small, 1975). For the vast majority of membrane proteins the site of synthesis is still unknown, although the alternatives are evidently limited.

In the case of membrane glycoproteins, neither the sites of synthesis of their polypeptides, nor the site (or sites) of their glycosylation are known at present. Work done on the membranes of enveloped viruses (which can serve as a simplified model for cellular membranes) indicates that the simple proteins of the envelope appear early in the plasmalemma of the infected cell, whereas the viral glycoproteins reach the same station with a lag of about 20 minutes, during which they appear to be associated with a composite membrane fraction derived, by cell fractionation, from infected cells (Compans, 1973; Morrison and Lodish, 1975; Atkinson et al., 1976). According to available evidence, the glycosylation of soluble, secretory proteins is a two-step operation: one step in the endoplasmic reticulum, and the second in the Golgi complex (cf. Leblond and Bennett, 1977), but the evidence concerning the sites of glycosylation of membrane proteins is limited and in part contradictory (Weiser, 1973; Porter and Bernacki, 1975; Waechter and Lennarz, 1976). One of the major problems stems from the fact that in the few cases for which we have adequate information (red blood cell membranes), membrane glycoproteins have an asymmetric distribution: the polypeptide segments to which the oligosaccharide chains are attached protrude some distance from the outer membrane surface (Steck, 1974). This asymmetry is assumed to be acquired at the time of assembly of the glycoproteins in the membrane, but the means by which asymmetric assembly is achieved remain unknown and are at present just a matter of speculation. Hypotheses and diagrams are published, but facts are few, and, because they are so few, diagrams and hypotheses continue to proliferate. A frequently discussed hypothesis assumes that newly synthesized membrane flows from the endoplasmic reticulum to the Golgi complex and finally to the plasmalemma, being modified in the process (Hirano, et al., 1972; Morré et al., 1974). The hypothesis is in part inspired by the flow of membrane containers that transport secretory proteins from the endoplasmic reticulum to the Golgi complex and from the latter to the plasmalemma (Palade, 1975; Jamieson and Palade, 1977), and is compatible with the increasing overlap of enzymic activities between the Golgi complex and the plasmalemma (Farquhar et al., 1974; Cheng and Farquhar, 1976), but there is little kinetic evidence at present to support the hypothesis (cf. Fleischer and Zambrano, 1974; Michaels and Leblond, 1976), and the biochemical information available puts severe restrictions on a simple general flow (Palade, 1976). If membrane flow exists, it must be limited to a few molecular species, including perhaps membrane glycolipids and glycoproteins. The membrane flow hypothesis does not explain asymmetric distribution; it only relegates it to an early unexplained event in the endoplasmic reticulum. Recently, the problem of asymmetric assembly has acquired vastly expanded dimensions. All membrane proteins (Singer and Nicolson, 1972) and probably all lipids (Zwaal et al., 1975) appear to be asymmetrically positioned in the plasmalemma and perhaps all other cellular membranes. The means by which asymmetric assembly is achieved represents a new but important aspect of membrane biogenesis that will undoubtedly receive considerable attention in the future.

Especially difficult to explain in terms of current knowledge is the transport of proteins from the cytoplasm to the inner structures (membranes and matrix) of mitochondria and chloroplasts; membranes are generally assumed to be impermeable to macromolecules, yet in this case most of the transported proteins cross two membranes. The four polypeptides of cytoplasmic origin that eventually find their way into the cytochrome aa_3 complex of the inner mitochondrial membrane are derived from a single cytoplasmic precursor of high molecular weight (Poyton and Kavanaugh, 1976), making the transport problem more difficult.

For the moment at least, the situation seems to be less confused when assembly itself is considered. In the cases so far studied, the cells appear to use a multistep mode of assembly, which, in logarithmically growing cells, imparts to the membrane a perpetually "unfinished" character, which is to be expected for structures that have the tendency of growing forever. A priori, one could argue that the multistep mode of assembly represents an advantage, since it gives the cell the opportunity of inserting new components seriatim, each of them at the most opportune time in the cell's metabolic cycle. Multistep assembly may have, however, a deeper meaning: it may be part of the mechanisms regulating membrane growth.

It is worthwhile pointing out that, in the cases so far studied (Schimke, 1975), the turnover of membrane proteins is asynchronous. This means that in removing molecules from its membranes the cell can deal (as in multistep assembly) with one component (or a group of components) at a time, rather than with all components at once.

Finally, perhaps the most important conclusion drawn from whatever we have learned so far is that membranes grow by expansion of preexisting membranes. There is no factual basis for the hypothesis—proposed some years ago (Green and Perdue, 1966)—that membranes are assembled de novo by polymerizing soluble subunits. If new molecules have always been inserted into preexisting membranes, it means that the latter have been transmitted as structural patterns without any discontinuity since the beginning of cellular life all the way down to our time in biological evolution. Expansion of preexisting membranes is compatible with the spatial and temporal continuity that characterizes cellular membranes and thus it satisfies the basic requirements for functional continuity. Membrane expansion is possible because the primary structure of the membrane is a bimolecular leaflet of polar lipids fluid at the temperature of the cell's environment. In this fluid bilayer, new lipid and protein molecules can be inserted (and, conversely, removed) with little structural and functional perturbation. The membranes can grow (and their components can turn over) while their function continues uninterrupted. In a solid bilayer the insertion of new molecules is expected to be a much slower process accompanied by more extensive perturbations in permeability. The key for membrane growth and, hence, for cell growth appears to be the fluid lipid bilayer. One can assume that it was selected very early in evolution because it makes possible growth without loss or interruption of function. The emerging picture would be incomplete without mentioning that another important process, that is, membrane fusion-fission, itself dependent on the fluidity of the bilayer, makes possible the distribution of the expanded membranes at each cell division, again without loss of function. The great biological merit of the fluid lipid bilayer is that it renders cellular membranes expandable partitions that can fuse and part without leaking.

REFERENCES

Armstrong, J. J., S. J. Surzyski, B. Moll, and R. P. Levine. 1971. Genetic transcription and translation specifying chloroplast components in *Chlamydomonas reinhardtii. Biochemistry* 10: 692-701.

Atkinson, P. H., S. A. Moyer, and D. F. Summers. 1976. Assembly of vesicular stomatitis virus glycoprotein and matrix protein into HeLa Cell plasma membranes. *J. Mol. Biol.* 102: 613-631.

Beck, D. P., and R. P. Levine. 1974. Synthesis of chloroplast membrane polypeptides during synchronous growth of *Chlamydomonas reinhardtii. J. Cell Biol.* 63: 759-772.

Beck, J., and R. P. Levine. 1973. Synthesis of chloroplast membrane lipids in *Chlamydomonas reinhardtii. J. Cell Biol.* 59: 20a.

Blobel, G., and B. Dobberstein. 1975. Transfer of proteins across membranes. I. Presence of proteolytically processed and unprocessed nascent immunoglobulin light chains on membrane-bound ribosomes of murine myeloma. II. Reconstitution of functional rough microsomes from heterologous components. *J. Cell Biol.* 67: 835-862.

Blobel, G., and D. D. Sabatini. 1971. Dissociation of mammalian polysomes into subunits by puromycin. *Proc. Natl. Acad. Sci. USA* 68: 390-394.

Bourguignon, L. Y. W., and G. E. Palade. 1976. Incorporation of polypeptides into thylakoid membranes of *Chlamydomonas reinhardtii. J. Cell Biol.* 69: 327-344.

Branton, D., and R. B. Park. 1967. Subunits in chloroplast lamellae. *J. Ultrastruct. Res.* 19: 283-303.

Bretcher, M. S. 1971. Major human erythrocyte glycoprotein spans the cell membrane. *Nature New Biol.* 231: 229-232.

Butow, R. A., W. F. Bennett, D. B. Finkelstein, and R. A. Kellems. 1975. Nuclearcytoplasmic interactions in the biogenesis of mitochondria in yeast. In A. Tzagoloff, ed., *Membrane biogenesis*, pp. 99-124. Plenum Press, New York.

Cheng, H., and M. G. Farquhar. 1976. Presence of adenylate cyclase activity in Golgi and other fractions from rat liver. I. Biochemical determination. II. Cytochemical localization within Golgi and ER membranes. *J. Cell Biol.* 70: 660-670, 671-684.

Chua, N. H., and P. Bennoun. 1975. Thylakoid membrane polypeptides of *Chlamydomonas reinhardtii:* wild type and mutant strains deficient in photosystem II reaction center. *Proc. Natl. Acad. Sci. USA* 72: 2175-2179.

Chua, N. H., G. Blobel, P. Siekevitz, and G. E. Palade. 1973. Attachment of chloroplast polysomes to thylakoid membranes in *Chlamydomonas reinhardtii. Proc. Natl. Acad. Sci. USA* 70: 1554-1558.

Chua, N. H., G. Blobel, P. Siekevitz, and G. E. Palade. 1976. Periodic variations in the ratio of free to thylakoid-bound chloroplast ribosomes during the cell cycle of *Chlamydomonas reinhardtii. J. Cell Biol.* 71: 497-514.

Chua, N. H., and N. W. Gillham. 1977. The sites of synthesis of the principal thylakoid membrane polypeptides in *Chlamydomonas reinhardtii. J. Cell Biol.* 74:441-452.

Chua, N. H., M. Karl, and P. Bennoun. 1975. A chlorophyll-protein complex lacking in photosystem I mutants of *Chlamydomonas reinhardtii. J. Cell Biol.* 67: 361-377.

Compans, R. W. 1973. Influenza virus proteins. II. Association with components of the cytoplasm. *Virology* 51: 56-70.

Cronan, J. E., and E. P. Gelmann. 1975. Physical properties of membrane lipids: biological relevance and regulation. *Bacter. Rev.* 39: 232-256.

Dales, S., and E. H. Mosbach. 1968. Vaccinia as a model of membrane biogenesis. *Virology* 35: 564-583.

Dallner, G., P. Siekevitz, and G. E. Palade. 1966. Biogenesis of endoplasmic reticulum membranes. I. Structural and chemical differentiation in developing rat hepatocyte. II. Synthesis of constitutive microsomal enzymes in developing rat hepatocyte. *J. Cell Biol.* 30: 73-96, 97-117.

De Petrocellis, B., P. Siekevitz, and G. E. Palade. 1970. Changes in chemical composition of thylakoid membranes during greening of the y-1 mutant of *Chlamydomonas reinhardtii*. *J. Cell Biol.* 44: 618-634.

Devillers-Thiery, A., T. Kindt, G. Scheele, and G. Blobel. 1975. Homology in animo-terminal sequence of precursors to pancreatic secretory proteins. *Proc. Nat. Acad. Sci. USA* 70: 1554-1558.

Ernster, L., and S. Orrenius. 1965. Substrate-induced synthesis of the hydroxylating enzyme system of liver microsomes. *Fed. Proc.* 24: 1190-1199.

Eytan, G., and I. Ohad. 1970. Biogenesis of chloroplast membranes. VI. Cooperation between cytoplasmic and chloroplast ribosomes in the synthesis of photosynthetic lamellar proteins during the greening process in a mutant of *Chlamydomonas reinhardtii* y-1. *J. Biol. Chem.* 245: 4297-4307.

Eytan, G., and I. Ohad. 1972. Biogenesis of chloroplast membranes. VII. The preservation of membrane homogeneity during development of the photosynthetic lamellar system in an algal mutant (*Chlamydomonas reinhardtii* y-1). *J. Biol. Chem.* 247: 112-121.

Farquhar, M. G., J. J. M. Bergeron, and G. E. Palade. 1974. Cytochemistry of Golgi fractions prepared from rat liver. *J. Cell Biol.* 60: 8-25.

Fleischer, B., and F. Zambrano. 1974. Golgi apparatus of rat kidney: preparation and role in sulfatide formation. *J. Biol. Chem.* 249: 5995-6003.

Goldberg, I., and I. Ohad. 1970. Biogenesis of chloroplast membranes. V. A radioautographic study of membrane growth in a mutant of *Chlamydomonas reinhardtii* y-1. *J. Cell Biol.* 44: 572-591.

Goodenough, U. W., and R. P. Levine. 1974. *Genetics*. Holt, Rinehart and Winston, Inc., New York.

Goodenough, U. W., and L. A. Staehelin. 1971. Structural differentiations of stacked and unstacked chloroplast membranes: freeze-etch electron microscopy of wild-type and mutant strains of *Chlamydomonas*. *J. Cell Biol.* 48: 594-619.

Green, D. E., and J. F. Perdue. 1966. Membranes as expressions of repeating units. *Proc. Natl. Acad. Sci. USA* 55: 1295-1302.

Hirano, H., B. Parkhouse, G. L. Nicolson, E. S. Lenox, and S. J. Singer. 1972. Distribution of saccharide residues on membrane fragments from a myeloma-cell homogenate: its implications for membrane biogenesis. *Proc. Natl. Acad. Sci. USA* 69: 2945-2949.

Hoober, J. K. 1970. Sites of synthesis of chloroplast membrane polypeptides in *Chlamydomonas reinhardtii* y-1. *J. Biol. Chem.* 245: 4327-4334.

Hoober, J. K. 1972. A major polypeptide of chloroplast membranes of *Chlamydomonas reinhardtii*: evidence for synthesis in the cytoplasm as a soluble component. *J. Cell Biol.* 52: 84-96.

Hoober, J. K., P. Siekevitz, and G. E. Palade. 1969. Formation of chloroplast membranes in *Chlamydomonas reinhardtii* y-1: effects of inhibitors of protein synthesis. *J. Biol. Chem.* 244: 2621-2631.

Ito, A., and R. Sato. 1968. Purification by means of detergents and properties of cytochrome b_5 from liver microsomes. *J. Biol. Chem.* 243: 4922-4923.

Ito, A., and R. Sato. 1969. Proteolytic dissection of smooth surfaced vesicles of liver microsomes. *J. Cell Biol.* 40:179-189.

Jamieson, J. D., and G. E. Palade. 1977. Production of secretory proteins in animal cells. In *Symposium volume of the 1st International Congress of Cell*

Biology, Boston, 1976. Rockefeller Univ. Press, New York (in press).

Kuriyama, Y., and D. J. L. Luck. 1973. Ribosomal RNA synthesis in mitochondria of *Neurospora crassa*. *J. Mol. Biol.* 73(4): 425-437.

Kuriyama, Y., T. Omura, P. Siekevitz, and G. E. Palade. 1969. Effects of phenobarbital on the synthesis and degradation of the protein components of rat liver microsomal membranes. *J. Biol. Chem.* 244: 2017-2026.

Leblond, C. P., and G. Bennett. 1977. The role of the Golgi apparatus in terminal glycosylation. In *Symposium volume of the 1st International Congress of Cell Biology*, Boston, 1976. Rockefeller Univ. Press, New York (in press).

Leskes, A., P. Siekevitz, and G. E. Palade. 1971. Differentiation of endoplasmic reticulum in hepatocytes. I. Glucose-6-phosphatase distribution *in situ*. II. Glucose-6-phosphatase distribution in rough microsomes. *J. Cell Biol.* 49: 264-287, 288-302.

Lodish, H. F. 1973. Biosynthesis of reticulocyte membrane proteins by membrane-free polyribosomes. *Proc. Natl. Acad. Sci. USA* 70:1526-1530.

Lodish, H. F., and B. Small. 1975. Membrane proteins synthesized by rabbit reticulocytes. *J. Cell Biol.* 65: 51-64.

Michaels, J., and C. P. Leblond. 1976. Transport of glycoprotein from Golgi apparatus to cell surface by means of "carrier vesicles" as shown by radioautography of mouse colonic epithelium after injection of ^{3}H-fucose. *J. Microscopie Biol. Cell.* 25:243-248.

Mindich, L. 1970. Membrane biosynthesis in *Bacillus Subtilis*. I. Isolation and properties of strains bearing mutations in glycerol metabolism. II. Integration of membrane proteins in the absence of lipid synthesis. *J. Mol. Biol.* 49: 415-432, 433-439.

Morré, D. J., T. W. Keenan, and C. M. Huang. 1974. Membrane flow and differentiation: origin of Golgi apparatus membranes from endoplasmic reticulum. In B. Ceccarelli et al., eds., *Advances in cytopharmacology*. Vol. 2. *Cytopharmacology of secretion*, pp. 107-126. New York: Raven Press.

Morrison, T. G., and H. F. Lodish. 1975. Sites of synthesis of membrane and nonmembrane proteins of vesicular stomatitis virus. *J. Biol. Chem.* 250: 6955-6962.

Ohad, I., P. Siekevitz, and G. E. Palade. 1967. Biogenesis of chloroplast membranes. I. Plastid differentiation in a dark-grown algal mutant (*Chlamydomonas reinhardtii*). II. Plastid differentiation during greening of a dark-grown algal mutant (*Chlamydomonas reinhardtii*). *J. Cell Biol.* 35: 521-552, 553-584.

Ojakian, G. K., and P. Satir. 1974. Particle movements in chloroplast membranes: quantitative measurements of membrane fluidity by the freeze-fracture technique. *Proc. Natl. Acad. Sci. USA* 71: 2052-2056.

Omura, T., and Y. Kuriyama. 1971. Role of rough and smooth microsomes in the biosynthesis of microsomal membranes. *J. Biochem. (Tokyo)* 69: 651-658.

Omura, T., P. Siekevitz, and G. E. Palade. 1967. Turnover of the constituents of the endoplasmic reticulum membranes of rat hepatocytes. *J. Biol. Chem.* 242: 2389-2396.

Palade, G. E. 1975. Intracellular aspects of the process of protein secretion. *Science* 189: 347-358.

Palade, G. E. 1976. Interactions among cellular membranes. In R. Passino, ed.,

Pontificiae Academiae Scientiarum Scripta Varia. Semaine d'étude sur le thème membranes biologiques et artificielles et la désalinisation de l'eau, pp. 85-109.

Porter, C. W., and R. J. Bernacki. 1975. Ultrastructural evidence for ectoglyco-syltransferase systems. *Nature* 256: 648-650.

Poyton, R. O., and J. Kavanagh. 1976. Regulation of mitochondrial protein synthesis by cytoplasmic proteins. *Proc. Natl. Acad. Sci. USA* 73: 3947-3951.

Ragnotti, G., G. R. Lawford, and P. N. Campbell. 1969. Biosynthesis of microsomal nicotinamide-adenine dinucleotide phosphate-cytochrome *c* reductase by membrane-bound and free polysomes from rat liver. *Biochem. J.* 112: 139-147.

Sager, R., and G. E. Palade. 1957. Structure and development of chloroplast in Chlamydomonas. I. The normal green cell. *J. Biophys. Biochem. Cytol.* 3: 463-488.

Sargent, J. R., and B. P. Vadlamudi. 1968. Characterization and biosynthesis of cytochrome b_5 in rat liver microsomes. *Biochem. J.* 107: 839-849.

Schatz, G., and T. L. Mason. 1974. The biosynthesis of mitochondrial proteins. *Ann. Rev. Biochem.* 43: 51-87.

Schimke, R. T. 1975. Turnover of membrane proteins in animal cells. In E. Korn, ed., *Methods in membrane biology* 3: 201-237.

Schor, S. 1971. Regular variations in chloroplast functional activity during the division cycle of *Chlamydomonas reinhardtii*. Ph.D. thesis. Rockefeller University.

Schor, S., P. Siekevitz, and G. E. Palade. 1970. Cyclic changes in thylakoid membrane of synchronized *Chlamydomonas reinhardtii*. *Proc. Natl. Acad. Sci. USA* 66: 174-180.

Segrest, J. P., I. Kahane, R. L. Jackson, and V. T. Marchesi. 1973. Major glycoprotein of the human erythrocyte membrane: evidence for an amphipathic molecular structure. *Arch. Biochem. Biophys.* 155: 167-183.

Singer, S. J., and Garth L. Nicolson. 1972. The fluid mosaic model of the structure of cell membranes. *Science* 175: 720-731.

Staehelin, L. A. 1976. Reversible particle movement associated with unstacking and restacking of chloroplast membranes *in vitro*. *J. Cell Biol.* 71: 333-355.

Stäubli, W., R. Hess, and E. R. Weibel. 1969. Correlated morphometric and biochemical studies on the liver cell. II. Effects of phenobarbital on rat hepatocytes. *J. Cell Biol.* 42: 92-112.

Steck, T. L. 1974. The organization of proteins in the human red blood cell membrane. *J. Cell Biol.* 62: 1-19.

Tillack, T. W., R. B. Scott, and V. T. Marchesi. 1972. The structure of erythrocyte membranes studied by freeze-etching. II. Localization of receptors for phytohemagglutinin and influenza virus to the intramembranous particles. *J. Exp. Med.* 135: 1209-1227.

Tomita, M., and V. T. Marchesi. 1975. Aminoacid sequence and oligosaccharide attachment sites of human erythrocyte glycophorin. *Proc. Natl. Acad. Sci. USA* 72: 2964-2968.

Waechter, C. J., and W. I. Lennarz. 1976. The role of polyprenol-linked sugars in glycoprotein synthesis. *Ann. Rev. Biochem.* 45: 95-112.

Weiser, M. M. 1973. Intestinal epithelial cell surface membrane glycoprotein synthesis. *J. Biol. Chem.* 248: 2542-2548.

Wirtz, K. W. A. 1974. Turnover of phospholipids between membranes. *Biochim. Biophys. Acta* 344: 95-117.

Zwaal, R. F. A., B. Roelafsen, P. Comifurius, and L. L. M. Van Deenen. 1975. Organization of phospholipids in human red cell membranes as detected by the action of various purified phospholipases. *Biochim. Biophys. Acta* 406: 83-96.

2   Lipid-Protein Associations

O. HAYES GRIFFTH AND PATRICIA C. JOST

The lipid regions of cell membranes have been implicated in the membrane permeability barrier, in the function of membrane proteins, in the lateral diffusion of protein components in the membrane, and in the events leading to cell-cell recognition and some cell-virus fusions. Very early, well before the thin cell membranes had been visualized by electron microscopy, Gortner and Grendel (1925) attempted to demonstrate that the red-cell membrane contained just enough lipid to form a continuous bilayer surrounding the cell. In turn, their experiments were based on even earlier work on the passive permeation of small hydrophobic molecules and on the lipid monolayer work of Langmuir (1917). By the time electron microscopy had demonstrated that fixed, imbedded cells and organelles were limited by a thin envelope (railroad tracks) that never exceeded 60-150Å in width, the idea of a continuous lipid bilayer was often assumed, and the protein was thought to coat the two sides of the bilayer, possibly as a β-pleated sheet (Robertson, 1960) to conform to the dimensions seen by electron microscopy. It was also recognized that membrane proteins were probably globular and, more importantly, at least some of them were very hydrophobic and required lipid for measurable enzyme activity.

A number of physical techniques clearly established the presence of the lipid bilayer structure in biological membranes (see Jain, 1972; Branton and Deamer, 1972; Rothfield, 1972; Finean, Coleman, and Michell, 1974). The current working model, the fluid mosaic model (Singer, 1971), is consistent with the data presently available for most membranes, such as membrane protein solubilities and lipid requirements of many membrane proteins for activity, passive permeability properties of membranes, and electron microscope dimensions. One schematic variation of this model is shown in Figure 2.1, where many membrane proteins are shown penetrating into the lipid bilayer. The membranes are asymmetric with respect to both lipid and protein. Many proteins are amphipathic, that is, have both hydrophobic regions surrounded by lipid and hydrophilic regions that extend into the aqueous compartments. Many membrane properties are known to involve the membrane proteins. These properties can be heavily dependent, in turn, on the presence of phospholipids, and may be specific for both the polar head group and the character of the acyl side chains of the phospholipids (Coleman, 1973; Fourcans and Jain, 1974). It is evident that an understanding of lipid-protein associations is central to the understanding of membrane function.

The general problem of lipid-protein associations is more easily stated than solved. While all of the sophisticated techniques developed for the characterization of the structure and function of water-soluble proteins are available to the membranologist, the problem is complicated by the nature of the binary solvent system. The functional membrane protein may, and probably does, have an active site exposed to the aqueous medium, but a two-dimensional lipid solvent is also present. Part of the amphipathic protein is surrounded by this solvent, which is heterogeneous in both the lipid polar head group and side chain composition. It is becoming increasingly evident from a variety of studies that this heterogeneous lipid matrix is not just an inert lipid support, but has a functional as well as a structural role.

Because of the complexity of even the simplest biological membranes, the experimental approach at the present state of knowledge is to deal with membrane proteins that are either isolated with accompanying phospholipids or reconstituted with defined lipids (see Fig. 2.2). A homogenous preparation of a membrane protein in a lipid environment is analogous to the isolation and preparation of a water-soluble protein, but the analogy can only be carried so far. There are many added complexities. For example, the lipid matrix forms a formidable barrier to the problems of obtaining three-dimensional crystals for x-ray crystallographic studies of the protein structure, and no such crystals have yet been obtained. Two-dimensional crystalline arrays have been obtained either by controlling the lipid-to-protein ratio, as seen in cytochrome oxidase (Seki, Hayashi, and Oda, 1970), or by taking advantage of the natural two-dimensional crystalline array that infrequently occurs, as in the purple membrane protein of *Halobacterium halobium* (Blaurock and Stoeckenius, 1971). In an ingenious series of experiments the structure of this purple membrane protein in its natural lipid environment has recently been determined to a 7 Å resolution (Henderson and Unwin, 1975). This is the first high-resolution structural information available on an amphipathic protein.

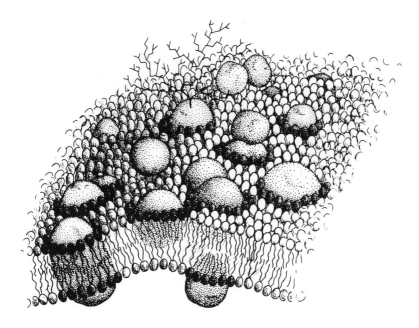

FIGURE 2.1. Cross-sectional diagram of a generalized membrane showing integral proteins imbedded in a phospholipid bilayer. Peripheral proteins are shown associated with the surface. The bushlike structure represents the carbohydrate of a glycoprotein.

MEMBRANE

HOMOGENEOUS
PROTEIN + LIPID

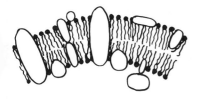

→

FIGURE 2.2. Diagram illustrating the isolation of a single protein complex and its associated phospholipids from a heterogeneous biological membrane. Minor contaminating proteins and some heterogeneity in lipid distribution are neglected.

When a lipid vesicle containing a single species of membrane protein in the bilayer is prepared, other experimental difficulties remain. The apparently simple measurements of enzyme activities may be haunted by unrecognized or uncontrollable aggregation in the assay mixture. Some preparations of lipid and protein form concentric closed compartments in aqueous dispersions (Chuang, Awasthi, and Crane, 1973), and many of the protein molecules can thus be shielded from molecules added in the assay. Membrane proteins are frequently isolated as complexes consisting of a large number of different polypeptides. There are also difficulties in separating kinetic and thermodynamic effects (for example, the monomeric solubility of phospholipids is vanishingly low, so that the competitive binding experiments in a homogeneous solvent are not practical). But, while many experimental difficulties are still to be overcome, one of the achievements in this field is the development of methods for the isolation and purification of many membrane proteins. Capitalizing on these developments, efforts have begun to probe the lipid-protein associations. In this chapter we discuss some of the examples reported thus far, and explore the structural and functional implications of lipid-protein associations.

The Hydrophobic Lipid-Protein Interface: Boundary Lipid

Assuming the current models of membrane structure to be substantially correct, some lipid molecules must be in direct contact with hydrophobic regions of the proteins that penetrate into or through the bilayer (see Fig. 2.1). The interface between the hydrophobic protein regions and those lipid molecules in direct protein contact may be thought of as a region of primary lipid-protein interaction. Some experimental data relating to this nearest neighbor lipid at the protein boundary (boundary lipid) have recently been obtained. We will review this evidence and discuss some of the physical and functional implications of this boundary region.

Cytochrome Oxidase

Initial studies characterizing boundary lipid involved membranous cytochrome oxidase isolated from beef heart mitochondria. Cytochrome oxidase is the terminal member of the mitochondrial transport chain (net chemical reaction, $4H^+ + O_2 + 4e^- \rightarrow 2H_2O$). The functional complex is large, about 200,000 molecular weight, and is composed of seven different polypeptide chains. Each complex contains two heme *a* molecules and two copper atoms. Cytochrome oxidase evidently spans the membrane, as it can be labeled with antibodies or chemical reagents from both sides of the mitochondrial membrane. (A detailed discussion of cytochrome oxidase isolated from yeast mitochondria is presented in this volume in the chapter by Gottfried Schatz.)

Cytochrome oxidase is isolated along with some associated lipids from the inner mitochondrial membrane. At appropriate lipid levels, closed vesicular structures form spontaneously (Sun et al., 1968; Chuang, Awasthi, and Crane, 1970). Typical electron micrographs of negatively stained preparations are shown in Figure 2.3a. These vesicles are not fragments of the inner mitochondrial membrane, but instead are a reasonably homogeneous system of lipids and the cytochrome oxidase complex. Cytochrome oxidase is by no means a fully characterized protein complex, but the crystalline patterns observed within a narrow range of lipid to protein ratios (Fig. 2.3b) indicates that the protein complex must have a well-defined structure.

Since the data to be discussed below are based on the motional characteristics of the lipids, it is useful to start with the motion of lipid chains in pure phospholipid vesicles, with no protein present. Most phospholipids spontaneously become closed bilayer-containing structures in water. These lipid bilayers have been used extensively to investigate many questions about the bilayer in the absence of protein. Among the well-characterized properties of such protein-free bilayers is the relative motion of the long lipid tails. The segmental motion of these lipid tails increases markedly near the center of the bilayer. This phenomenon is largely responsible for the dip in the center of the electron density profile from x-ray diffraction of phospholipid bilayers (Levine and Wilkins, 1971) and biological membranes (Caspar and Kirschner, 1971; Wilkins, Blaurock, and Engelman, 1971), and can be seen in the narrowing of the ESR lines as the spin reporter group is positioned near the center of the bilayer (Jost et al., 1971; Hubbell and McConnell, 1971), and by NMR (Seiter and Chan, 1973; Seelig and Seelig, 1974). The labeling experiment uses the spin label analogs of natural lipids or portions of such lipids (for example, the fatty acyl chains) at low concentrations. The behavior of the bulk unlabeled phospholipids is then deduced from the observable behavior of the reporter groups. Qualitatively, the width of the three lines becomes very narrow with motion that is rapid on the time scale of the technique. As motion becomes slower, the three lines become broader, overlap, and finally assume the line shape characteristic of immobilization. (Subtler differences in line shape are discussed in detail in Berliner, 1976.)

Lipid-protein associations in cytochrome oxidase have been examined by incorporating lipid spin labels into vesicles containing the integral membrane protein (Jost et al., 1973a). At very low lipid-to-protein ratios the spectrum (Fig. 2.4a) shows a broad line shape, indicating extremely hindered motion and resembling the general line shape of an immobilized spin label. As the lipid content increases, the lines become sharper, increasing in amplitude (for a constant number of spin labels). At much higher lipid content the line shape resembles that seen in bilayers without protein (Fig. 2.4e). At intermediate lipid levels (Fig. 2.4b, c, d) the line

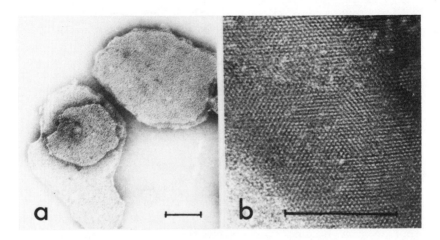

FIGURE 2.3 Electron micrographs of membranous cytochrome oxidase isolated from beef heart mitochondria by the method of Sun et al., 1968, showing (*a*) vesicular structures and (*b*) high-magnification micrograph of a crystalline region in the same sample. Bar indicates $0.1\,\mu$ in both photographs. The sample contained 0.49 mg phospholipid/mg protein and was negatively stained with phosphotungstic acid, pH 5.

shape is complex. The protein clearly immobilizes some lipid, since no immobilization is seen when protein is absent. At the other end of the scale, spectra resemble that of fluid bilayer (Fig. 2.4*e*). This observation and additional data on the effects of dehydration and orientation (Jost et al., 1973b) establish that at high lipid concentrations the cytochrome oxidase vesicles contain fluid bilayer regions.

The samples of intermediate lipid content are analyzed in terms of two spectral components, one arising from the protein-immobilized lipid, the other from the lipids in the bilayer. Spectral subtraction establishes the proportion of the total absorption due to the immobilized component, with no assumptions needed regarding the influence of the protein on the motion of the lipids in the bilayer. However, for simplicity, it is easier to present the results visually as the sum of two components, and the results of the two methods are in agreement. In the latter approach, spectra (such as Fig. 2.4*b*, *c*, *d*, left column) are approximated by summing varying proportions of the two components (immobilized and fluid bilayer) and adjusting the ratios by visual fit with the experimental spectra. The results are shown in the right column of Figure 2.4. The agreement between the experimental and summed spectra shows that major spectral features are accounted for by assuming that there are different proportions of the two components, rather than a continuous variation in a single component. If the assumption is made that the partitioning of the

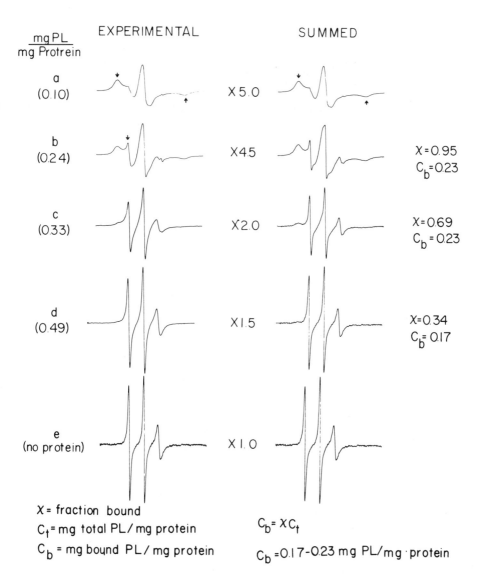

FIGURE 2.4. Electron spin resonance spectra of 16-doxylstearic acid diffused into cytochrome oxidase samples of different lipid contents (as indicated at the left). The summed spectra on the right are visual fits to the experimental (see text). The factors in the center column indicate relative gain, for example, X 5.0 indicates that the experimental spectrum has been amplified by a factor of five relative to the spectrum of the fluid bilayer (e). When the spectra are multiplied by these factors, all spectra reflect the same concentration of the spin label.

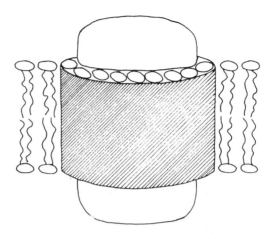

FIGURE 2.5. Diagrammatic sketch of a single cytochrome oxidase complex and associated lipid. The hydrophobic surface of the protein is surrounded by an immobilized boundary layer of lipid. Beyond the boundary lipid is the phospholipid bilayer. The orientation and disposition of the boundary lipid tails are deliberately obscured and are probably irregular.

lipid spin label between the bilayer and boundary region is reflecting the distribution of the unlabeled phospholipids, the amount of bound lipid can be calculated from the known lipid to protein ratio. The amount of bound lipid, C_b, is $C_b = C_t X$ where C_t is the total mg phospholipid per mg protein determined by phosphate and protein analysis and X is the fraction of the total absorption contributed by the bound component. Although spectral line shapes differ markedly with lipid content of the sample, the contribution of the bound component, C_b, remains relatively constant (Fig. 2.4), even in this data treatment that ignores second-order effects and is uncorrected for nonexchangeable lipid. Cytochrome oxidase thus appears to require a relatively constant number of lipids for maximum occupancy of protein-bound sites. This number roughly corresponds to the number of phospholipids required to coat the periphery of the cytochrome oxidase, assuming dimensions consistent with the electron micrographs (Vanderkooi et al., 1972); however, this calculated number can vary by a factor of two, depending on whether one chain or two chains per phosphate are in contact with the protein. This approximate correspondence, plus the fact that the spin label exchanges between the fluid bilayer and protein-bound sites, has led to the working hypothesis that there is a boundary layer of immobilized lipid associated with the protein surface, as shown in Figure 2.5. This figure emphasizes the two lipid regions, boundary and bilayer, but exaggerates the demarcation between the two lipid regions and neglects any protein-protein contacts.

These experiments show that the lipids at the protein-lipid interface are

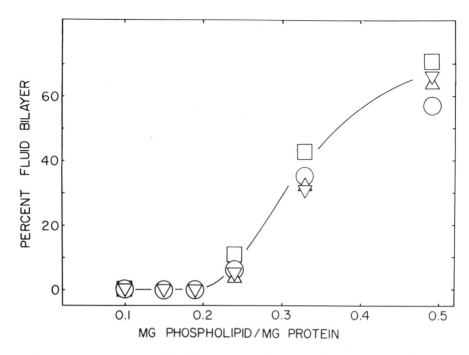

FIGURE 2.6. The percent fluid bilayer in cytochrome oxidase samples as a function of phospholipid content, arrived at by four different spin-labeling procedures. All methods are in substantial agreement. The two sets of points ∇ and Δ represent two spectral summations of the 16-doxylstearic acid data; ○ is the estimate of fluid bilayer based on difference spectra, using the same data. A separate estimate is based on the difference spectra obtained from samples labeled with 3-doxyl-5a-androstan-17β-ol, symbolized by □. (From Jost et al., 1973c.)

immobilized. They do not give a value for the binding constant, which could be relatively low in the equilibrium between bound lipid and fluid bilayer. However, the change in free energy proceeding from the fluid bilayer to the bound sites is negative. As the lipid content of the samples increases, the bound sites are filled first until the maximum occupancy in the boundary region is achieved, then excess lipid forms fluid bilayer of increasing extent (Fig. 2.6). Exchange of lipid between these two regions occurs, and the minimum residence time on the protein must be longer than $\sim 10^{-5}$ to 10^{-6} second in order to be detected in these experiments, and may well be much longer.

Other biochemical and biophysical data are consistent with the idea that a certain minimum amount of exchangeable lipid must play a special role in the cytochrome oxidase system. When lipid is removed from cytochrome oxidase, the activity (measured by cytochrome c oxidation or oxygen uptake) drops off precipitously and the protein aggregates. With successive lipid extractions using cold aqueous acetone, the majority of the lipid is removed with the first 30 minutes extraction. The remaining 0.2-0.3 mg phospholipid/mg protein is less easily removed and requires repeated extractions (Griffith et al., 1973). Maximum recoverable enzyme activity, measured after adding back mitochondrial phospholipids to lipid-depleted cytochrome oxidase solubilized in detergent, is achieved at approximately 0.2-0.4 mg phospholipid/mg protein (cf. Tzagoloff and MacLennan, 1965; Chuang, Awasthi, and Crane, 1973; Jost et al., 1973c; Yu, Yu, and King, 1975), as shown in Figure 2.7. Furthermore, electron micrographs of negatively stained cytochrome oxidase preparations of varying phospholipid content show some vesicular structures only above 0.2-0.3 mg phospholipid/mg protein. Below this range, amorphous aggregates are seen (Chuang, Awasthi, and Crane, 1970; Jost et al., 1973b).

A summary of the cytochrome oxidase system as a function of lipid content is shown in the phase diagram of Figure 2.8. The dark line gives the relationship between the phospholipid content and the calculated membrane area per protein complex. Below roughly 0.2 mg phospholipid/mg protein all lipid is immobilized and the samples are amorphous and aggregated. As the lipid content is increased, fluid bilayer forms. In the narrow range indicated by C crystallinity is often observed in electron micrographs (Seki, Hayashi, and Oda, 1970; Vanderkooi et al., 1972; see Fig. 2.3). In a cytochrome oxidase sample with a lipid-to-protein weight ratio of 0.4-0.5, approximately 30-40 percent of the total membrane surface area is calculated to be occupied by lipids in fluid bilayer, with the protein and its associated boundary lipid taking up the rest of the area.

Properties of the Lipid-Protein Interface

The two main properties of boundary lipid that have been discussed so far are that it is immobilized and that it exchanges with the fluid bilayer. Additional evidence that the immobilized lipid is in direct contact with protein comes from polarity measurements. The average polarity sensed by the spin labels is quite different when the boundary and bilayer regions are compared. Based on spectral measurements at low temperatures, where the effects of motion are negligible, the splittings indicate that the lipid spin labels in the boundary see a more polar environment than when they are near the center of the bilayer. This presumably reflects hydrogen bonding of the immobilized label to the polypeptide chain. The same polarity differences are also seen in lipid binding to the hydrophobic regions of the liver microsomal protein, cytochrome b_5 (Dehlinger, Jost, and Griffith, 1974). In addition, there is indirect evidence that the packing arrangement of lipids in contact with the proteins is irregular. The bound lipid does not appear to be oriented when samples are macroscopically ordered as multilayers, even though there is

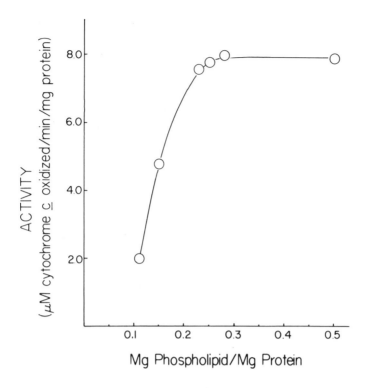

FIGURE 2.7. Cytochrome oxidase activity as a function of total phospholipid present. (From Jost et al., 1973c.)

a preferential orientation of the lipids in the fluid bilayer component of these same samples as judged by spectral anisotropy (Jost et al., 1973b).

Another interesting question is that of lipid specificity. It has been shown, for example, that cardiolipin is required for the marked stimulation of respiration by uncouplers when cytochrome oxidase is reconstituted with defined phospholipids (Racker, 1972). Under some assay conditions, the activation of cytochrome oxidase in the oxidation of reduced cytochrome c by molecular oxygen is affected by the kind of phospholipid added (Yu, Yu, and King, 1975). Small differences are seen in the occupancy of sites in the boundary region by a phospholipid spin label (16-doxylphosphatidylcholine) when compared to the fatty-acid spin label (16-doxylstearic acid), although the difference is only on the order of 10-15 percent (Jost, Nadakavukaren, and Griffith, 1977), which does not exceed by very much the propagated experimental errors. While it is clear that many interactions are nonspecific, some lipid specificity is apparently important in the function of integral membrane proteins.

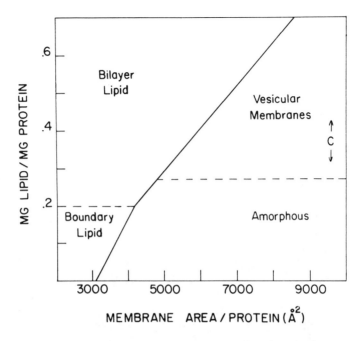

FIGURE 2.8. Correlation of various phases observed in the cytochrome oxidase system. The area of the protein is roughly 3100 Å. (From Jost et al., 1973c.)

Sarcoplasmic Reticulum ATPase

Lipid-protein association has been examined in the calcium transport protein, sarcoplasmic reticulum ATPase. This membrane protein and its associated phospholipid can be dispersed in cholate without loss of activity as measured by ATP hydrolysis. Codisperal of the ATPase-lipid complex with large excesses of a defined phospholipid in cholate allows equilibration of the lipid pools. In this way a complex is prepared in which one defined lipid predominates. The activity of the protein is dependent on the nature of the lipid chains present (Warren et al., 1974a). Activity of the ATPase at 37°C is significantly reduced when the added phosphatidylcholine contains saturated C_{14} side chains and reversibly restored when the side chains are $C_{18:1}$. The longer chain length with a *cis* double bond confers functional properties that are almost absent when the protein associates with the shorter saturated chains, even though both lipids are well above their chain melting temperatures, T_m. There is no evidence as yet that the level of activity is strongly dependent on the polar head group.

Excess lipid can be stripped from the ATPase-lipid complex in cholate by centrifugation through detergent-free sucrose. The amount of lipid re-

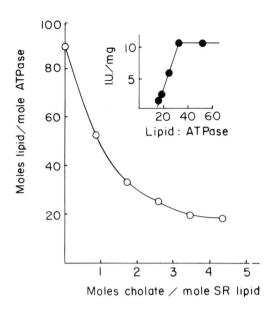

FIGURE 2.9. The relationship between cholate concentration and the amount of lipid associated with sarcoplasmic reticulum (SR) ATPase, when purified by centrifugation through sucrose. The inset shows the ATPase activity at 37° as a function of the amount of phospholipid in the purified complexes. (From Warren et al., 1974a.)

maining in association with the protein is dependent on the amount of cholate initially present, so that detergent-free preparations of various lipid/protein ratios can be prepared (Fig. 2.9). Regardless of the structure of the lipid present, maximum activity requires a minimum of 26-33 moles of phospholipid per mole of protein. These data indicate that about 30 lipid molecules may be in direct contact with the protein, forming a lipid annulus (Fig. 2.10). Activity is decreased proportional to the decrease in the number of molecules forming the annulus, down to a minimum of 15. Below this number the protein suffers irreversible loss of activity (Warren et al., 1974b). If the ATPase is stripped of the annular phospholipid in the presence of cholesterol, enzymic activity decreases proportional to the number of cholesterol molecules present in the annulus. Since cholesterol in the presence of sufficient phospholipid to fill the annulus is not inhibitory, apparently cholesterol is normally excluded from the annulus (Warren et al., 1975).

The phenomenon of boundary lipid as characterized by the spin-labeling experiment is also seen with the sarcoplasmic reticulum ATPase. From two somewhat different spin-labeling experiments, the boundary lipid of sarcoplasmic reticulum ATPase has been estimated as about 20 moles phospholipid (Nakamura and Ohnishi, 1975) or 15-18 moles per

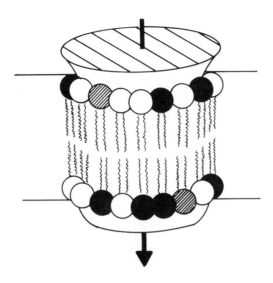

FIGURE 2.10. Model proposed for the organization of the native lipids associated with the calcium transport protein, sarcoplasmic reticulum ATPase. (Reproduced from Warren et al., 1974b.)

mole of ATPase (Fig. 2.11). The operational definitions of annular lipid and boundary lipid are different, but considering the propagated experimental errors of the various methods used, quantitatively they appear to be very similar. A reasonable assumption is that these are two different names for the same lipid domain, but it is possible, for example, that the annulus may consist of boundary lipid plus some less immobilized lipid (see halo lipid discussion below).

Other Systems

Evidence for a heterogeneous lipid environment in membranes has been obtained from studies on rat liver microsomal membranes (Stier and Sackmann, 1973). In this experiment, the spin label 5-doxylstearic acid was diffused into the heterogeneous membrane preparation and the rate of decay of the ESR signal was measured as a function of temperature. The rate of signal loss is greatly increased by the addition of NADPH, and Stier and Sackmann attribute the signal loss to enzymatic reduction by the cytochrome P_{450}-cytochrome P_{450} reductase hydroxylating enzyme system. There is a break in the Arrhenius plot at about 32°C. Below 32°C the rate of signal disappearance proceeds slowly, whereas above 32°C the rate increases. Control experiments ruled out phase changes in enzymatic activity and in the bulk lipids. From these data, it

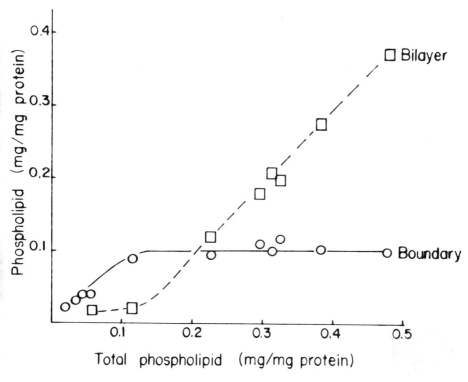

FIGURE 2.11. Estimates of the phospholipid in the two lipid domains, bound and bilayer, in samples of purified sarcoplasmic reticulum ATPase as a function of the total lipid-to-protein weight ratio (P.C. Jost, D.H. MacLennan, and O.H. Griffith, unpublished data).

was suggested that a quasi-crystalline environment surrounded the protein, but underwent a phase transition at 32°C, becoming more fluid above that temperature (Stier and Sackmann, 1973). There are also fluorescence measurements that are interpreted as indicating that a certain amount of lipid in a bacterial membrane is not in the fluid bilayer state (Träuble and Overath, 1973).

Boundary lipid refers to lipid immobilized on the surface of the protein; the bilayer beyond is fluid. When appropriate mixtures of lipids are cooled, fluid and ordered solid phases can coexist in the plane of the membrane. There is little evidence that the boundary lipid acts as a nucleation site for the formation of the solid phases. In fact, in the case of glycophorin, a transmembranous red-cell glycoprotein, the protein is excluded from the solid phase regions and is seen to be confined almost entirely to the regions of fluid phase, as seen by freeze-fracture electron microscopy (Grant and McConnell, 1974).

Captive Lipid

In addition to boundary lipids, one can envision another type of bound lipid, captive lipid, that does not exchange with the bilayer. This lipid could be deeply imbedded in or totally surrounded by the polypeptide chains of the protein complex. The spin-labeling experiments would not detect such captive lipid nor, for that matter, any lipid that associates with a high binding constant and that has a structure not resembling the spin-labeled lipids examined thus far. In the cytochrome oxidase complex, the two molecules of heme a, containing the porphyrin ring attached to an isoprenoid 17-carbon chain, probably are captive lipids. In addition, cytochrome oxidase cannot be freed of several molecules of a very negative phospholipid, cardiolipin, except by procedures that result in irreversible loss of activity (Awasthi et al, 1971). These few molecules, which are also shielded from the action of phospholipase, may fall in this class of captive lipids.

There is one demonstrated case of a specialized captive lipid, bacterio-chlorophyll a. The arrangement of this lipid in a protein complex from *Chlorobium limicola* has recently been determined by x-ray crystallography (Fenna and Matthews, 1975).

In the green photosynthetic bacterium *Chlorobium limicola*, the major light-harvesting pigment is bacteriochlorophyll c, which is located in the cigar-shaped chlorobium vesicle beneath the cell membrane. This antenna chlorophyll is intrinsic to the chlorobium vesicle. In addition, a small amount of water-soluble bacteriochlorophyll a (BChl a) protein complex can be extracted in aqueous salt solutions without the use of detergents (Olson and Romano, 1962). This BChl a protein complex therefore fits the operational definition of a peripheral membrane protein. The protein complex has a molecular weight of about 150,000 and it includes 21 molecules of noncovalently bound BChl a (Fig. 2.12) (Thornber and Olson, 1968). It is functionally implicated in the transfer of excitation energy from the light-harvesting antenna pigment to the photochemical reaction centers (Sybesma and Vrendenberg, 1964).

The x-ray crystallographic analysis shows that the protein complex consists of three identical subunits, each containing a core of seven BChl a molecules. One subunit is shown in Figure 2.13. The structure is a distorted hollow cylinder (silk bag) with the seven bacteriochlorophylls packed inside. The protein area facing the aqueous solvent is composed of 15 strands of β sheet. The side of the polypeptide in contact with an adjacent protein subunit is more complex and includes four short lengths of a helix (Fenna and Matthews, 1975). The packing of the seven BChl a molecules does not have the regularity one would expect for a microcrystal within a protein cavity, nor does it represent a pool of lipid. The 20-carbon phytol chain is clearly visible in the electron-density map, indicating the lipid motion is severely hindered. The phytol chains are

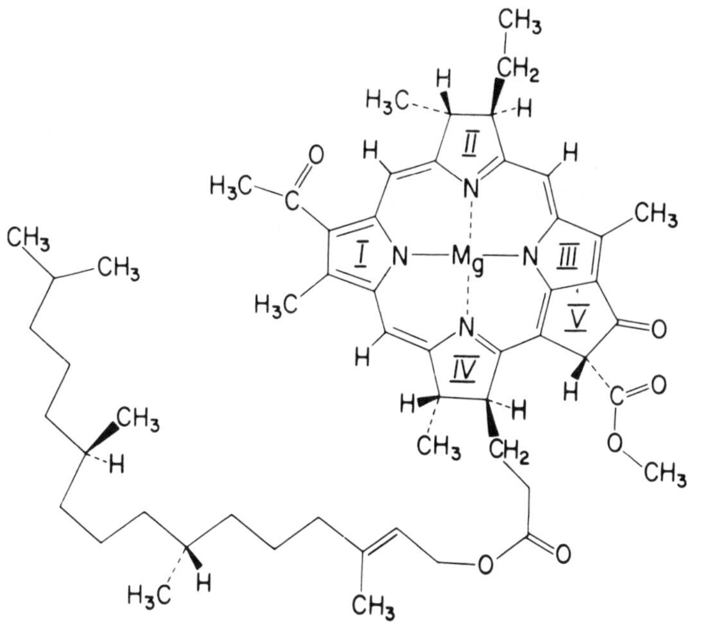

FIGURE 2.12. Structure of bacteriochlorophyll *a*. (From Fenna and Matthews, 1975.)

bent backward about the chlorin head groups so that they lie sandwiched between five of the chlorin rings on one side and the remaining two rings and the β-sheet protein wall on the other. The hydrocarbon chains are in an irregular extended conformation, with occasional bends. For example, the chain of one BChl *a* is bent into a U-shaped loop (1). Without the amino acid sequence it is difficult to deduce the atomic detail from the 2.8 Å resolution electron-density map. The β structure provides an inherent asymmetry across the plane of the sheet and the inwardly directed side chains are probably hydrophobic, forming close interactions with the phytol chain segments, whereas the polar residues face outward and interact with the aqueous surroundings (Fenna and Matthews, 1975).

The bacteriochlorophyll *a* protein complex is a specialized structure. It does nevertheless provide a clear example of captive lipid in a hydrophobic enivironment, one of the types of lipid-protein associations thought to be present in many protein complexes (for example, the lipid heme *a* and nonexchangeable cardiolipin of cytochrome oxidase). It also shows the lipid is immobilized and suggests a considerable degree of contact between lipid methylene side chains and the hydrophobic amino acid residues, resulting in an irregular packing of the lipid chains. These observations are relevant to boundary lipid, where instead of being sur-

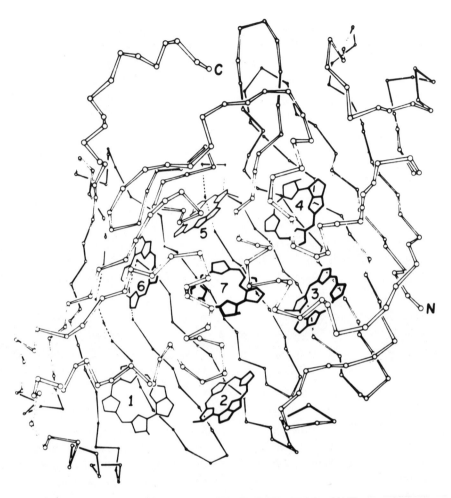

FIGURE 2.13. Diagram of the polypeptide backbone of one subunit of the BChl a-protein complex from *Chlorobium limicola*. The positions of the seven chlorin rings are shown, but the phytol chains are omitted for clarity. (From Fenna and Matthews, 1975.)

rounded by protein the lipid is at a protein-lipid interface free to exchange with the fluid phospholipid bilayer (Jost et al., 1977).

Before leaving the topic of chlorophyll-protein associations, we note that there is evidence for the association of the light-harvesting antenna chlorophyll *a* with integral membrane proteins of plant thylakoids (Thornber, 1975), and it has even been postulated recently that the antenna chlorophyll may exist as boundary lipid with the chlorin head groups imbedded into the hydrophobic integral membrane protein complex (Anderson, 1975). Athough interesting, the latter suggestion is speculation at this point and must await experimental test.

Lipid Association with Peripheral Proteins

In contrast to the behavior of integral proteins, peripheral membrane proteins are removed by dilute salt solutions and are evidently bound electrostatically to the membrane surface. Of the many peripheral proteins reported, the most fully studied is mitochondrial cytochrome c, and we discuss it as a model peripheral protein. The crystal structure of this small protein (MW $\sim$ 13,400) reveals that it is globular with a remarkably assymetric charge distribution (Fig. 2.14). The lysine residues are all on the outside and are clustered into two positively charged regions separated by a zone of negative charge, yielding a net positive charge of $+8$ at neutral pH (Dickerson, 1972). Cytochrome c is a member of the mitochondrial electron transport chain, which is essential for energy transduction. In its normal functional role it transfers electrons from one integral protein complex (cytochrome c_1) to another (cytochrome oxidase). Cytochrome c is also known to bind to bilayers containing negatively charged lipids such as cardiolipin of the inner mitochondrial membrane, and this may be important to the function of cytochrome c (Green and Fleischer, 1963).

The association of cytochrome c with vesicles of negatively charged lipids is readily demonstrated by centrifuging the phospholipid dispersion in the presence of the reddish cytochrome c. The normally clear pellet is red, and the complex can be further purified by column chromatography (Green and Fleischer, 1963). Studies using x-ray diffraction, optical spectroscopy, protein spin labeling, and fluorescence spectroscopy are consistent with the idea that the complex consists of layers of cytochrome c bound to the phospholipid bilayers (Gulik-Krzywicki et al., 1969; Kimelberg et al., 1970; Steinemann and Lauger, 1971; Blaurock, 1973; Letellier and Schechter, 1973; Vanderkooi, Erecińska, and Chance, 1973a, b). Spin labels attached to various positions along the lipid chains and also steroid spin labels report only small differences in the molecular motion before and after addition of oxidized cytochrome c, indicating very little if any penetration of the protein into the bilayer (Van and Griffith, 1975). However, there are suggestions that *reduced* cytochrome c may have some hydrophobic interaction with the bilayer (Azzi, Fleischer, and Chance, 1969; Letellier and Schechter, 1973).

Cytochrome c has recently been shown to induce lateral phase separations in mixtures of cardiolipin and a steroid spin label, as illustrated in Figure 2.15 (Birrell and Griffith, 1976). In the absence of cytochrome c the two lipids are randomly mixed. Through electrostatic interactions the cytochrome c sequesters the cardiolipin into patches, leaving behind

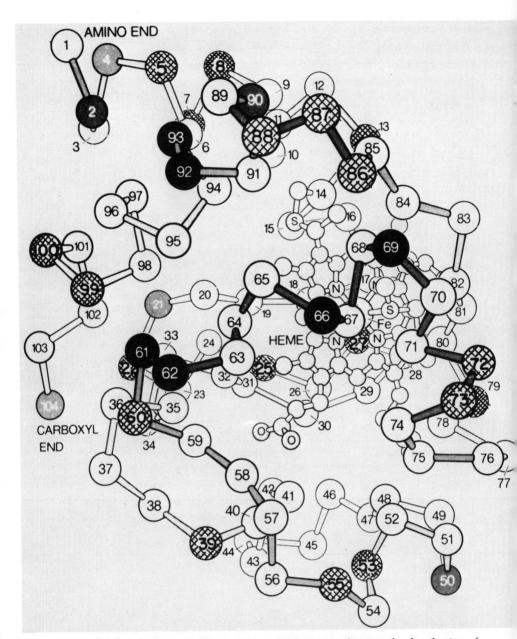

FIGURE 2.14. Structure of horse heart cytochrome c, showing the distribution of the 19 lysines (cross-hatched). These positively charged residues are distributed in two regions of the molecule and a number of the acidic side chains (gray and black) are clustered in an intermediate zone. (Adapted from Dickerson, 1972.)

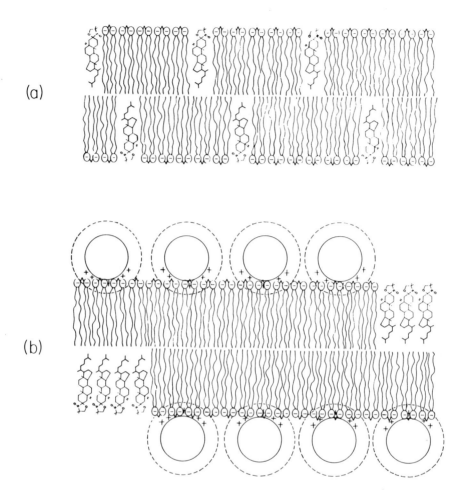

(a)

(b)

FIGURE 2.15. Schematic diagram indicating the distribution of lipids in the absence (*a*) and presence (*b*) of cytochrome *c*. The lipid mixture contains a negatively charged phospholipid, diphosphatidylglycerol, and a steroid spin label, 3-doxyl-5a-cholestane. (From Birrell and Griffith, 1976.)

pools of the steroid spin label. In contrast, cytochrome *c* does not cause a phase separation in egg lecithin mixtures with the same steroid spin label. There is no evidence that cytochrome *c* can cause lipid phase separations in the inner mitochondrial membrane. It seems more plausible at this point that there is a nonrandom distribution of cardiolipin in the membrane with cardiolipin-rich regions about the integral electron transport proteins. The cardiolipin could then attract the cytochrome *c* through electrostatic interactions and confine it to the regions of its cytochrome b-c_1 and cytochrome oxidase binding sites, increasing the efficiency of electron transport.

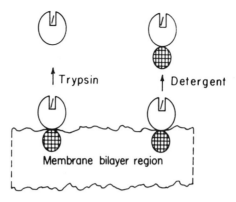

FIGURE 2.16. The relationship between cytochrome b_5 molecules that have been trypsin cleaved and detergent extracted, respectively. The cross-hatched region indicates the hydrophobic tail anchored in the membrane. (From Dehlinger, Jost, and Griffith, 1974.)

Enzymes with Hydrophobic Tails

Some membrane proteins consist of a water-soluble functional unit anchored by a relatively small hydrophobic segment to the phospholipid bilayer. Cytochrome b_5 is representative of this type. It is an electron transport protein found in the endoplasmic reticulum of mammalian cells. Cytochrome b_5 is a single polypeptide chain consisting of two parts, as shown in Figure 2.16. The hydrophilic part containing the heme and active site has a molecular weight of about 11,000 and is cleaved from the microsomes by trypsin. The intact cytochrome b_5 can be freed from the membrane with detergents and has a molecular weight of about 17,000 (Ito and Sato, 1968; Spatz and Strittmatter, 1971). The amino acid sequence of the water-soluble fragment is known for six mammalian species (Nobrega and Ozols, 1971) and the hydrophobic fragment has been partially sequenced (Ozols, 1974).

The three-dimensional structure of the water-soluble part of cyto-chrome b_5 (oxidized form) obtained from calf liver microsomes has been determined by x-ray diffraction to 2.0 Å resolution (Mathews, Argos, and Levine, 1971), and the structure of the reduced form has recently been determined to 2.8 Å (Argos and Mathews, 1975). The water-soluble fragment is globular, with a large percentage of its surface covered with hydrophilic amino acid side chains, and the heme group is in a hydro-phobic cleft. Thus, the structure of the hydrophilic fragment of cyto-chrome b_5 resembles structures of typical water-soluble proteins. There is a sequence similarity between cytochrome b_5 and hemoglobin near the active site (Ozols and Strittmatter, 1967).

The membranous segment of cytochrome b_5, comprising about one-

fourth of the total protein, is rich in hydrophobic amino acids, supporting the idea that it anchors the protein in the membrane (Ozols, 1974).
The water-soluble fragment does not bind to membranes but the intact
detergent-extracted cytochrome b_5 can bind to microsomes (Strittmatter,
Rogers, and Spatz, 1972; Enomoto and Sato, 1973) and to egg lecithin
vesicles (Rogers and Strittmatter, 1975). The detergent-extracted cytochrome b_5 molecule immobilizes lipid spin labels, whereas the water-
soluble fragment does not. The lipid association with the hydrophobic
fragment is evidently very similar to boundary lipid observed in cytochrome oxidase. The lipid in contact with the protein is exchangeable and
immobilized, even toward the end of the 18-carbon chain lipid, in
marked contrast with the motion of lipids in the fluid phospholipid
bilayer (Dehlinger, Jost, and Griffith, 1974). The amphipathic nature of
cytochrome b_5 has also been examined in binding experiments with
deoxycholate, Triton X-100, and lecithins (Robinson and Tanford, 1975;
Visser, Robinson, and Tanford, 1975).

Protein Influence on the Bilayer (Halo)

The cytochrome oxidase experiments show that there are two lipid envi-
ronments in the reconstituted cytochrome oxidase-phospholipid membrane. Immobilized boundary lipid and fluid bilayer are both present.
Thus a membrane can contain a large amount of integral membrane
protein without abolishing the fluid bilayer. As a first-order approximation, the presence of the protein has little effect on the bilayer regions.
This is consistent with the idea that the protein is globular with a well
defined three-dimensional structure, and does not have polypeptide
chains extending randomly out into the plane of the bilayer.

Nevertheless, if lipid in the boundary is immobilized by the protein,
intuitively one would expect that the nearby *bilayer* lipids would be influenced by the complex of protein-boundary lipid. That is, a gradient of
motion in the plane of the bilayer radiates outward from the protein-lipid
complex, so that the bilayer nearest the boundary lipid would show some
hindered motion, and this hindrance would decrease as a function of the
distance between the boundary and the instantaneous position of a single
lipid molecule in the bilayer. Another way of saying much the same thing
would be to consider that the packing of the lipids in the bilayer would be
a function of the radial distance from the nearest boundary. For convenience we can think of "perturbed" bilayer near the boundary and
"unperturbed" bilayer farther away. While this kind of radial distribution is difficult to determine quantitatively (as by some radial distribution function describing a motion parameter), there is an experimental
way to approach the question of whether or not such a gradient exists
in the bilayer surrounding the solvated protein. If a halo of perturbed bilayer surrounds the boundary region, and the effect dies off with distance

from the protein, then the *average* motion of lipid tails over all the bilayer should be a function of the ratio of lipid to protein. At higher lipid levels the extent of the bilayer increases (see Fig. 2.6 or 2.11) and the parameter measuring motion has a larger component due to unperturbed bilayer at greater distances from the protein. At lower lipid levels, the average motion parameter is heavily weighted by the fact that a larger percentage of the bilayer lipid molecules are involved in the perturbed bilayer, the bilayer "halo" (see the diagrammatic sketches in Figs. 2.17 and 18, where the halo bilayer is shown around each protein).

Advantage can be taken of the fact that the line width in the ESR spectrum is a function of the motion. The difference spectrum (that is, the spectrum remaining after removal of the immobilized component that is used to define boundary lipid) is characteristic of the motional range seen in bilayers and has been shown to represent lipid bilayer by several criteria (Jost et al., 1973b). So the question can be asked: Does the line width of the difference spectrum (bilayer spectrum) decrease as the amount of lipid is increased? It is clear from the difference spectra shown on the right in Figure 2.17 that the bilayers of the two samples differ. In this presentation the line heights are inversely proportional to the square root of the average line width, since the difference spectra have been scaled to reflect the same concentration (that is, to give the same value for the integrated absorption spectrum). The bilayer (non-boundary) lipid in the first sample is $\sim$ 25 percent larger, and the average line width in the bilayer is decreased by $\sim$20 percent. The simplest way to account for these results is depicted schematically in the left side of Figure 2.17. The influence of the protein is felt by the lipids in the bilayer beyond the boundary, but this effect falls off with distance. The available evidence supports the notion that the bilayer around the boundary forms a region of slightly more hindered motion than the bilayer farther out from the boundary. This region of slightly restricted motion, the halo, may be only a physical consequence of insertion of the solvated protein in the lipid bilayer. Or, to enter the realm of conjecture, the bilayer halo surrounding the boundary lipid may have some as yet unknown chemical and functional properties that distinguish it from the rest of the bilayer.

Summary

Several types of lipid-protein associations occur in biological membranes. Some of these are illustrated in Figure 2.18. Surrounding the hydrophobic regions of the integral proteins, such as cytochrome oxidase and sarcoplasmic reticulum ATPase, there appears to be a boundary layer (or annulus) of lipid that is characterized by marked restriction of lipid chain motion. The interaction is hydrophobic; the lipid exchanges with the bilayer lipids and is required for activity. We speculate that one

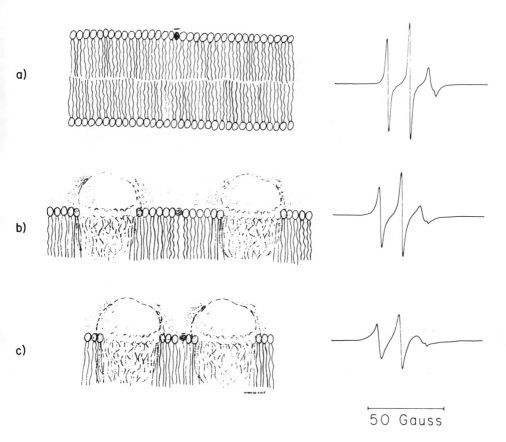

a)

b)

c)

50 Gauss

FIGURE 2.17. The influence of cytochrome oxidase on motion in the bilayer. The top ESR spectrum (*a*) is for 16-doxylstearic acid intercalated in liposomes of mitochondrial lipid; the center spectrum (*b*) and bottom spectrum (*c*) are for bilayer components (difference spectra) derived by subtracting out the bound components. All three spectra are normalized to reflect the same amount of spin label, so that the decrease in line height indicates an increase in the average line width, indicative of somewhat slower average motion in the bilayer. The interpretation that this reflects a halo of perturbed bilayer surrounding the boundary lipid-protein complex is indicated by the shadowy halos in the sketch. The hazy outlines of the protein and boundary lipid indicate that this subtraction has been performed.

function of this boundary lipid is to solvate the protein and thus to prevent indiscriminate protein aggregation in the plane of the bilayer.

Captive lipid represents a second type of hydrophobic lipid-protein association. This lipid is immobilized but is not rapidly exchangeable with the bilayer. Captive lipid is evidently highly specific. It is deeply imbedded in, or surrounded by, polypeptide chains, and attempts to remove it irreversibly abolish activity. Examples of captive lipid are the

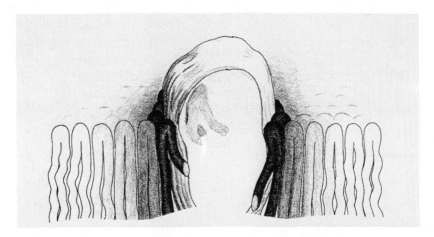

FIGURE 2.18. Summary model, showing three lipid-protein associations. The integral protein penetrates deeply into the phospholipid bilayer and is solvated by the immobilized but exchangeable boundary lipid. Beyond the boundary layer is the halo or region of influence of the protein on the bilayer, and, finally, the relatively unperturbed phospholipid bilayer beyond the halo. The possibility of non-exchangeable captive lipid is indicated by the lipid molecule sketched within the protein complex.

bacteriochlorophyll *a* molecules within the BChl *a* protein complex of *Chlorobium limicola* and heme *a*, and perhaps a few molecules of cardiolipin in the cytochrome oxidase complex.

Another type of lipid-protein association can occur between peripheral proteins and the phospholipid bilayer. This interaction is largely electrostatic at the lamellar interface, and can be charge-specific. An example is cytochrome *c*, which, in addition to having important integral protein binding sites, associates with bilayer regions containing negatively charged phospholipids.

The lipid chain mobility in bilayers is preserved when both peripheral and integral proteins are present. Peripheral proteins such as cytochrome *c* have very little influence on the segmental motion of the lipids. Integral proteins immobilize the lipids in contact with the protein, but the influence of the protein drops off rapidly with distance. The perturbed bilayer near the boundary can be thought of as a halo of bilayer with motional and possibly compositional differences from the bulk bilayer phase. These lipid-protein associations are important to the function of membrane proteins and may help to explain the variety of lipids found in biological membranes.

REFERENCES

Anderson, J. M. 1975. Possible location of chlorophyll within chloroplast membranes. *Nature* 253:536-537.

Argos, P., and F. S. Mathews. 1975. The structure of ferrocytochrome b_5 at 2.8 Å resolution. *J. Biol. Chem.* 250:747-751.

Azzi, A., S. Fleischer, and B. Chance. 1961. Cytochrome *c* phospholipid interaction: structural transitions associated with valency changes. *Biochem. Biophys. Res. Comm.* 36:322-327.

Awasthi, Y. C., T. F. Chuang, T. W. Keenan, and F. L. Crane. 1971. Tightly bound cardiolipin in cytochrome oxidase. *Biochim. Biophys. Acta* 226: 42-52.

Berliner, L. J., ed. 1976. *Spin labeling: theory and applications.* Academic Press, New York.

Birrell, G. B., and O. H. Griffith. 1976. Cytochrome *c*—induced lateral phase separation in a diphosphatidylglycerol-steroid spin label model membrane. *Biochemistry,* in press.

Blaurock, A. E. 1973. The structure of a lipid-cytochrome *c* membrane. *Biophys. J.* 13:290-298.

Blaurock, A. E., and W. Stoeckenius. 1971. Structure of the purple membrane. *Nature New Biol.* 233:152-155.

Branton, D., and D. W. Deamer. 1972. *Membrane structure.* Springer-Verlag, New York.

Caspar, D. L. D., and D. A. Kirschner. 1971. Myelin membrane structure at 10 Å resolution. *Nature New Biol.* 231:46-52.

Chuang, T. F., Y. C. Awasthi, and F. L. Crane. 1970. Model mosaic membrane: cytochrome oxidase. *Proc. Indiana Acad. Sci. 1969* 79:110-120.

Chuang, T. F., Y. C. Awasthi, and F. L. Crane. 1973. A model mosaic membrane: association of phospholipids and cytochrome oxidase. *Bioenergetics* 5:27-72.

Coleman, R. 1973. Membrane-bound enzymes and membrane ultrastructure. *Biochim. Biophys. Acta* 300:1-30.

Dehlinger, P. J., P. C. Jost, and O. H. Griffith. 1974. Lipid binding to the amphipathic membrane protein cytochrome b_5 . *Proc. Natl. Acad. Sci. USA* 71: 2280-2284.

Dickerson, R. 1972. The structure and history of an ancient protein. *Sci. Amer.* 226:58-72.

Enomoto, K., and R. Sato. 1973. Incorporation *in vitro* of purified cytochrome b_5 into liver microsomal membranes. *Biochem. Biophys. Res. Comm.* 51: 1-7.

Fenna, R. E., and B. W. Matthews. 1975. Chlorophyll arrangement in a bacteriochlorophyll protein from *Chlorobium limicola. Nature* 258:573-577.

Finean, J. B., R. Coleman, and R. H. Michell. 1974. *Membranes and their cellular function.* John Wiley and Sons, New York.

Fourcans, B., and M. K. Jain. 1974. Role of phospholipids in transport and enzyme reactions. *Adv. Lipid Res.* 12:147-226.

Gortner, E., and F. Grendel. 1925. On biomolecular layers of lipids on the chromocytes of the blood. *J. Exp. Med.* 41:439-443.

Grant, C. W. M., and H. M. McConnell. 1974. Glycophorin in lipid bilayers. *Proc. Nat. Acad. Sci. USA* 71:4653-4657.

Green, D. E., and S. Fleischer. 1963. The role of lipids in mitochondrial electron transfer and oxidative phosphorylation. *Biochim. Biophys. Acta* 70: 554-582.

Griffith, O. H., P. C. Jost, R. A. Capaldi, and G. Vanderkooi. 1973. Boundary lipid and fluid bilayer regions in cytochrome oxidase model membranes. *Ann. N.Y. Acad. Sci.* 222:561-573.

Gulik-Krzywicki, T., E. Schechter, V. Luzzati, and M. Faure. 1969. Interactions of proteins and lipids: structure and polymorphism of protein-lipid-water phases. *Nature* 223:1116-1121.

Henderson, R., and P. N. T. Unwin. 1975. Three dimensional model of purple membrane obtained by electron microscopy. *Nature* 257:28-32.

Hubbell, W. L., and H. M. McConnell. 1971. Molecular motion of spin-labeled phospholipids and membranes. *J. Am. Chem. Soc.* 93:314-326.

Ito, A., and R. Sato. 1968. Purification by means of detergents and properties of cytochrome b_5 from liver microsomes. *J. Biol. Chem.* 243:4922-4930.

Jain, M. K. 1972. *The bimolecular lipid membrane: a system.* Van Nostrand Reinhold Co., New York.

Jost, P. C., R. A. Capaldi, G. Vanderkooi, and O. H. Griffith. 1973c. Lipid-protein and lipid-lipid interactions in cytochrome oxidase model membranes. *J. Supramol. Struc.* 1:269-280.

Jost, P. C., O. H. Griffith, R. A. Capaldi, and G. Vanderkooi. 1973a. Evidence for boundary lipid in membranes. *Proc. Nat. Acad. Sci. USA* 70: 480-484.

Jost, P. C., O. H. Griffith, R. A. Capaldi, and G. Vanderkooi. 1973b. Identification and extent of fluid bilayer regions in membranous cytochrome oxidase. *Biochim. Biophys. Acta* 311:141-152.

Jost, P. C., L. J. Libertini, V. C. Hebert, and O. H. Griffith. 1971. Lipid spin labels in lecithin multilayers: a study of motion along fatty acid chains. *J. Mol. Biol.* 59:77-98.

Jost, P. C., K. K. Nadakavukaren, and O. H. Griffith. 1977. Phosphatidylcholine exchange between the boundary lipid and bilayer domains in cytochrome oxidase-containing membranes. *Biochemistry* 16:3110-3114.

Kimelberg, H. K., C. P. Lee, A. Claude, and E. Mrena. 1970. Interactions of cytochrome *c* with phospholipid membranes. I. Binding of cytochrome *c* to phospholipid liquid crystals. *J. Membrane Biol.* 2:235-251.

Langmuir, I. 1917. The constitution and fundamental properties of solids and liquids. II. Liquids. *J. Am. Chem. Soc.* 39:1848-1906.

Letellier, L., and E. Schechter. 1973. Correlations between structure and spectroscopic properties in membrane model system: fluorescence and circular dichroism of the cytochrome *c*-cardiolipin system, *Eur. J. Biochem.* 40:507-12.

Levine, Y. K., and M. H. F. Wilkins. 1971. Structure of oriented lipid bilayers. *Nature New Biol.* 230:69-72.

Mathews, F. S., P. Argos, and M. Levine. 1971. The structure of cytochrome b_5 at 2.0 Å resolution. *Cold Spring Harbor Symposium* 36:387-393.

Nakamura, M., and S. Ohnishi. 1975. Organization of lipids in sarcoplasmic reticulum membrane and Ca^{2+}-dependent ATPase activity. *J. Biochem.* 78:1039-1045.

Nobrega, F. G., and J. Ozols. 1971. Amino acid sequences of tryptic peptides of cytochromes b_5 from microsomes of human, monkey, porcine and chicken liver. *J. Biol. Chem.* 246:1706-1717.

Olson, J. M., and C. A. Romano. 1962. A new chlorophyll from green bacteria. *Biochim. Biophys. Acta* 59:726-728.

Ozols, J. 1974. Cytochrome b^5 from microsomal membranes of equine, bovine, and porcine livers: isolation and properties of preparations containing the membranous segment. *Biochemistry* 13:426-433.

Ozols, J., and P. Strittmatter. 1967. The homology between cytochrome b_5, hemoglobin and myoglobin. *Proc. Nat. Acad. Sci. USA* 58:264-267.

Racker, E. 1972. Reconstitution of cytochrome oxidase vesicles and conferral of sensitivity to energy transfer inhibitors. *J. Membrane Biol.* 10:221-235.

Robertson, J. D. 1960. The molecular structure and contact relationships of cell membranes. *Progr. Biophys. Biophys. Chem.* 10:344-418.

Robinson, N. C., and C. Tanford. 1975. The binding of deoxycholate, Triton X-100, sodium dodecyl sulfate and phosphatidylcholine vesicles to cytochrome b_5. *Biochemistry* 14:369-377.

Rogers, M. J., and P. Strittmatter. 1975. The interaction of NADH-cytochrome b_5 reductase and cytochrome b_5 bound to egg lecithin liposomes. *J. Biol. Chem.* 250:5713-5718.

Rothfield, L., ed. 1972. *Structure and function of biological membranes.* Academic Press, New York.

Seelig, A., and J. Seelig. 1974. The dynamic structure of fatty acyl chains in a phospholipid bilayer measured by deuterium magnetic resonance. *Biochemistry* 13:4839-4845.

Seiter, C. H. A., and S. I. Chan. 1973. Molecular motion in lipid bilayers: a nuclear magnetic resonance linewidth study. *J. Am. Chem. Soc.* 95:7541-7553.

Seki, S., H. Hayashi, and T. Oda. 1970. Studies on cytochrome oxidase. I. Fine structure of cytochrome oxidase-rich submitochondrial membrane. *Arch. Biochem. Biophys.* 138:110-121.

Singer, S. J. 1971. The molecular organization of biological membranes. In L. I. Rothfield, ed., *The structure and function of biological membranes*, pp. 145-222. Academic Press, New York.

Spatz, L., and P. Strittmatter. 1971. A form of cytochrome b_5 that contains an additional hydrophobic sequence of 40 amino acid residues. *Proc. Nat. Acad. Sci. USA* 68:1042-1046.

Steinemann, A., and P. Läuger. 1971. Interaction of cytochrome c with phospholipid monolayers and bilayer membranes. *J. Membrane Biol.* 4:74-86.

Stier, A., and E. Sackmann. 1973. Spin labels and enzyme substrates: heterogeneous lipid distribution in liver microsomal membranes. *Biochim. Biophys. Acta* 311:400-408.

Strittmatter, P., M. J. Rogers, and L. Spatz. 1972. The binding of cytochrome b_5 to liver microsomes. *J. Biol. Chem.* 247:7188-7194.

Sun, F. F., K. S. Prezbindowski, F. L. Crane, and E. E. Jacobs. 1968. Physical state of cytochrome oxidase: relationship between membrane formation and ionic strength. *Biochim. Biophys. Acta* 153:804-818.

Sybesma, C., and W. J. Vrendenberg. 1964. Kinetics of light-induced cytochrome oxidation and P_{840} bleaching in green photosynthetic bacteria under various conditions. *Biochim. Biophys. Acta* 88: 205-207.

Thornber, J. P. 1975. Chlorophyll-proteins: light-harvesting and reaction center components of plants. *Ann. R. Plant Physiol.* 26:127-158.

Thornber, J. P., and J. M. Olson. 1968. Chemical composition of a crystalline bacteriochlorophyll-protein complex isolated from green bacterium *Chloropseudomonas ethylicum. Biochemstry* 7:2242-2249.

Träuble, H., and P. Overath. 1973. The structure of *Escherichia coli* membranes studied by fluorescence measurements of lipid phase transitions. *Biochim. Biophys. Acta* 307:491-512.

Tzagoloff, A., and D. H. MacLennan. 1965. Studies of the electron transfer system. LXIV. Role of phospholipids in cytochrome oxidase. *Biochim. Biophys. Acta* 99:476-485.

Van, S. P., and O. H. Griffith. 1975. Bilayer structure in phospholipid-cytochrome *c* model membranes. *J. Membrane Biol.* 20:155-170.

Vanderkooi, G., A. E. Senior, R. A. Capaldi, and H. Hayashi. 1972. Biological membrane structure. III. The lattice structure of membranous cytochrome oxidase. *Biochim. Biophys. Acta* 274:38-48.

Vanderkooi, J., M. Erecińska, and B. Chance. 1973a. Cytochrome *c* interaction with membranes. I. Use of a fluorescent chromophore in the study of cytochrome *c* interaction with artificial and mitochondrial membranes. *Arch. Biochem. Biophys.* 154:219-229.

Vanderkooi, J., M. Erecińska, and B. Chance. 1973b. Cytochrome *c* interaction with membranes. II. Comparative study of the interaction of *c* cytochromes with the mitochondrial membrane. *Arch. Biochem. Biophys.* 157:531-540.

Visser, L., N. C. Robinson, and C. Tanford. 1975. The two-domain structure of cytochrome b_5 in deoxycholate solution. *Biochemistry* 14:1194-1199.

Warren, G. B., N. J. M. Birdsall, A. G. Lee, and J. C. Metcalfe. 1974b. Lipid substitution: the investigation of functional complexes of single species of phospholipid and a purified calcium transport protein. In G. F. Azzone, ed., *Membrane proteins in transport and phosphorylation*, pp. 1-12. North Holland Publishing Company, Amsterdam, The Netherlands.

Warren, G. B., M. D. Houslay, J. C. Metcalfe, and N. J. M. Birdsall. 1975. Cholesterol is excluded from the phospholipid annulus surrounding an active calcium transport protein. *Nature* 255:684-687.

Warren, G. B., P. A. Toon, N. J. M. Birdsall, A. G. Lee, and J. C. Metcalfe. 1974a. Reversible lipid titrations of the activity of pure adenosine triphosphatase-lipid complexes. *Biochemistry* 13:5501-5507.

Wilkins, M. H. F., A. E. Blaurock, and D. M. Engelman. 1971. X-ray diffraction from membrane dispersions: bilayer structure in membranes. *Nature New Biol.* 230:72-76.

Yu, C., L. Yu, and T. E. King. 1975. Studies on cytochrome oxidase: interactions of the ctyochrome oxidase protein with phospholipids and cytochrome *c*. *J. Biol. Chem.* 250:1383-1392.

3 Molecular Architecture of Oligomeric Membrane Proteins

FERNANDO CABRAL, WALTER BIRCHMEIER,
CARMEN E. KOHLER, AND GOTTFRIED SCHATZ

Biological membranes are asymmetric. It is now well established that the major membrane components, such as proteins, lipids, and carbohydrates, are not randomly distributed, but specifically oriented within the membrane. The techniques that have been employed to demonstrate "sidedness" of membranes have included electron microscopy, chemical analysis of freeze-cleaved membrane halves, and determination of accessibility of membrane components to membrane-impermeable molecules, such as enzymes, antibodies, or small hydrophilic "probes" capable of covalent attachment to membrane components (for review see Singer, 1974; Carraway, 1975). Taken together, these studies have greatly enhanced our knowledge of the molecular architecture of biological membranes and have raised intriguing questions about membrane biogenesis. More importantly, however, they have provided a new frame of reference within which to understand the function of membranes in living cells. While preceding generations of membranologists were preoccupied with the permeability properties or the electron microscopic appearance of membranes, the interest of present researchers is focused on the three-dimensional arrangement of membrane molecules. This trend, like the preceding trends, will undoubtedly run its course and be superseded by others. However, at the moment we are becoming aware that most, if not all, membrane functions are vectorial and that asymmetry is one of the major raisons d'être of biological membranes.

In studying the arrangement of membrane molecules, the biochemist must ask the following questions:

1. How many different types of molecules are present in a given functionally defined membrane component?

2. How are these molecules arranged within that component and the membrane?
3. To what extent is the biological function explained by the chemical properties of the constituent molecules?
4. How is the functional component assembled in the living cell?

A particularly attractive system for approaching these questions is the mitochondrial inner membrane. It carries out a variety of well-defined vectorial functions (see Racker, 1970 for review) and is generally one of the best studied of all biological membranes. Since it is synthesized by a close cooperation of mitochondrial and nuclear genes (Schatz and Mason, 1974), it has also been a popular system for investigating membrane biogenesis. The two functional components that have been especially well characterized are cytochrome c oxidase and the mitochondrial ATPase complex. Cytochrome c oxidase is the terminal enzyme of the mitochondrial respiratory chain and participates in the generation of an electrochemical potential across the mitochondrial inner membrane during substrate oxidation. The mitochondrial ATPase complex is the enzymic unit that synthesizes ATP from ADP and P_i, probably at the expense of the electrochemical potential.

During the past eight years, our laboratory has studied the chemical structure and the biosynthesis of cytochrome oxidase and the ATPase complex. Similar efforts have been under way in other laboratories (for review see Schatz, 1970; Tzagoloff et al., 1973; Schatz and Mason, 1974). Both functional components share certain chemical features and are synthesized similarly. The present survey will therefore be limited to cytochrome oxidase with the tacit understanding that most of the general principles also apply to the ATPase complex and perhaps to other mitochondrial inner membrane complexes as well.

The Polypeptide Composition of Cytochrome Oxidase

Purified cytochrome oxidase from yeast, *Neurospora,* and bovine heart contains seven different polypeptides, two molecules of noncovalently bound heme a, two copper atoms, and variable amounts of lipid (Mason and Schatz, 1973; Tzagoloff et al., 1973; Weiss et al., 1975a; Downer et al., 1976). The polypeptide composition of yeast cytochrome oxidase is illustrated in Figure 3.1. The same subunit composition is observed if cytochrome oxidase is immunoprecipitated from a crude lysate of whole mitochondria (Mason and Schatz, 1973).

The individual polypeptides have been isolated after dissociating the enzyme with guanidine HCl, formic acid, or dodecyl sulfate, and have been characterized with respect to molecular weight, amino acid composition, and antigenic specificity. The three large subunits are very hydrophobic, whereas the four smaller subunits are more hydrophilic (Poyton and Schatz, 1975a). There is no detectable immunological cross-reaction

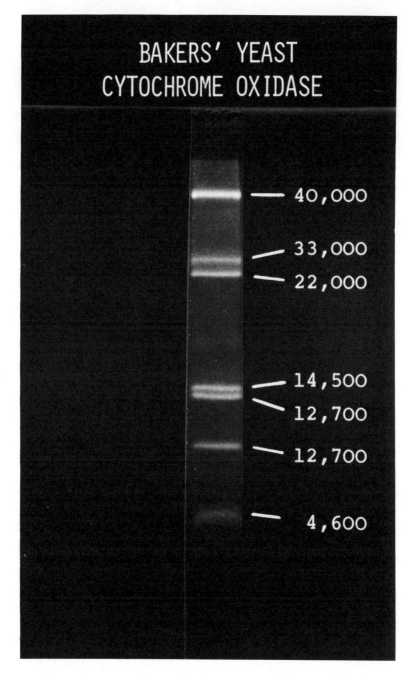

FIGURE 3.1. The seven subunits of cytochrome oxidase from *Saccharomyces cerevisiae*. The enzyme was electrophoresed in the presence of dodecyl sulfate on 10-20 percent polyacrylamide step gels (Cabral, 1976). The photograph is of the Coomassie-blue-stained gel slab. The apparent molecular weights are indicated on the right.

between the subunits. Antisera directed against the holoenzyme, against subunit II, or against mixtures of two or more of the small subunits IV-VII inhibit the activity of purified cytochrome oxidase (Poyton and Schatz, 1975b); this suggests that all seven subunits represent a structural and functional unit. Work on the amino acid sequence of subunits IV, VI, and VII is currently under way in collaboration with A. Tsugita. A major impediment to these studies was the microheterogeneity of cytochrome oxidase subunits isolated from commercial baker's yeast. Consequently, sequence work is now performed on cytochrome oxidase subunits isolated from the haploid laboratory strain D273-10B (Sherman, 1965).

Orientation of Cytochrome Oxidase Polypeptides in the Mitochondrial Inner Membrane

In order to explore the arrangement of the subunits in intact yeast cytochrome oxidase, the reactivity of each subunit with a variety of "surface probes" was tested with the isolated enzyme and with mitochondrially bound enzyme. The surface probes included iodination with lactoperoxidase and coupling with $[^{35}S]$-p-diazonium benzene sulfonate. In addition, external subunits were identified by linking them to bovine serum albumin carrying a covalently bound isocyanate group (Eytan and Schatz, 1975). In the membrane-bound enzyme, subunit I was almost completely inaccessible and subunit II was only partly accessible. All other subunits were accessible to at least one of the surface probes. Similar results were obtained with the solubilized enzyme except that the differences between the individual subunits were less clear-cut. These data suggested that the two largest subunits are localized in the interior of the enzyme and that they are genuine components of cytochrome oxidase. However, these experiments did not reveal which of the exposed subunits may be situated on the outer side or the inner side of the mitochondrial inner membrane. In order to approach this question, one has to compare the accessibility of a given subunit in intact mitochondria and in sonically prepared submitochondrial particles in which the "sidedness" of the mitochondrial inner membrane has been reversed. Since we were unsuccessful in preparing yeast mitochondria that had completely retained the native "sidedness" of their inner membrane, we concluded that a possible asymmetric arrangement of cytochrome oxidase across the mitochondrial inner membrane could be better studied in bovine heart mitochondria, which can be isolated in a sufficiently intact state to permit experiments on membrane asymmetry. These mitochondria can also be inverted by sonic irradiation (Lee and Ernster, 1966). Cytochrome oxidase from bovine heart has a subunit composition similar to that of the yeast enzyme except that subunits II and III co-migrate on the commonly used

SDS-polyacrylamide gels (Downer et al., 1976).[1] Exposed subunits on the outer side of the mitochondrial inner membrane were identified by labeling intact mitochondria with [^{35}S]-p-diazonium benzene sulfonate. Exposed subunits on the inner side of the inner membrane were identified by labeling sonically-prepared submitochondrial particles. Since sonic irradiation led to a rearrangement of cytochrome oxidase in a large fraction of the resulting submitochondrial particles, an immunochemical procedure was developed for isolating particles with a low content of displaced cytochrome oxidase (Eytan et al., 1975).

With intact mitochondria, subunits II, III, VI, and VII were labeled,[2] whereas in purified submitochondrial particles most of the label was in subunit IV. We concluded, therefore, that the arrangement of cytochrome oxidase in the mitochondrial inner membrane was transmembraneous and asymmetric; subunits II (or III, or both), VI, and VII are situated on the outer side, subunit IV is situated on the inner side, and subunits I and V are buried in the interior of the membrane. This arrangement is illustrated in Figure 3.2.

Obviously, the model depicted in Figure 3.2 is only a very crude approximation. We are currently attempting to refine the model by crosslinking adjacent subunits with bifunctional reagents (cf. Wang and Richards, 1974) and by analyzing two-dimensional cytochrome oxidase crystals (Vanderkooi, 1974) with the aid of optically filtered electron micrographs as well as electron diffraction.

Which Cytochrome Oxidase Polypeptide Binds Cytochrome c?

Cytochrome c, the substrate of cytochrome oxidase, is situated on the outer side of the mitochondrial inner membrane (Schneider et al., 1972). In order to identify cytochrome c-binding subunits of yeast cytochrome oxidase, the purified enzyme was reacted with yeast iso-1-cytochrome c whose single free sulfhydryl residue at position 107 had been activated with 5,5'-dithiobis(2-nitrobenzoate) (Fig. 3.3a). The resulting cytochrome c derivative appeared to function as an "affinity label" since it rapidly inactivated cytochrome oxidase (Fig. 3.3b). Inactivation could be competitively prevented by underivatized cytochrome c and could be re-

---

[1]Subunits II and III can, however, be separated on SDS-polyacrylamide gels containing either 8 M urea (Downer et al., 1976) or very high ($\sim$ 20 percent) concentrations of acrylamide (Cabral and Schatz, 1976). In order to differentiate between subunits II and III of the bovine heart enzyme, the surface-labeling experiments with bovine heart cytochrome oxidase will have to be repeated with these more highly resolving gel systems.

[2]The numbering of the subunits of bovine heart cytochrome oxidase is according to apparent molecular weight. It does not imply that a given subunit of bovine heart cytochrome oxidase is homologous to the yeast-enzyme subunit designated by the same number.

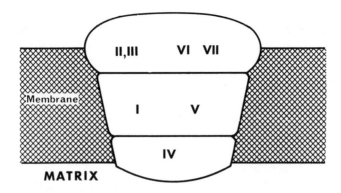

FIGURE 3.2. Schematic illustration of the asymmetric arrangement of cytochrome oxidase subunits in the inner membrane of bovine heart mitochondria. The size and shape of the various subunits (or groups of subunits) is displayed arbitrarily. See note 2 for the numbering of the subunits. (Adapted from Eytan et al., 1975.)

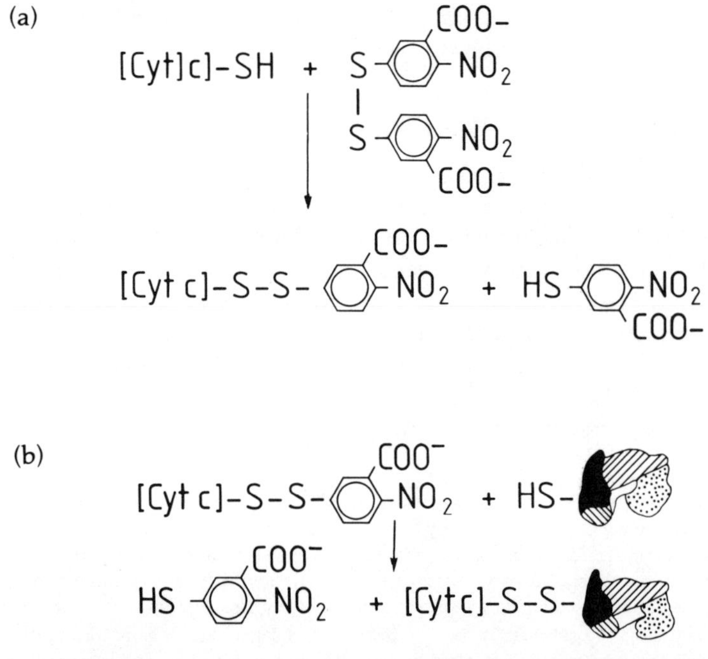

(a)

$$[Cyt\ c]-SH \quad + \quad \underset{COO-}{\overset{COO-}{S-\bigcirc-NO_2}}$$
$$\downarrow$$
$$[Cyt\ c]-S-S-\bigcirc-NO_2 \quad + \quad HS-\bigcirc-NO_2$$

(b)

$$[Cyt\ c]-S-S-\bigcirc-NO_2 \quad + \quad HS-\blacksquare$$
$$\downarrow$$
$$HS-\bigcirc-NO_2 \quad + \quad [Cyt\ c]-S-S-\blacksquare$$

FIGURE 3.3. "Affinity-labeling" of yeast cytochrome oxidase with activated iso-1-cytochrome c: (a) activation of iso-1-cytochrome c; (b) cross-linking of activated iso-1-cytochrome c to cytochrome oxidase (Birchmeier et al., 1976).

versed by excess sulfhydryl compounds. When the affinity labeled oxidase was analyzed by dodecyl sulfate polyacrylamide gel electrophoresis in the absence of sulfhydryl compounds, a new band of apparent molecular weight 38,000 was observed (Fig. 3.4). Upon a second electrophoresis in the presence of excess sulfhydryl compounds, this band was split into approximately equimolar amounts of subunit III and iso-1-cytochrome c (not shown). Similar results were obtained when underivatized iso-1-cytochrome c was cross-linked to the oxidase by oxidative disulfide bridge formation in the presence of o-phenanthroline and Cu^{2+}. Control experiments indicated that this cross-link was not simply an artefact caused by the presence of an unusually reactive sulfhydryl group on subunit III (Birchmeier et al., 1976).

Since cytochrome c is a fairly large molecule that may cover more than one exposed cytochrome oxidase subunit, these results do not unequivocally identify subunit III as the substrate-binding site (for additional uncertainties, see also Phan and Mahler, 1976; Ferguson-Miller et al., 1976; Vanderkooi et al., 1973). They do suggest, however, that subunit III is, at the very least, situated close to the actual cytochrome c binding site and, by implication, on the same side of the mitochondrial inner membrane as cytochrome c itself. The cross-linking studies thus provide independent support for the surface-labeling studies mentioned in the preceding section.

Biosynthesis of Cytochrome Oxidase

It is now well documented that cytochrome oxidase is made by the cooperation of two genetic systems. The three large, hydrophobic polypeptides are synthesized on mitochondrial ribosomes and the four small polypeptides are synthesized on cytoplasmic ribosomes (for review see Tzagoloff et al., 1973; Schatz and Mason, 1974). Accumulation of subunits I and II within the mitochondria is controlled by oxygen (Mason and Schatz, 1973).

In order to study the assembly of cytochrome oxidase in vivo, Ebner et al. (1973a, b) isolated numerous nuclear yeast mutants that were lacking cytochrome oxidase. More recently, similar mutants have been isolated by others (Ono, 1974; Tzagoloff et al., 1975; Solioz, 1976). When the mutants isolated by Ebner et al. were analyzed immunochemically for residual cytochrome oxidase subunits by immunoprecipitation of radioactively labeled mitochondrial extracts, all of the mutants were found to lack at least one of the mitochondrially made subunits (Ebner et al., 1973b). Within the resolution limits of the electrophoretic methods then available, all mutants appeared to contain a normal set of cytoplasmically made cytochrome oxidase subunits. These results indicated clearly that the mutation of a nuclear gene could affect the synthesis or the accumulation of a mitochondrially made cytochrome oxidase sub-

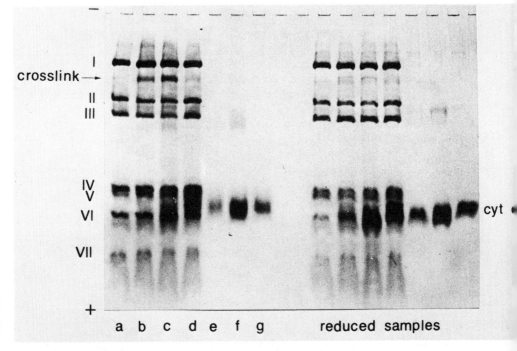

FIGURE 3.4. Cross-linking of cytochrome oxidase to thionitrobenzoate-cytochrome c: (a) cytochrome oxidase alone, (b) cytochrome oxidase plus 1.0 equivalent of modified yeast cytochrome c, (c) cytochrome oxidase plus 2.5 equivalents modified yeast cytochrome c, (d) cytochrome oxidase plus 1.0 equivalent of (thiol-free) horse-heart cytochrome c that had been treated with 5,5'-dithiobis (2-nitrobenzoate) in the same way as the yeast cytochrome c. Samples e-g correspond to b-d except that cytochrome oxidase was omitted. The unlettered samples on the right correspond to a-g except that they were treated with 0.3 M 2-mercaptoethanol before electrophoresis.

unit. Since one of these mutants (pet 494-1) could be suppressed by a nuclearly coded amber suppressor known to affect translation on cytoplasmic ribosomes, Ono et al. (1975) concluded that the nuclear mutation had affected a cytoplasmically made protein that governed the accumulation of the mitochondrially made subunit III.

In order to exploit fully the potential of yeast mutants lacking cytochrome oxidase, it seemed necessary to develop improved methods for detecting subtle changes in the polypeptide composition of mitochondrial inner membranes without the need for cumbersome immunoprecipitation methods. The following high-resolution methods are currently employed in our laboratory:

1. Concentrated polyacrylamide step gels (10-20 percent acrylamide) offering excellent resolution of small polypeptides (Cabral and Schatz, 1976; see Fig. 3.1).

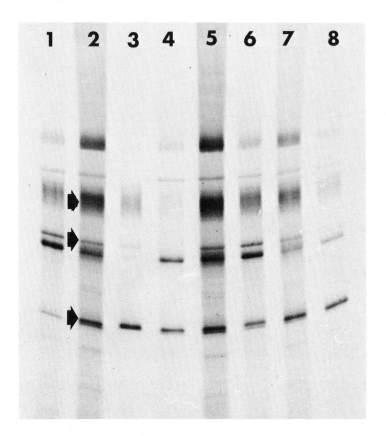

FIGURE 3.5. High-resolution analysis of mitochondrial translation products in nuclear cytochrome-oxidase-less yeast mutants. The samples were dissociated in dodecyl sulfate and electrophoresed in 10-15 percent polyacrylamide gradient gels (Douglas and Butow, 1976): (1) Pet 494-1 mitochondria labeled in vivo with $^{35}SO_4$ in the presence of cycloheximide; (2 and 5) wild-type mitochondria treated as in 1; (3) immunoprecipitate obtained by treating a mitochondrial lysate from wild-type cells grown in $^{35}SO_4$ with (unlabeled) antiserum to holo-cytochrome oxidase; (4) pet E-11 mitochondria treated as in 1; (6) mixture of 1 and 3; (7) mixture of 2 and 3; (8) mixture of 4 and 3. The arrows mark the positions of the three mitochondrially made cytochrome oxidase subunits. Note that, in wild-type cells, subunit III is only barely separated from a slightly faster moving mitochondrial translation product that is still present in mutant pet 494-1. These two closely adjacent bands can, however, be distinguished by the mixing experiments shown in slots 6-8.

2. Exponential acrylamide gradient gels (10-15 percent acrylamide) affording high resolution of larger polypeptides (Douglas and Butow, 1976). The application of these gels to mitochondria from yeast cells labeled in the presence of cycloheximide, coupled with radioautography, resolves at least one dozen distinct mitochondrial translation products

TABLE 3.1. Cytochrome oxidase subunits in yeast mutants lacking cytochrome oxidase.

	Mutant		
Subunit	Pet 494-1	Pet E11-1	ϱ^-
I	+	−	−
II	+	±	−
III	−	+	−
IV	±	±	+
V	+	+	+
VI	+	+	+
VII	+	+	+

Note: See text and Figure 3.5 for further details.

without the need for immunoprecipitation methods (Fig. 3.5, slots 2 and 5).

3. Two-dimensional separation methods combining isoelectrofocusing in the presence of urea and nonionic detergent with dodecyl sulfate-acrylamide gel electrophoresis (O'Farrell, 1975; Ames and Nikaido, 1976; Cabral and Schatz, 1976).

Analysis of cytochrome-oxidase-less mutants by the various methods has so far yielded the following results (see Fig. 3.5):

1. The pet 494-1 mutation leads to a complete *loss* of subunit III. (Earlier results had only shown that this subunit was no longer *integrated* with the other subunits, cf. Ebner et al., 1973b.) While all four of the cytoplasmically made subunits are still present, the level of subunit IV is drastically lowered (Table 3.1).

2. Mutant pet E11-1 has lost subunits I and II but has retained subunit III. (Subunit III is obviously not bound to the remaining cytoplasmically made subunits since it can no longer be immunocoprecipitated with these latter subunits; see Ebner et al., 1973b.)

3. A cytoplasmic petite mutant lacking mitochondrial protein synthesis has lost subunits I, II, and III but has retained all four cytoplasmically made subunits. (Earlier immunoprecipitation experiments had failed to detect the presence of subunit VII, probably because this subunit is not firmly bound to the other remaining subunits IV-VI).

We are now screening a large number of mutants and their revertants by two-dimensional analysis of radioactively labeled mitochondria (see below). In this manner we hope to identify the structural genes for the various cytochrome oxidase subunits. Once a structural mutation has been identified, nearest-neighbor relationships between the various sub-

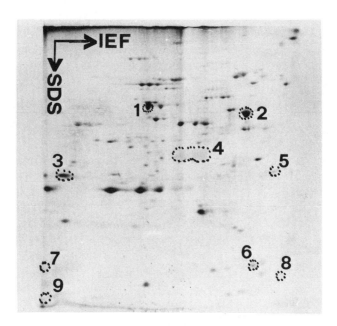

FIGURE 3.6. Two-dimensional analysis of yeast submitochondrial particles. Wild-type yeast cells (D273-10B) were grown in $^{35}SO_4$ and converted to spheroplasts. The spheroplasts were lysed in the presence of the protease inhibitor phenylmethyl sulfonyl fluoride, and the mitochondria were isolated and purified on a Metrizamide density gradient. They were then converted to submitochondrial particles by sonication in the presence of protease inhibitor and the submitochondrial particles were isolated by centrifugation. The particles were dissociated as described in the text. (1) a subunit of F_1 ; (2) β subunit of F_1 : (3) γ subunit of F_1; (4) cytochrome oxidase subunit I; (5) cytochrome oxidase subunit II; (6) cytochrome oxidase subunit IV; (7) cytochrome oxidase subunit V; (8) cytochrome oxidase subunit VI; (9) cytochrome oxidase subunit VII. Because the intensity of the various polypeptide spots is so different, it is usually not possible to display all of them adequately on a single radioautogram. As the radioautogram shown here was designed to visualize the major spots, some of the minor ones (for example, 8 and 9) may not be clearly visible. For unknown reasons, cytochrome oxidase subunit I appears as a broad but characteristic "cloud" in this two-dimensional system. Subunit III cannot be clearly displayed in this system.

units can be uncovered by second-site reversion experiments analogous to those applied in phage systems (Jarvick and Botstein, 1975).

Toward a Two-Dimensional Polypeptide Map of the Mitochondrial Inner Membrane

We are convinced that progress in elucidating mitochondrial membrane assembly is now critically dependent on the ability to screen the polypep-

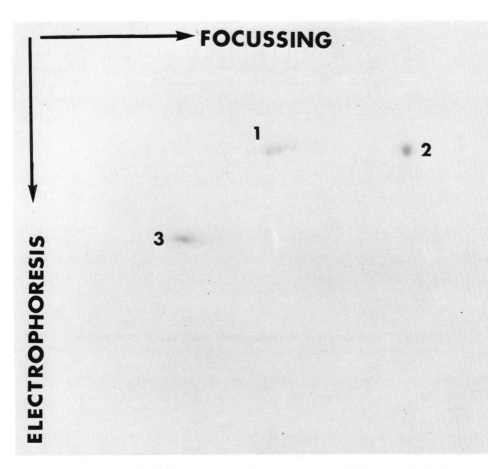

FOCUSSING

ELECTROPHORESIS

1

2

3

FIGURE 3.7. Two-dimensional analysis of yeast F_1 containing only the three large (a, β, γ) subunits. The purified enzyme was iodinated with $^{125}I_2$ in the presence of chloramine T, mixed with unlabeled yeast mitochondria, and subjected to two-dimensional analysis as described in the text and in the legend of Figure 3.6. The gel slab was stained for protein (not shown) and then subjected to radioautography. This figure depicts the radioautogram: (1) a subunit (some charge heterogeneity is evident); (2) β subunit; (3) γ subunit. Superimposition of the radioautogram onto the stained gel slab identified the three major F_1 subunits (cf. Fig. 3.6).

tide patterns of a large number of samples and to detect missense mutations in individual membrane proteins without having to isolate these proteins. Because the mitochondrial inner membrane contains so many different polypeptides, this requirement is best met by two-dimensional separation techniques that separate polypeptides by charge as well as by size. The method employed in our laboratory is as follows:

1. Mitochondria or submitochondrial particles are first dissociated in

dodecyl sulfate and subjected to isoelectric focusing in the presence of urea and Triton X-100 in cylindrical polyacrylamide gels.

2. The isoelectric focusing gels are then placed on top of gel slabs containing an exponential (10-15 percent) gradient of polyacrylamide and electrophoresed in the presence of dodecyl sulfate.

3. The gel slabs are stained with Coomassie Blue; if radioactive material is being used, the gels are subsequently dried and subjected to radioautography.

4. In order to identify polypeptide spots, unlabeled mitochondria are mixed with uniformly labeled cytochrome oxidase, cytochrome oxidase subunits, or F_1-subunits and carried through steps (1) through (3). By superimposing the radioautograms onto the stained gel slabs, individual polypeptides can be "placed on the map."

Figures 3.6-3.8 illustrate this approach. It can be seen that nine important membrane polypeptides have already been identified. Since so many

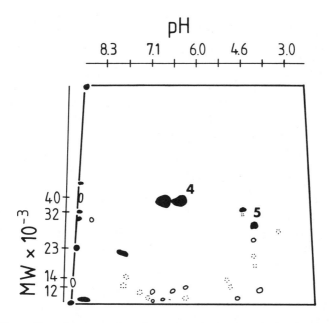

FIGURE 3.8. Two-dimensional analysis of total mitochondrial translation products. Wild-type yeast cells were labeled with $^{35}SO_4$ in the presence of cycloheximide, the mitochondria were processed as in Figure 3.6, and the gel slab was subjected to radioautography. Note the large number of species resolved by this system. Since these unusually hydrophobic polypeptides cannot be as satisfactorily separated by isoelectric focusing as the bulk of the mitochondrial membrane proteins can, the radioactive spots showed extensive horizontal "streaking." For this reason, this figure depicts a tracing of the spots: (4) cytochrome oxidase subunit I; (5) cytochrome oxidase subunit II.

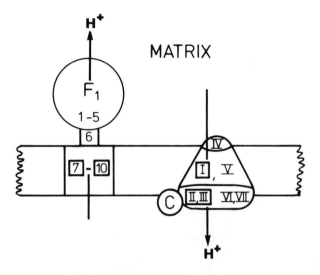

FIGURE 3.9. Schematic view of the molecular architecture of cytochrome oxidase and the ATPase complex in the mitochondrial inner membrane. Some aspects of this drawing are still speculative and no attempt is made to draw the components to size. (F_1) The F_1-ATPase (coupling factor 1) contains five dissimilar subunits $(a\text{-}\varepsilon)$, most of these in two to three copies. F_1 may also contain a small inhibitor protein, which is not shown here. The stalk section (6) may be identical with the "oligomycin-sensitivity conferring protein" (MacLennan and Tzagoloff, 1968). The "membrane-sector" $(7\text{-}10)$ consists of two to four hydrophobic polypeptides synthesized by mitochondria. One of these polypeptides appears to be the binding site for oligomycin and dicyclohexylcarbodiimide (Tzagoloff et al., 1973). The numbering of cytochrome oxidase subunits is explained in the text and Figure 3.1. The arrows depict the proton flow catalyzed by each of the units during oxidative phosphorylation.

functionally defined mitochondrial membrane polypeptides can be purified, we should soon have a usable polypeptide map of the mitochondrial inner membrane.

The Function of Mitochondrially Made Polypeptides

It is now known that mitochondria synthesize three subunits of cytochrome oxidase, two to four subunits of the ATPase complex, and one to two subunits of the cytochrome bc_1 complex (Schatz and Mason, 1974; Weiss et al., 1975b). The role of these extremely hydrophobic polypeptides is still an enigma. One possible clue may be the observation that these polypeptides ensure the tight binding of their cytoplasmically made partner proteins to the mitochondrial inner membrane (Schatz, 1968; Ebner et al., 1973 b). To a certain degree, the mitochondrially made polypeptides may thus have a "structural" fraction. It may also be significant that all three of the functional units known to contain mitochondrially made subunits are involved in the generation or the utilization of a proton gradient and that at least two of these units (and probably all three of them) span the mitochondrial inner membrane (Fig. 3.9). It is tempting to speculate that these hydrophobic polypeptides function as proton pores. Since an excess of such pores would uncouple mitochondria, this may explain why mitochondria stop synthesizing polypeptides if the supply of cytoplasmically made "partner proteins" becomes limiting (Schatz and Mason, 1974).

ACKNOWLEDGMENTS
These studies were supported by grant GM 16320 from the U.S. Public Health Service, grant GB-40541 X from the U.S. National Science Foundation, and grant 3.2350.74 from the Swiss National Science Foundation. We wish to thank Urs Müller and Henri Homberger for excellent technical assistance.

REFERENCES
Ames, G. F.-L., and K. Nikaido. 1976. Two-dimensional gel electrophoresis of membrane proteins. *Biochemistry* 15:616-623.
Birchmeier, W., C. E. Kohler, and G. Schatz. 1976. Interaction of integral and peripheral membrane proteins: affinity labeling of yeast cytochrome oxidase by modified yeast cytochrome *c. Proc. Natl. Acad. Sci. USA* 73:4334-4338.
Cabral, F. 1976. Unpublished.
Cabral, F., and G. Schatz. 1976. *Methods Enzymol.*, in press.
Carraway, K. L. 1975. Covalent labeling of membranes. *Biochim. Biophys. Acta* 415:379-410.
Douglas, M. G., and R. A. Butow. 1976. Variant forms of mitochondrial translation products in yeast: evidence for location of determinants on mitochondrial DNA. *Proc. Natl. Acad. Sci. USA* 73: 1083-1086.
Downer, N. W., N. C. Robinson, and R. A. Capaldi. 1976. Characterization of a seventh different subunit of beef heart cytochrome *c* oxidase: similarities between the beef heart enzyme and that from other species. *Biochemistry* 15:2930-2936
Ebner, E., L. Mennucci, and G. Schatz. 1973a. Mitochondrial assembly in respiration-deficient mutants of *Saccharomyces cerevisiae*. I. Mitochondrial protein synthesis in nuclear petite mutants. *J. Biol. Chem.* 248:5360-5368.
Ebner, E., T. L. Mason, and G. Schatz. 1973b. Mitochondrial assembly in respiration-deficient mutants of *Saccharomyces cerevisiae*. II. Effect of nuclear and extrachromosomal mutations on the formation of cytochrome oxidase. *J. Biol. Chem.* 248:5369-5378.
Eytan, G. D., R. C. Carroll, G. Schatz, and E. Racker. 1975. Arrangement of the subunits in solubilized and membrane-bound cytochrome *c* oxidase from bovine heart. *J. Biol. Chem.* 250:8598-8603.
Eytan, G. D., and G. Schatz. 1975. Cytochrome *c* oxidase from baker's yeast. V. Arrangement of the subunits in the isolated and membrane-bound enzyme. *J. Biol. Chem.* 250:767-774.
Ferguson-Miller, S., D. L. Brautigan, and E. Margoliash. 1976. Correlation of the

kinetics of electron transfer activity of various eukaryotic cytochromes c with binding to mitochondrial cytochrome c oxidase. *J. Biol. Chem.* 251:1104-1115.

Jarvick, J., and D. Botstein. 1975. Conditional-lethal mutations that suppress genetic defects in morphogenesis by altering structural proteins. *Proc. Natl. Acad. Sci. USA* 72:2738-2742.

Lee, C.-P., and L. Ernster. 1966. The energy-linked nicotinamide nucleotide transhydrogenase reaction: its characteristics and its use as a tool for the study of oxidative phosphorylation. In J. M. Tager, S. Papa, E. Quagliariello, and E. C. Slater, eds., *Regulation of metabolic process in mitochondria*, pp. 218-234. Elsevier, Amsterdam.

MacLennan, D. H., and A. Tzagoloff. 1968. Studies on the mitochondrial adenosine triphosphatase system. IV. Purification and characterization of the oligomycin sensitivity conferring protein. *Biochemistry* 7:1603-1610.

Mason, T. L., and G. Schatz. 1973. Cytochrome c oxidase of baker's yeast. II. Site of translation of the protein components. *J. Biol. Chem.* 248:1355-1360.

O'Farrell, P. H. 1975. High-resolution two-dimensional electrophoresis of proteins. *J. Biol. Chem.* 250:4007-4021.

Ono, B. 1974. Unpublished.

Ono, B., G. Fink, and G. Schatz. 1975. Mitochondrial assembly in respiration-deficient mutants of *Saccharomyces cerevisiae*. IV. Effects of nuclear amber suppressors on the accumulation of a mitochondrially made subunit of cytochrome c oxidase. *J. Biol. Chem.* 250:775-782.

Phan, S. H., and H. R. Mahler. 1976. Studies on cytochrome oxidase: preliminary characterization of an enzyme containing only four subunits. *J. Biol. Chem.* 251:270-276.

Poyton, R. O., and G. Schatz. 1975a. Cytochrome c oxidase of baker's yeast. III. Physical characterization of the isolated subunits and chemical evidence for two different classes of polypeptides. *J. Biol. Chem.* 250:572-761.

Poyton, R. O., and G. Schatz. 1975b. Cytochrome c oxidase of baker's yeast. IV. Immunological evidence for the participation of a mitochondrially synthesized subunit in enzymic activity. *J. Biol. Chem.* 250:762-766.

Racker, E., ed. 1970. *Membranes of mitochondria and chloroplasts*. Van Nostrand Reinhold, New York.

Schatz, G. 1968. Impaired binding of mitochondrial adenosine triphosphatase in the cytoplasmic "petite" mutant of *Saccharomyces cerevisiae*. *J. Biol. Chem.* 243:2192-2199.

Schatz, G. 1970. The biogenesis of mitochondria. In E. Racker, ed., *Membranes of mitochondria and chloroplasts*, pp. 251-314. Van Nostrand Reinhold, New York.

Schatz, G., and T. L. Mason. 1974. The biosynthesis of mitochondrial proteins. *Annu. Rev. Biochem.* 43:51-87.

Schneider, D. L., Y. Kagawa, and E. Racker. 1972. Chemical modification of the inner mitochondrial membrane. *J. Biol. Chem.* 247: 4074-4079.

Sherman, F. 1965. The genetic control of the cytochrome system yeast. In *Mécanismes de régulation des activités cellulaires chez les microorganismes*, pp. 465-479. Edition du CNRS, Paris.

Singer, S. J. 1974. The molecular organization of membranes. *Annu. Rev. Biochem.* 43:805-834.

Solioz, M. 1976. Unpublished.

Tzagoloff, A., A. Akai, and R. B. Needleman. 1975. Assembly of the mitochondrial membrane system: isolation of nuclear and cytoplasmic mutants of *Saccharomyces cerevisiae* with specific defects in mitochondrial function. *J. Bacteriol.* 122:826-831.

Tzagoloff, A., M. S. Rubin, and M. Sierra. 1973. Biosynthesis of mitochondrial enzymes. *Biochim. Biophys. Acta* 301:71-104.

Vanderkooi, G. 1974. Organization of proteins in membranes with special reference to the cytochrome oxidase system. *Biochim. Biophys. Acta* 344:307-344.

Vanderkooi, J., M. Erecińska, and B. Chance. 1973. Cytochrome *c* interaction with membranes. II. Comparative study of the interaction of *c* cytochromes with the mitochondrial membrane. *Arch. Biochem. Biophys.* 157:531-540.

Wang, K., and F. M. Richards. 1974. An approach to nearest neighbor analysis of membrane proteins: application to the human erythrocyte membrane of a method employing cleavable cross-linkages. *J. Biol. Chem.* 249:8005-8018.

Weiss, H., A. J. Schwab, and S. Werner. 1975a. Biogenesis of cytochrome oxidase and cytochrome *b* in *Neurospora crassa*. In A. Tzagoloff, ed., *Membrane biogenesis*, pp. 125-152. Plenum, New York.

Weiss, H., B. Ziganke, and H. J. Kolb. 1975b. Protein composition and site of translation of mitochondrial cytochrome *b*. In E. Quagliariello, S. Papa, F. Palmieri, E. C. Slater, and N. Siliprandi, eds., *Electron transfer chains and oxidative phosphorylation*, pp. 15-22. North Holland, Amsterdam.

4   Transmembrane Compositional Asymmetry of Lipids in Bilayers and Biomembranes

THOMAS E. THOMPSON

The basic hypothesis underlying much of today's research on biological membranes was formulated by Singer and Nicolson in 1972. This construct, known as the fluid mosaic hypothesis, contains two essential elements. The first of these, a derivative of the model of Danielli and Davson (1935), requires that the lipid component of the membrane be a bilayer in structure and contribute to the membrane its basic barrier properties. The simplest expression of this set of properties is the impermeability of the membrane to small, water-soluble molecules. Although the lipid molecules are confined to the bilayer, they are free to exhibit a variety of motional modes such as vibration, rotation, and translation. The second element of the fluid mosaic hypothesis deals with the disposition of the protein components of the membrane. These components are immersed to varying degrees in the lipid bilayer. Some may be only superficially associated with the polar faces of the bilayer, others are embedded in its hydrophobic core, while still others may completely span the bilayer. The fluid nature of the bilayer permits the protein components to move in the bilayer in both rotational and translational modes. These diffusional motions of the individual protein components may give rise to time-dependent patterns in the compositional mosaic.

Although the basic physiological properties of biological membranes must derive primarily from protein components, there is much evidence to suggest that these functions may be markedly influenced by lipid composition. In general, the relations between physiological functions and lipid composition have been formulated in terms of bilayer fluidity and lipid-protein (as well as protein-protein) interactions. These interactions are in turn strongly modulated by the spatial distribution of the component lipids, both within the plane of the bilayer and between the two opposing monolayers that comprise it.

Much of the support for the existence of the lipid bilayer as an essential structural element in biological membranes is based on the study of model systems in the form of liposome dispersions (Bangham et al., 1965) and the Mueller-Rudin bilayer (Mueller et al., 1962). Although these simple systems have provided and will continue to provide much important information directly relevant to the problems posed by biological membranes, it has recently become clear that these model systems and the bilayers of biological membranes differ in one important respect. Whereas the usual model bilayer systems are compositionally symmetric in a transmembrane sense, the biological bilayer may not be. (Recently, two general methods for the formation of compositionally assymetric bilayers of the Mueller-Rudin type have been described by Montal and Mueller, 1972; Montal, 1974; and Michaels and Dennis, 1973.)

The transbilayer lipid compositional asymmetry recently reported for a variety of biological membranes, and discussed below, raises three interesting and important questions:

1. What are the special properties of asymmetric bilayers that derive directly from the compositional asymmetry?
2. How is this asymmetry of lipid composition across the bilayer generated?
3. How is the asymmetry, once generated, maintained?

At the present time these questions have no ready answers. They will, however, serve as a backdrop for the following discussion of the information that is available about compositionally asymmetric bilayers. Let us begin by outlining the information available for biological membranes.

Lipid Asymmetry in Biological Membranes

The principal lipids of biological membranes are phospholipids, cholesterol, and glycosphingolipids. There is considerable evidence that the glycosphingolipids, which are a minor component usually confined to the mammalian plasma membrane, are asymmetrically localized on the external surface of this membrane (Wallach, 1975). Recently evidence has been presented in several membrane systems that the phospholipid components may also be distributed between the opposing membrane faces in a highly asymmetric fashion. The situation with regard to cholesterol is less clear. This is in part due to the difficulty in identifying the cholesterol molecule, which is relatively inert chemically, and in part may be attributed to the relative ease with which cholesterol moves between various membrane systems (Cooper et al., 1975; Inbar and Shinitzky, 1974; Werb and Cohen, 1971), presumably because of its micellar form in aqueous systems (Haberland and Reynolds, 1973).

Phospholipids

Bretscher (1972b) first drew attention to transmembrane phospholipid asymmetry in his studies of the erythrocyte membrane. His initial observations, based in part on the chemical reactivity of the primary amine lipids, suggested that the choline lipids, sphingomyelin and phosphatidylcholine, were largely confined to the outer surface of this membrane while phosphatidylethanolamine and serine were located on the cytoplasmic surface. Although this observation generated considerable controversy at the time it was proposed, it has been thoroughly substantiated by subsequent work (Bretscher, 1972a; Gordesky et al., 1973; Verkleij et al., 1973; Renooij et al., 1974; Bloj and Zilversmit, 1976). This transmembrane phospholipid compositional asymmetry in the human erythrocyte membrane is clearly illustrated in Figure 4.1, taken from Verkleij et al. (1973). These results, which are based on the susceptibility of the various phospholipid components to hydrolysis by several phospholipases, show that although the total phospholipid is about equally distributed between the two sides of the membrane, there is a marked asymmetry in composition. A generally similar result for rat erythrocyte membranes has been recently reported by Renooij et al. (1976), with the principal difference, compared to the human RBC, being that about twice as much phosphatidylcholine is found on the cytoplasmic face.

Transmembrane compositional asymmetries of phospholipid components for other membrane systems have been reported quite recently. Rothman and co-workers (1976) have shown a marked asymmetry in the phospholipid composition in the membrane of influenza A virus grown in Maden-Darby bovine kidney cells. These data are summarized in Figure 4.2. In this membrane system the compositional asymmetry is quite different from that displayed by the human erythrocyte. The viral membrane shows a high content of sphingomyelin, together with phosphatidylethanolamine and serine on the cytoplasmic face, while phosphatidylcholine is about equally distributed between the two surfaces. It is also striking that the total phospholipid is quite asymmetrically distributed, with about twice as much on the cytoplasmic face as on the outside surface. This asymmetry is in part accounted for by the presence of glycosphingolipid on the outside face, not shown in Figure 4.2.

The phospholipid distribution in the membrane of the bovine rod outer segment disc has been found by Smith and co-workers (1977) to be asymmetrical. Using the chemical reactivity of the amino group of phosphatidylethanolamine and serine, these workers have found that 85 percent of these two lipids, which constitute 52 percent of the total phospho-

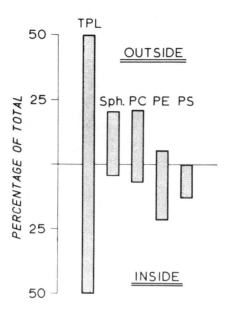

FIGURE 4.1. The distribution of phospholipids between inside and outside surfaces of the human erythrocyte membrane (Verkleij et al., 1973): (*TPL*) total phospholipid, (*Sph*) sphingomyelin, (*PC*) phosphatidylcholine, (*PE*) phosphatidylethanolamine, (*PS*) phosphatidylserine.

lipid, are localized on the outer surface of the disc (the cytoplasmic face). Assuming the total phospholipid concentration to be equal on both faces of the membrane, the remaining phospholipid, phosphatidylcholine, must be predominantly localized on the cisternal surface of the disc.

Rothman and Kennedy (1977) have recently reported that phosphatidylethanolamine, which comprises 69 percent of the plasma membrane of *Bacillus megaterium*, is distributed so that 33 percent is on the external surface and 68 percent on the cytoplasmic face of this membrane. Since this lipid together with phosphatidylglycerol comprises 99 percent of the total phospholipid of the membrane, it is apparent that phosphatidylglycerol is largely confined to the external face, if the total distribution of phospholipids is equally apportioned between the two membrane surfaces.

To date, information on lipid asymmetry is confined to the five systems discussed above. These five membranes all exhibit a marked asymmetry. In addition, it is important to note that there are no reports of any membrane that does not show an asymmetrical distribution of lipids.

Although the rat and human erythrocyte membranes have a qualitatively similar distribution of phospholipids, the dissimilarity of distribution exhibited among the four types of membranes is striking. There is no clearly discernible pattern to be seen. This is particularly interesting

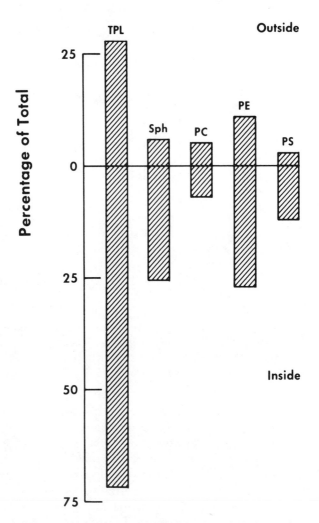

FIGURE 4.2. The distribution of phospholipids between inside and outside surfaces of influenza A virus membrane. The abbreviations are the same as in Figure 4.1. (From the data of Rothman et al., 1976.)

since, although the membrane sources span a wide range of cell types, all are either plasma membranes or derivatives thereof. Thus, the viral membrane is initially a special section of the host cell plasma membrane that envelops the virion during the budding process. In the instances that have been examined, the total lipid composition of the viral membrane is quite similar, but not identical, to the composition of the host cell plasma membrane (Lenard and Compans, 1974). There is considerable evidence to suggest that the discs of the mammalian rod outer segment arise as invaginations of the rod plasma membrane, which subsequently pinch off to form the isolated discs (Young, 1967). It is tempting to generalize that a marked asymmetry of lipid composition may be expected to be the rule in plasma membranes, and perhaps in other membranes as well. It is clear, however, that the five systems discussed above do not provide a sound basis for this speculation without additional data on a variety of membranes.

Cholesterol

The localization of cholesterol in the erythrocyte membrane has been widely studied by self-exchange of labeled cholesterol and by net depletion of cholesterol. The self-exchange method is used to determine the size of the readily exchangeable pool of membrane cholesterol. The fact that this pool is smaller in size than the total membrane cholesterol content is interpreted to mean that the exchangeable pool is located on the external surface of the membrane while the nonexchangeable pool is on the cytoplasmic surface. Self-exchange studies of this type carried out in erythrocyte membranes have given a wide variety of results ranging from an exchangeable pool size of 35 to 100 percent of the total membrane cholesterol (Basford et al., 1964; Bruckdorfer and Green, 1967; Bell and Schwartz, 1971; Bjornson et al., 1975). The reasons for this lack of agreement are unknown.

The net depletion technique assesses the fraction of the total erythrocyte membrane cholesterol that can be removed during incubation of the membranes with a suitable acceptor such as phospholipid vesicles. By inference, the removable cholesterol is taken to be located on the external surface of the membrane and hence accessible to the acceptor system. Results obtained by this method range from 35 to 75 percent removable (Mason, Barenholz, and Thompson, unpublished observation; Gottlieb, 1976). Although the basis for these discrepant results is unknown, there is some evidence that the apparent size of the removable pool may be a function of the rate of depletion (H. Ginsburg, unpublished observation). In any event it is clear that, unlike the self-exchange experiment, net depletion may be expected to cause alterations in structure, which may in turn affect the apparent pool size.

The self-exchange of cholesterol between phospholipid vesicles and influenza A virus has been used by Lenard and Rothman (1976) to estimate the size of the rapidly exchangeable pool in this viral membrane. They conclude that in this system the total cholesterol is approximately equally distributed between the two membrane surfaces. The weakness of this conclusion is the assumption connecting *exchangeability* with *localization* of the cholesterol. Clearly independent verification of this assumption is required.

Maintenance of Compositional Asymmetry

The large compositional asymmetries exhibited by the five membrane systems discussed above give rise to very large transbilayer concentration gradients of the various lipid components. In general, it would be expected that these gradients would drive the transmembrane diffusion of components with the eventual elimination of all concentration gradients. The apparent stability of such large compositional differences in the erythrocyte and viral membranes, which are associated with systems that are essentially biosynthetically incompetent, creates an interesting and important problem.

Transbilayer Migration of Lipids in Bilayer Vesicles

The kinetics of transbilayer movement of lipid components was first examined by Kornberg and McConnell (1971), using an electron spin resonance probe in a dispersion of single-walled phospholipid vesicles (Huang, 1969). The results of their experiments suggested that the half-time for transbilayer equilibration of a spin-labeled phospholipid derivative was as short as 6.5 hours at 30°C. Although this half-time is large compared to the equilibration half-time for lipid concentration gradients in the plane of the bilayer (Lee, 1975), it is much too short to account for the long period of stability exhibited by the erythrocyte and viral membranes.

Because of the importance of the design of the Kornberg-McConnell experiment, which has been subsequently used by a number of workers employing other probes, the basic protocol will be described in detail. The experiment is based on creation of a concentration gradient of spin-labeled derivative across the bilayer, and then determination of the time rate of decay of this gradient. This is accomplished by preparation of single-walled phosphatidylcholine vesicles containing a small amount of spin-labeled derivative in which the nitroxide moiety is covalently linked to the nitrogen of the phosphatidylcholine. Initially the vesicles contain spin label in both bilayer faces. The asymmetry in concentration is then established by reduction of the nitroxide group on the outer surface of the bilayer by adding ascorbate to the external aqueous phase at 0°C. Because, under these conditions, the bilayer is impermeable to ascorbate, the spin label on the interior surface of the bilayer remains intact. The rate of transbilayer migration of spin-labeled phospholipid from interior to exterior bilayer surface can then be assessed by measuring the amplitude of the ESR signal in aliquots of the system that have been reacted with ascorbate at successive time intervals. Thus, as transbilayer migration occurs, there is an increase in the fraction of the ESR signal that is sensitive to ascorbate added to the external aqueous phase.

Recently, Roseman et al. (1975) have reported the results of experiments, similar in principle to the Kornberg-McConnell experiment, designed to establish the rate of transbilayer exchange of phosphatidylethanolamine for phosphatidylcholine and a closely related amidine derivative of phosphatidylethanolamine, N-acetimidoylphosphatidylethanolamine. This exchange rate has direct bearing on the stability of the transmembrane compositional asymmetry found in the erythrocyte.

In order to measure the net transfer of phosphatidylethanolamine in phospholipid vesicles, it is necessary to prepare vesicles having a nonequilibrium distribution of this phospholipid across the bilayer. This was accomplished by reacting a phosphatidylcholine vesicle preparation, (Litman, 1973) containing 0.1 mole fraction of phosphatidylethanolamine, with the reagent isethionyl acetimidate hydrochloride, as shown in Figure 4.3 (Whiteley and Berg, 1974). This reagent, to which the bilayer is impermeable (H. G. Smith and B. J. Litman, unpublished results), converted 90 percent of the external phosphatidylethanolamine molecules to their amidine derivative, leaving the mole ratio of unreacted phosphatidylethanolamine to phosphatidylcholine on the outside surface of the vesicle bilayer much lower than that on the inside surface. Equilibration of phosphatidylethanolamine across the bilayer was then measured as a function of time by monitoring the appearance of phosphatidylethanolamine on the outside surface, utilizing the reaction of the amino group with 2,4,6-trinitrobenzenesulfonic acid (Litman, 1973). The amidine derivative of phosphatidylethanolamine does not react with this reagent. The design of this experiment is illustrated in Figure 4.4.

Amidination of the phosphatidylethanolamine in the external surface of the vesicle bilayer, which initially was composed of an equilibrium distribution of phosphatidylcholine and phosphatidylethanolamine, generates a three-component system of some complexity. In order to understand the results fully, it is necessary to describe the chemical-species gradients present at zero time in this system. The ratios of the concentrations in the opposite faces of the bilayer of each lipid species, both before and after amidination, can be calculated as follows. If the ratio of the amount of a specific lipid in the outer bilayer face to the total amount of that lipid in the vesicle is x, then the ratio of the amount of that lipid in the outer face to the amount in the inner face is $x/(1-x)$. The ratio of the concentrations of that lipid in the two faces is then obtained by multiplying $x/(1-x)$ by R_i^2/R_o^2 where R_i and R_o are the radii of the inner and outer bilayer faces, respectively. The outer radius of the vesicle determined by autocorrelation light scattering is 112 Å (J. Goll, Y. Barenholz, F. Carlson, and B. Litman, manuscript in preparation). Assuming that the bilayer thickness is 40 Å (Thompson et al., 1974) then $R_i^2/R_o^2 =$ 0.41. Before amidination, $x = 0.71$ for phosphatidylethanolamine. Employing the value $x = 0.71$, the ratio of the concentration of phosphatidylethanolamine in the outer to that in the inner bilayer surface is 1.01. Although the ratio of the amounts of phosphatidylcholine in the two

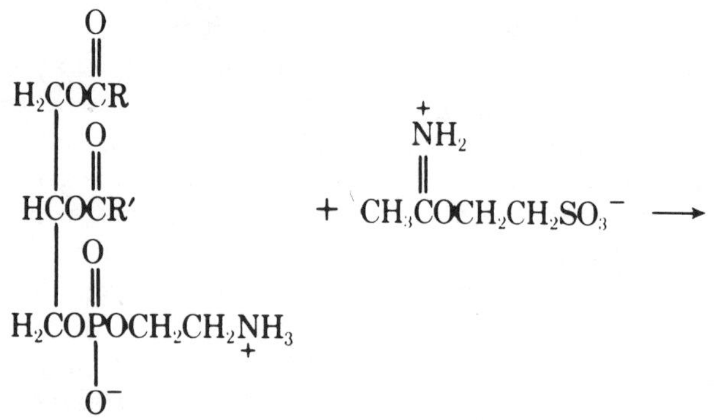

$$
\begin{array}{c}
\underset{\displaystyle H_2COCR}{\overset{\displaystyle \overset{O}{\|}}{}} \\
\underset{\displaystyle HCOCR'}{\overset{\displaystyle \overset{O}{\|}}{}} \\
\underset{\displaystyle H_2COPOCH_2CH_2NH_3}{\overset{\displaystyle \overset{O}{\|}}{}} \\
\underset{\displaystyle O^-}{}
\end{array}
\quad + \quad
\underset{\displaystyle CH_3COCH_2CH_2SO_3^-}{\overset{\displaystyle \overset{+}{NH_2}}{\overset{\|}{}}}
\quad \longrightarrow
$$

phosphatidylethanolamine   IAI

I   II

$$
\begin{array}{c}
\overset{O}{\|}\\
H_2COCR\\
\overset{O}{\|}\\
HCOCR'\\
\overset{O}{\|}\\
H_2COPOCH_2CH_2\overset{H}{N}\!\!=\!\!\underset{\overset{+}{NH_2}}{\overset{\|}{C}}CH_3\\
|\\
O^-
\end{array}
\qquad + \quad HOCH_2CH_2SO_3^-
$$

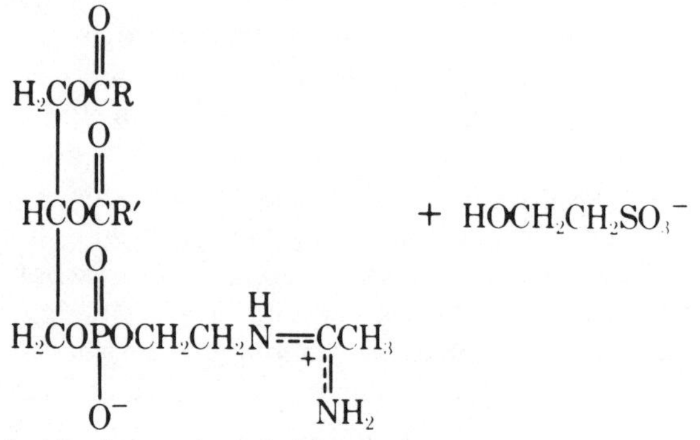

N-acetimidoylphosphatidylethanolamine

III

FIGURE 4.3.

faces of the bilayer has not been determined, if equal packing densities in both surfaces are assumed, then the ratio of the concentration of phosphatidylcholine in the outer to that in the inner face is 0.99. Thus, in the initial configuration of the system before amidination there are essentially no concentration gradients of lipid species existing across the bilayer. However, this situation is altered by amidination of 90 percent of the external phosphatidylethanolamine. Now x will be 0.2 for phosphatidylethanolamine, and hence, for this component, the ratio of concentration in the outer face to that in the inner face is 0.10. The corresponding concentration ratio for the amidine derivative is infinite, while the transbilayer concentration ratio of phosphatidylcholine remains 1.0. Assuming that the total number of lipid molecules in each bilayer face remains constant, two exchange processes could result in the appearance of external phosphatidylethanolamine. (1) The exchange of internal phos-

1. Cosonicate PC/PE
   Mole Ratio 9 : 1 ———▸ Symmetric Vesicles
2. Amidination External PE

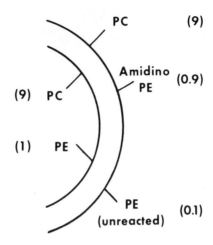

PC   (9)

Amidino   (0.9)
  PE

(9) PC

(1)   PE

PE   (0.1)
(unreacted)

$$\frac{\text{Internal } [\text{PE}]}{\text{External } [\text{PE}]} = 10.$$

FIGURE 4.4. The design of the experiment to measure transbilayer lipid migration based on the establishment of concentration asymmetries by chemical modification of phosphatidylethanolamine.

phatidylethanolamine for the amidine derivative. This process would cause the decay of the gradients of both of these species and hence would be expected to occur spontaneously. (2) The exchange of internal phosphatidylethanolamine for external phosphatidylcholine. In this case, decay of the transbilayer gradient of phosphatidylethanolamine would necessarily be coupled to the generation of a transbilayer gradient of phosphatidylcholine. Nonetheless, this process would also be spontaneous, since, in order to achieve equilibrium, the relative concentrations of all exchangeable phospholipids must reach the same value on each side of the bilayer. Since the data presented in Figure 4.5 clearly show that no new phosphatidylethanolamine appeared on the external surface of the vesicle over a period of 12 days, it can be concluded that neither of these processes occurred. These results, of course, say nothing about self-exchange or the exchange of phosphatidylcholine for the amidine deriva-

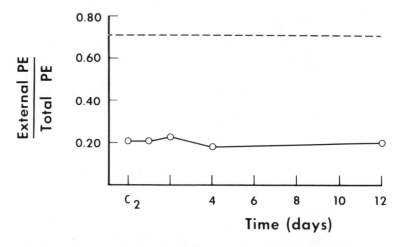

FIGURE 4.5. Observed ratio of external to total phosphatidylethanolamine as a function of time (————) and predicted ratio if equilibrium were achieved (———). (From Roseman et al., 1975; reproduced by permission of the American Chemical Society.)

tive. A conservative estimate of the precision of the measurements is .10 percent. On this basis, the estimated half-time for phosphatidylethanolamine exchange by either process in small, single-lamellar vesicles must be greater than 80 days at 22°C.

Assuming that the long half-time for phospholipid exchange observed for this particular system obtains in all bilayer systems under a variety of conditions, maintenance of the very large transmembrane compositional asymmetries in the biological membranes discussed above can be accounted for.

A different type of experiment designed to determine the self-exchange of a phospholipid component across the bilayer has been introduced by Johnson and co-workers (1975) and Rothman and Dawidowicz (1975). In this approach a compositional asymmetry of [^{14}C]phospholipid in the vesicle bilayer is generated using phospholipid exchange protein prepared from beef liver (Wirtz et al., 1972; Kamp et al., 1973). This protein, which has an absolute specificity for phosphatidylcholine, causes the exchange of this phospholipid between bilayer systems without net transfer. A similar protein prepared from beef heart catalyzes the exchange of phosphatidylcholine primarily, but also sphingomyelin to a lesser extent (Johnson and Zilversmit, 1975; Bloj and Zilversmit, 1976). By using a donor bilayer system containing [^{14}C]phospholipid, the acceptor vesicle bilayer can be enriched with this tracer molecule. If the acceptor vesicle system is now separated from the donor system by some means, such as centrifugation or chromatography, the acceptor vesicles can, after a suitable time lapse, be in turn used as the donor for a new unlabeled acceptor system. The exchange-out kinetics are expected to reflect the existence of two pools of [^{14}C]phospholipid: one rapidly exchangeable, a second more slowly exchangeable. By inference, the rapidly exchangeable pool is taken to be on the outer surface of the donor bilayer and the slowly exchangeable pool on the inner surface. The kinetics of slow exchangeability are then assumed to be the kinetics of transbilayer migration of the [^{14}C]phospholipid. Using this technique with egg phosphatidylcholine vesicles, Johnson and co-workers (1975) and Rothman and Dawidowicz (1975) were unable to detect any transmembrane migration of [^{14}C]phosphatidylcholine. The error limits on their experiments set a lower limit to the half-time for this process of 4 days and 11 days, respectively, at 37°C.

The obvious limitation with this type of experiment is the assumption that the two pools identified by their rate constants are in fact located on the opposite faces of the vesicle bilayer. This limitation can only be overcome by an independent method of establishing the transmembrane localization of the two pools of exchangeable label.

Determination of the kinetics of transbilayer lipid migration can be carried out by an entirely different procedure utilizing NMR and paramagnetic shift reagents to distinguish between phospholipid molecules in the outer and inner bilayer faces (Shaw et al., 1977). The basic experiment can be carried out in several variations. The simplest of these has been used to estimate the self-exchange of phosphatidylcholine across the bilayer of small, single-lamellar vesicles. The design of the experiment is as follows: The transbilayer compositional asymmetry is generated by using the beef liver exchange protein. By this means phosphatidylcholine from normal erythrocyte ghost membranes is incorporated into the outer bilayer surface of vesicles formed from egg phosphatidylcholine that has the N-methyl protons completely replaced by deuterium. Since the N-methyl deuterons do not give rise to an NMR signal in the spectrometer operating at 100 mHz, only the resonance line of the N-methyl protons exchanged into the outer surface of the vesicle is seen. The fact that these protons are associated with externally located phosphatidylcholines can be established by the introduction into the vesicle dispersion of an impermeant paramagnetic ion such as Pr^{3+}. When this is done, all of the N-methyl proton signal is shifted down field. If aliquots of this asymmetric vesicle system are then examined in the NMR specastrometer as a function of time, both in the absence and presence of Pr^{3+}, the appearance of internalized phosphatidylcholine with protonated N-methyl groups can be readily detected. The sensitivity of this experiment to transbilayer migration is high compared to the double exchange experiment utilized by Johnson and co-workers (1975) and by Rothman and Dawidowicz (1975). Results in our laboratory indicate that the half-time for the exchange migration in this vesicle system at 20°C is about 26 days (Shaw et al., 1977).

The results obtained for single-bilayer vesicle systems are summarized in Table 4.1. On the basis of these limited data it appears that exchange of one chemical species for another, which is the biologically interesting process, is substantially slower than self-exchange. It is clear that the half-time determined with ESR spin-label reduction by ascorbate is in disagreement with the other determinations. The limitations of the spin-label approach are discussed more fully below.

Experiments designed to measure the transbilayer migration of cholesterol in single-lamellar phospholipid vesicles have been carried out utilizing [^{14}C]cholesterol, following the procedures outlined above (Poznansky and Lange, 1977). Somewhat earlier Smith and Green (1974) reported a study of the kinetics of this process using the quenching of a fluorescent cholesterol analogue, sterol sterophenol, in an experiment of the same type as that introduced by Kornberg and McConnell (1971). Huang and co-workers (1970), in an ESR spin-label study involving the incorporation of a racemic mixture of a and β thiocholesterol into phospholipid vesicles, estimated the vesicle content of thiocholesterol by titration with 5,5'-dithiobis (2-nitrobenzoic acid), and found the kinetics of the titration to be biphasic. If the slow phase is identified with the transbilayer migration of thiocholesterol from inner to outer surface of the vesicle bilayer, then a half-time for this process can be calculated for this cholesterol analogue.

The half-times obtained in these three studies are summarized in Table 4.2. It is immediately apparent that the half-time determined in the self-exchange experiment (> 6 days) is very much longer than the half-times determined in the experiments utilizing cholesterol analogues (42 and 70 minutes). The explanation for this large difference is probably the compositional differences in the three systems. Thus, in the case of the self-exchange experiment, the vesicle phospholipid is saturated dipalmitoyl-phosphatidylcholine, while in the analogue experiments it is egg phosphatidylcholine, a mixture of closely related molecules with an average degree of unsaturation of about 0.8 double bonds per acyl chain. In addition, it is quite obvious that neither a, β-thiocholesterol nor the sterol sterophenol are in fact cholesterol. As a third point of compositional difference, the cholesterol or cholesterol analogue to phosphatidylcholine ratios are quite different in the three experiments. It is tempting to believe that a half-time greater than 6 days is the most reliable estimate. Short apparent half-times can be caused by oxidation or hydrolysis of the component lipids, resulting in a loss of vesicle integrity, possibly followed by fusion (Roseman et al., 1975). Additional studies may, however, show that the transbilayer migration of cholesterol is a strong function of bilayer composition.

It is clear from the preceding discussion that the transbilayer migration in simple bilayer systems is a relatively slow process. The most probable half-times for chemical exchange of phospholipids are of the order of several months and for self-exchange a few weeks. The situation for

TABLE 4.1. Half-times for transbilayer migration of phospholipids and phospholipid analogs in single-bilayer vesicles.

System	Method	$t_{1/2}$	Molecule studied[a]	Temp (°C)
1. Egg phosphatidylcholine vesicles	Exchange protein	Not detected; >11 days[b]	PC	37
		> 4 days[c]	PC	37
	Exchange protein plus shift reagent	26 days[d]	PC	22
2. Egg phosphatidylcholine vesicles	Spin label reduction	6.5 hours[e]	S-PC	30
3. Egg phosphatidylcholine/ phosphatidylethanolamine/ amidinoyl-phosphatidylethanolamine	Chemical titration	>80 days[f]	PE	22

[a]Abbreviations: *PC*, phosphatidylcholine; *S-PC*, spin-labeled phosphatidylcholine; *PE*, phosphatidylethanolamine.
[b]Rothman and Dawidowicz, 1975.
[c]Johnson et al., 1975.
[d]Shaw et al., 1977.
[e]Kornberg and McConnell, 1971.
[f]Roseman et al., 1975.

cholesterol migration is less clear; the data that are apparently most reliable set the half-time in excess of 6 days. More extensive studies in this area are required. This is particularly true if such studies are to form an adequate basis for the interpretation of analogous data obtained in biological membrane system.

Transbilayer Migrations in Biological Membranes

A limited number of studies have been carried out on several different biological membranes in order to estimate the half-times for transbilayer migration of lipid components. These studies have utilized certain of the methods outlined in the previous section, that is, spin-label probe reduction (McNamee and McConnell, 1973; Grant and McConnell, 1973; Rousselet et al., 1976a, b), phospholipase digestion (Renooij et al., 1976;

TABLE 4.2. Half-times for transbilayer migration of cholesterol and cholesterol analogs in single-bilayer phosphatidylcholine vesicles.

System	Method	$t_{1/2}$	Composition[a]	Temp (°C)
1. Dipalmitoyl-phospha-tidylcholine vesicles	Exchange of labeled cholesterol	> 6 days[b]	0.9	37
2. Egg phospha-tidylcholine liposome	Quenching of fluo-rescent choles-terol analogue	70 min[c]	0.56-0.71	30
3. Egg phospha-tidylcholine vesicles	Thiocholesterol chemical titration	42 min[d]	0.1	22

[a]Mole ratio of cholesterol or analog to phosphatidylcholine.
[b]Poznansky and Lange, 1976.
[c]Smith and Green, 1974.
[d]Huang et al., 1970.

Rothman et al., 1976), and exchange of isotopically labeled membrane components either with or without exchange proteins (Bloj and Zilver-smit, 1976; Rothman et al., 1976; Lenard and Rothman, 1976; Rousselet et al., 1976b). In the case of the spin-label probes, the probe has usually been introduced into the biological membrane by "fusion" of probe-containing phospholipid vesicles (Scandella et al., 1972).

The results of these studies are summarized in Table 4.3. It is apparent that, although the number of systems examined is small, the membranes are derived from a wide range of cell types and include membranes for which compositional asymmetry data exist. Examination of Table 4.3 shows that the half-times obtained in the same membrane system using the ESR-ascorbate method are much shorter than the values obtained by other methods. This is particularly well illustrated in systems 1 and 7. Rousselet and co-workers (1976a, b) examined erythrocyte and mito-chondrial inner membranes using ascorbate reduction and reduction by endogenous components of these systems. Their data make it clear that ascorbate reduction gives smaller half-times by far. This situation is reminiscent of that for single-bilayer vesicles. The results summarized in Table 4.1 show that ascorbate reduction of the spin-label gives markedly shorter half-times than do other methods. The data summarized in Tables 4.1 and 4.3 suggest that the results obtained with ascorbate should be viewed with caution until the origin of the discrepancy is understood (Rousselet et al., 1976a, b).

If the data obtained by ascorbate reduction of spin-labeled phospho-lipid are set aside, the results obtained by other spin-label studies show

TABLE 4.3. Half-times for transbilayer migration of component lipids and lipid analogs in biological membranes.

System	Method	$t_{1/2}$	Molecule studied[a]	Temp (°C)
1. Human erythrocytes	Spin label reduction	+ ascorbate; 7 hr[b]	S-PC	37
		+ ascorbate; 25 min[c]	S-PC	37
		− ascorbate; not detected[b]	S-PC	
2. Rat erythrocytes	Phospholipase digestion of ^{31}P-labeled external surface lipid	4.5 hr[d]	PC	37
3. Human erythrocyte ghosts	Exchange protein	2.3 hr[e]	PC	37
4. Inside-out vesicles from human erythrocyte ghosts	Exchange protein	5.3 hr[e]	PC	37
5. Vesicles from electric organ of *Electrophoros electricus*	Spin label reduction	3.8-7 min[f]	S-PC	15
6. *Acholeplasma laidlawii*	Spin label reduction	+ ascorbate; > 1 min[g]	S-PC	0
7. Inner mitochondrial membrane	Spin label reduction	+ ascorbate; 4 hr[h]	S-PC	22
		− ascorbate; 25 hr[h]	S-PC	22
8. Influenza A virus	Exchange protein	> 10 days[i]	PC	32
	Phospholipase digestion of external surface lipid	> 30 days[i]	SPM	37
	Spontaneous exchange	13 days[j]	Chol	37

[a]Abbreviations: *S-PC*, spin-labeled phosphatidylcholine; *PC*, phosphatidylcholine; *SPM*, sphingomyelin; *Chol*, cholesterol.

[b]Rousselet et al., 1976a.

[c]Cited by McNamee and McConnell, 1973.

[d]Renooij et al., 1976.

[e]Bloj and Zilversmit, 1976.

[f]McNamee and McConnell, 1973.

[g]Grant and McConnell, 1973.

[h]Rousselet et al., 1976b.

[i]Rothman et al., 1976.

[j]Lenard and Rothman, 1976.

rather long half-times for transmembrane migration for both the erythrocyte and the mitochondrial membranes. With mitochondrial membranes $t_{1/2}$ is 25 hours at 22°C, and, although Rousselet et al. (1976a) state they were unable to detect migration, it can be estimated from their data that $t_{1/2}$ for the erythrocyte membrane must be greater than 24 hours at 37°C. It is important to note that in studies utilizing spin-label derivatives it is not clear whether the measured half-times refer to self-exchange across the bilayer of the spin-label and its reduction product or to the exchange of spin-label for some intrinsic membrane lipid.

In contrast to these spin-label results, the studies that measure self-exchange for the erythrocyte membranes (Table 4.3, systems 2, 3, and 4) give half-times that are considerably smaller. These values are also much smaller than the reliable value for the mitochondrial inner membrane determined by spin-label reduction (25 hours) shown in Table 4.3, system 7. It is interesting to note that the human erythrocyte ghost preparations (Table 4.3, systems 3 and 4) give self-exchange half-times similar to that obtained for the rat erythroctye. In contrast to this situation, the self-exchange data for influenza A virus yield half-times that are very long (phosphatidylcholine, >10 days; sphingomyelin, >30 days; and cholesterol, 13 days).

Although the available data for simple bilayers, summarized in Tables 4.1 and 4.2, and the data for biological membranes, summarized in Table 4.3, are not extensive, it is certain that transbilayer migration of lipids does occur in some systems. The time constant for the process, however, is probably not less than hours and may be as long as months. The data also strongly suggest that the half-time may be markedly dependent on the type of membrane, the conditions, and the design of the experiment. It would seem that the process of prime biological interest is the net transbilayer migration of a specific chemical species. The most satisfactory design for this type of study requires that net movement of an intrinsic membrane lipid be determined under conditions that permit the unambiguous localization of this component on one side or the other of the membrane.

The available data for the erythrocyte and influenza A virus, together with the data on simple bilayers, suggest that the large transmembrane compositional asymmetries in these systems are in fact stable for times that are long in biological terms. The reason for this stability is unknown, but it could rest in the large energy required to transfer the polar portion of a membrane lipid from its aqueous environment on the surface of the bilayer to the highly nonpolar central core of this structure. It is not difficult, however, to imagine that other processes could also operate to effect transbilayer migration. For example, a local disruption of bilayer structure caused by a different type of molecule, such as a membrane protein or a lysophosphatide, might give rise to a migration route with a much lower activation energy. Taupin and co-workers (1975) have recently suggested that the spontaneous formation of self-sealing micro-holes in bilayers may give rise to transbilayer migration. The existence of such a fluctuation phenomenon in bilayers has been discussed by Litster (1975). In addition to these intrinsic processes there are many possible mechanisms, introduced as methodological artifacts, that can give rise to apparent transbilayer migration of lipid components.

The physiological significance of lipid compositional asymmetries in biological membranes is not immediately obvious. In contrast, the physiological significance of the absolute transmembrane protein compositional asymmetries reflect the functional relations between the aqueous compartments separated by the membrane. For example, the metabolic interrelations between a mitochondrion and its surrounding cytoplasm depend critically on the asymmetrical localization of the protein components of the mitochondrial inner membrane (Fessenden-Raden and Racker, 1971). It is not difficult to imagine, however, that such marked asymmetries in protein composition give rise to large differences in the free energies of the two opposing membrane surfaces. This energetically unfavorable situation could be relieved in two ways: one, by the bending of the membrane, the other, by generating an opposing compositional asymmetry of lipid components. That a relation does in fact exist between curvature and compositional asymmetry has been shown for simple two-component lipid bilayers (Thompson et al., 1974). Thus, lipid compositional asymmetry in biological membranes may be required for the maintenance of a stable membrane shape when the membrane must, for functional reasons, have an absolute asymmetry of protein composition. In addition, the manipulation of transmembrane lipid compositional asymmetry by the cell may be important in the alteration of membrane curvature to modulate function. Whether or not this is the basis for lipid compositional asymmetry in biological membrane can only be decided by further experimentation.

ACKNOWLEDGMENTS

The author wishes to thank Drs. Mark Roseman, John Lenard, and James Rothman for many helpful and stimulating discussions. This work was supported in part by National Institutes of Health grant GM-14628.

REFERENCES

Bangham, A., M. M. Standish, and J. C. Wadkins. 1965. Diffusion of univalent ions across the lamellae of swollen phospholipids. *J. Mol. Biol.* 13:238-252.
Basford, J. M., J. Glover, and C. Green. 1964. Exchange of cholesterol between human β-lipoproteins and erythrocytes. *Biochim. Biophys. Acta* 84:764-766.
Bell, F. P., and C. J. Schwartz. 1971. Exchangeability of cholesterol between

swine serum lipoproteins and erythrocytes, *in vitro. Biochim. Biophys. Acta* 231:553-557.

Bjornson, L. K., C. Gniewkowski, and H. J. Kayden. 1975. Comparison of exchange of L-tocopherol and free cholesterol between rat plasma lipoproteins and erythrocytes. *J. Lipid Res.* 16:39-53.

Bloj, B., and D. B. Zilversmit. 1976. Asymmetry and transposition rates of phosphatidyl choline in rat erythrocyte ghosts. *Biochemistry* 15:1277-1283.

Bretscher, M. S. 1972a. Phosphatidyl-ethanolamine: differential labelling in intact cells and cell ghosts of human erythrocites by a membrane-impermeable reagent. *J. Mol. Biol.* 71:523-528.

Bretscher, M. S. 1972b. Asymmetrical lipid bilayer structure for biological membranes. *Nature New Biol.* 236:11-12.

Bruckdorfer, K. R., and C. Green. 1967. The exchange of unesterified cholesterol between human low-density lipoproteins and rat erythrocyte 'ghosts'. *Biochem. J.* 104:270-277.

Cooper, R. A., E. C. Arner, J. S. Wiley, and S. J. Shattel. 1975. Modification of red cell membrane structure by cholesterol-rich lipid dispersions. *J. Clin. Invest.* 55:115-126.

Danielli, J., and H. Davson. 1935. A contribution to the theory of permeability of thin films. *J. Cell. Comp. Physiol.* 5:495-508.

Fessenden-Raden, J. M., and E. Racker. 1971. *Structural and functional organization of mitochondrial membranes.* In L. I. Rothfield, ed., *Structure and function of biological membranes,* pp. 401-438. Academic Press, New York.

Gordesky, S. E., and G. V. Marinetti. 1973. The asymmetric arrangement of phospholipids in the human erythrocyte membrane. *Biochem. Biophys. Res. Commun.* 50:1027-1031.

Gottlieb, M. 1976. The limited depletion of cholesterol from erythrocyte membranes on treatment with incubated plasma. *Biochim. Biophys. Acta* 433:333-343.

Grant, C. W. M., and H. M. McConnell. 1973. Fusion of phospholipid vesicles with viable *Acholeplasma laid lawii. Proc. Natl. Acad. Sci. USA* 70:1238-1240.

Haberland, M. E., and J. A. Reynolds. 1973. Self-association of cholesterol in aqueous solution. *Proc. Natl. Acad. Sci. USA* 70:2313-2316.

Huang, C. 1969. Studies on phosphotidylcholine vesicles: formation and physical characteristics. *Biochemistry* 8:344-352.

Huang, C., J. P. Charlton, C. I. Shyr, and T. E. Thompson. 1970. Studies on phosphotidylcholine vesicles with thiocholesterol and a thiocholesterol-linked spin label incorporated in the vesicle wall. *Biochemistry* 9:3422-3426.

Inbar, M., and M. Shinitzky. 1974. Increase of cholesterol level in the surface membrane of lymphoma cells and its inhibitory effect on ascites tumor development. *Proc. Natl. Acad. USA* 71:2128-2130.

Johnson, L. W., M. E. Hughes, and D. B. Zilversmit. 1975. Use of phospholipid exchange protein to measure inside-outside transposition in phosphatidyl choline liposomes. *Biochim. Biophys. Acta* 375:176-185.

Johnson, L. W., and D. B. Zilversmit. 1975. Catalytic properties of phospholipid exchange protein from bovine heart. *Biochim. Biophys. Acta* 375:165-175.

Kamp, H. H., K. W. A. Wirtz, and L. L. M. van Deenen. 1973. Some properties of phosphatidyl choline exchange protein purified from beef liver. *Biochim.*

Biophys. Acta 318:313-325.

Kornberg, R. D., and H. M. McConnell. 1971. Inside-outside transitions of phospholipids in vesicle membranes. Biochemistry 10:1111-1120.

Lee, A. G. 1975. Functional properties of biological membranes: a physical-chemical approach. Prog. Biophys. Mol. Biol. 29:3-56.

Lenard, J., and R. W. Compans. 1974. The membrane structure of lipid-containing viruses. Biochim. Biophys. Acta 344:51-94.

Lenard, J., and J. E. Rothman. 1976. Transbilayer distribution and movement of cholesterol and phospholipid in the membrane of influenza virus. Proc. Natl. Acad. USA 73:391-395.

Litman, B. J. 1973. Lipid model membranes: characterization of mixed phospholipid vesicles. Biochemistry 12:2545-2554.

Litster, J. D. 1975. Stability of lipid bilayers and red blood cell membranes. Physics Letters 53A:193-194.

McNamee, M. G., and H. M. McConnell. 1973. Transmembrane potentials and phospholipid flip-flop in excitable membrane vesicles. Biochemistry 12:2951-2958.

Michaels, D. W., and D. Dennis. 1973. Asymmetric phospholipid bilayer membranes: formation and electrical characterization. Biochem. Biophys. Res. Commun. 51:357-363.

Montal, M. 1974. Formation of bimolecular membranes from lipid monolayers. Methods Enzymol. 32:545.

Montal, M., and P. Mueller. 1972. Formation of bimolecular membranes from lipid monolayers and a study of their electrical properties. Proc. Natl. Acad. Sci. 69:3561-3566.

Mueller, P., D. O. Rudin, H. T. Tien, and W. C. Wescott. 1962. Reconstitution of excitable cell membrane structure in vitro. Circulation 26:1167-1171.

Poznansky, M., and Y. Lange. 1977. Transbilayer movement of cholesterol in dispalmitoyl lecithin-cholesterol vesicles. Nature 259:420-421.

Renooij, W., L. M. G. van Golde, R. F. A. Zwaal, B. Roelofsen, and L. L. M. van Deenen. 1974. Preferential incorporation of fatty acids at the inside of human erythrocyte membranes. Biochim. Biophys. Acta 363:287-292.

Renooij, W., L. M. G. van Golde, R. F. A. Zwaal, and L. L. M. van Deenen. 1976. Topological asymmetry of phospholipid metabolism in rat erythrocyte membranes. Eur. J. Biochem. 61:53-58.

Roseman, M., B. J. Litman, and T. E. Thompson. 1975. Transbilayer exchange of phosphatidyl ethanolamine for phosphatidyl choline and N-acetimidoyl-phosphatidyl ethanolamine in single-walled bilayer vesicles. Biochemistry 14:4826-4830.

Rothman, J. E., and E. A. Dawidowicz. 1975. Asymmetric exchange of vesicle phospholipids catalyzed by the phosphatidylcholine exchange protein: measurement of inside-outside transitions. Biochemistry 14:2809-2816.

Rothman, J. E., and E. P. Kennedy. 1977. Asymmetrical distribution of phospholipids in the membrane of Bacillus megaterium. J. Mol. Biol. 110:603-618.

Rothman, J. E., D. K. Tsai, E. A. Dawidowicz, and J. Lenard. 1976. Transbilayer phospholipid asymmetry and its maintenance in the membrane of influenza virus. Biochemistry 15:2361-2370.

Rousselet, A., C. Guthmann, J. Matricon, A. Bienvenue, and P. E. Devaux.

1976a. Study of the transverse diffusion of spin-labeled phospholipids in biological membranes. I. Human blood cells. *Biochim. Biophys. Acta* 426:357-371.

Rousselet, A., A. Colbeau, P. M. Vignais, and P. F. Devaux. 1976b. Study of the transverse diffusion of spin-labeled phospholipids in biological membranes. II. Inner mitochondrial membrane of rat liver: use of phosphatidyl choline exchange protein. *Biochim. Biophys. Acta* 426:372-384.

Scandella, C. J., P. Devaux, and H. M. McConnell. 1972. Rapid lateral diffusion of phospholipids in rabbit sarcoplasmic reticulum. *Proc. Natl. Acad. Sci. USA* 69:2056-2060.

Shaw, J. M., W. C. Hutton, B. R. Lentz, and T. E. Thompson. 1977. Proton NMR study of the decay of transbilayer compositional asymmetry generated by a phosphatidylcholine exchange protein. *Biochemistry* 16:4156-4163: in press.

Singer, S. J., and G. L. Nicolson. 1972. The fluid mosaic model of the structure of cell membranes. *Science* 175:720-731.

Smith, H. G., R. Fager, and B. J. Litman. 1977. Light-activated calcium release from sonicated bovine retinal rod outer segment disks. *Biochemistry* 16:1399-1405.

Smith, R. J. M., and C. Green. 1974. The rate of cholesterol 'flip-flop' in lipid bilayers and its relation to membrane sterol pools. *FEBS Letters* 42:108-111.

Taupin, C., M. Dvolaitzky, and C. Sauterey. 1975. Osmotic pressure induced pores in phospholipid vesicles. *Biochemistry* 14:4771-4775.

Thompson, T. E., C. Huang, and B. J. Litman. 1974. Bilayers and biomembranes: compositional asymmetries induced by surface curvatures. In A. A. Moscona, ed., *Cell surface in development*, pp. 1-16. John Wiley and Sons, New York.

Verkleij, A. J., R. F. A. Zwaal, B. Roelofsen, P. Comfurius, D. Kostelijn, and L. L. M. van Deenen. 1973. The asymmetric distribution of phospholipids in the human red cell membrane. *Biochim. Biophys. Acta* 323:178-193.

Wallach, D. F. H. 1975. *Membrane molecular biology of neoplastic cells*, pp. 183-215. Elsevier Scientific Publishing Co., Amsterdam.

Werb, Z., and Z. A. Cohen. 1971. Cholesterol metabolism in the macrophage. *J. Exp. Med.* 134:1545-1569.

Whiteley, N. W., and H. C. Berg. 1974. Amidination of the outer and inner surfaces of the human erythrocyte membrane. *J. Mol. Biol.* 87:541-561.

Wirtz, K. W. A., H. H. Kamp, and L. L. M. van Deenen. 1972. Isolation of a protein from beef liver which specifically stimulates the exchange of phosphatidyl choline. *Biochim. Biophys. Acta* 274:606-617.

Young, R. W. 1967. The renewal of photo receptor cell outer segments. *J. Cell Biol.* 33:61-72.

5 Ligand-Binding Properties of Membrane-Bound Cholinergic Receptors of *Torpedo Marmorata*

JONATHAN B. COHEN

Communication between neurons or between neurons and muscle cells depends upon the existence of specific receptors in the target cell plasma membrane. These receptors have a dual function: to recognize (bind) the appropriate neurotransmitter and to translate that binding into an appropriate intracellular signal. The binding of acetylcholine (AcCh) to its postsynaptic receptor (AcChR) on vertebrate skeletal muscle fibers leads to an increased permeability of the plasma membrane to inorganic cations, primarily Na and K (Fatt and Katz, 1951; delCastillo and Katz, 1954; Takeuchi and Takeuchi, 1960). This control of permeability by cholinergic ligands exists in the absence of intracellular energy stores or ion concentration gradients. Hence, the energetics of ligand binding is utilized to control the changes of membrane structure necessarily involved in the modification of membrane permeability (Nachmansohn, 1959). In order to describe the action of AcCh, it is necessary to identify the membrane structures involved in the binding of AcCh, in the recognition of cations, and in the transport of cations across the postsynaptic membrane. Beyond a description of these structures, it is necessary to describe how they are functionally related.

The study of the mechanism of permeability control can be approached at several different levels: electrophysiological studies of the permeability response of intact tissues; biochemical and biophysical studies of the structure and function of isolated membrane vesicles containing nicotinic cholinergic receptors; chemical characterization of isolated receptors in solution; analysis of the functional properties of isolated receptors reconstituted in artificial membranes. It is the purpose of this report to discuss in some detail one facet of these researches: the structural and functional properties of receptor-rich membranes isolated from *Torpedo* electric tissue. However, in order to put these studies in perspective, I will first review briefly general features of the cholinergic response. More extensive reviews have appeared recently, some emphasizing electrophysiological studies in vertebrates (Rang, 1975; Gage, 1976), and others emphasizing the study of cholinergic mechanisms in electric tissue (Karlin, 1974; Potter, 1974; Changeux, 1975; Cohen and Changeux, 1975).

Progress in the identification of the molecular structures involved in permeability control has been possible largely for two reasons: (1) There exist polypeptide toxins of molecular weight 7,000-8,000, the a-neurotoxins isolated from the venom of cobra and related species, that bind with high affinity and high specificity to the AcCh binding site of nicotinic cholinergic receptors (Lee and Chang, 1966; Lee, 1972). The use of radioactively labeled neurotoxins makes it possible to quantify the numbers and distribution of receptors in intact tissue, in tissue homogenates, or in detergent solutions. (2) The electric organs of strongly electric fish (*Torpedo, Electrophorus*), which are evolved from skeletal muscle, provide a uniquely rich source of cholinergic synapses. The two electric organs of an average *T. marmorata* provide 300-400 grams of tissue containing 1 nmole a-toxin sites per gram of tissue. For comparison, one of the richest sources of AcChR from vertebrate sources would be denervated skeletal muscle such as a rat hemidiaphragm, which weighs about one gram and contains 50 pmoles a-toxin sites (Potter, 1974).

The drugs that modify the cholinergic postsynaptic response serve as tools to probe the permeability control mechanism. The action of AcCh can be blocked by specific, potent competitive antagonists such as tubocurarine. Analysis of the concentration dependence of the AcCh response function (whether muscle contraction or membrane depolarization) in the presence of tubocurarine suggests that tubocurarine acts by binding to the same site as AcCh without causing a change of membrane permeability. Agonists such as carbamylcholine act as AcCh does to increase membrane permeability, and their effects are blocked by the antagonists. Noncompetitive antagonists appear to modify the cholinergic response by acting at a site other than the AcCh binding site. High agonist concentrations cannot overcome the effect of these compounds, suggesting that there is no competitive interaction between them and the agonists.

Electrophysiological studies define the magnitude and time course of the cholinergic permeability response. Under normal circumstances, the postsynaptic response is rapid, occurring in milliseconds. After its release from the nerve terminal, the concentration of AcCh near the postsynaptic receptors is significant for no more than several milliseconds, by which time losses due to diffusion or enzymatic hydrolysis are dominant. Although it is not known whether one or several molecules of AcCh or other agonist must be bound to cause a single elementary conductance event, that elementary event must be characterized as an aqueous channel or pore having a lifetime of several milliseconds (Katz and Miledi, 1972; Anderson and Stevens, 1973; Neher and Sakman, 1976). The permeability response is described in terms of channel formation because the magnitude of ion transport (40,000 ions per millisecond per conductance unit) is comparable to that associated with the antibiotic gramicidin, a compound known to create discrete channels in model membranes (Haydon and Hladky, 1972).

Studies have also been made of the cholinergic permeability response when agonists are present for extended periods of time. When a tissue is exposed to AcCh or other agonists for seconds or longer, there is no steady-state conductance response. Rather, there is a progressive loss of tissue responsiveness, even though the agonist concentration remains constant. Katz and Thesleff (1957) noted that the phenomenon, known as desensitization, suggested the existence of a form of the receptor that binds AcCh and agonists with high affinity but that is no longer associated with a permeability response. Even when there is appreciable desensitization, however, the characteristics of the remaining elementary conductance events are not modified. Rather, the number of channels that open decreases. This time dependence of the cholinergic response suggests that the ligand-binding properties of cholinergic receptors may be more complicated than a simple bimolecular association, and that the observed equilibrium binding properties may not be closely related to the ligand binding directly involved in permeability control.

It is possible to determine the ligand-binding properties of cholinergic receptors in vitro—either isolated receptors or receptors in membrane fragments. The properties of isolated receptors can be determined because it is possible to extract active receptors (that is, receptors binding cholinergic ligands) from the membranes, but only through the use of neutral or anionic detergents. It is now possible to isolate and purify mg quantities of AcChR in detergent solution. It is not the purpose of this chapter to review either the techniques of receptor purification or the characterization of the purified AcChR, but interested readers can find references in reviews (Karlin, 1974; Potter, 1974; Changeux, 1975) or in several recent reports (Eldefrawi, Eldefrawi, and Shamoo, 1975; Changeux et al., 1976; Karlin et al., 1976; Raftery et al., 1976). Since it is possible to isolate the AcChR, it should be possible to determine whether that macromolecule also contains the structural components necessary for permeability control. Many attempts have been made to reincorporate detergent-solubilized AcChR into artificial membrane systems to study its permeability control properties (Hazelbauer and Changeux, 1974; Michaelson and Raftery, 1974; Eldefrawi, Eldefrawi, and Shamoo, 1975; Karlin et al., 1976). Unfortunately, the results to date are not conclusive, and it still remains only a reasonable hypothesis that the AcChR (molecular weight about 300,000 daltons) contains the permeability control elements as well as the ligand-binding function. Perhaps the difficulty of reconstituting this system is related to the fact that low levels of detergent (10^{-4} percent w/v) act as noncompetitive blocking agents in vivo (Bartels and Rosenberry, 1972; Brisson, Devaux, and Changeux, 1975).

It is possible to avoid the use of detergents and to study the mechanism of permeability control by AcCh by studying the ligand-binding properties and permeability control functions of isolated membranes containing cholinergic receptors. Kasai and Changeux (1971) demonstrated that cholinergic ligands control the permeability properties of membrane vesicles isolated from *Electrophorus* electric tissue. Although those membrane vesicles do contain functional cholinergic receptors, they only comprise several percent of the membrane protein. They remain a useful preparation for studying cholinergic mechanisms (see for example, Hess et al., 1975), but it is also possible to study the functional properties of membrane fragments containing the high densities of cholinergic receptors characteristic of the postsynaptic membranes underlying the nerve terminal.

Isolation and Characterization of *Torpedo* Receptor-Rich Membranes

Torpedo electric tissue is a richer source of postsynaptic membranes than is *Electrophorus* electric tissue. About 50 percent of the ventral surface of each *Torpedo* electroplacque is subsynaptic (Sheridan, 1965; Israel, Gautron, and Lesbats, 1970), while for *Electrophorus* only 2-3 percent of each caudal electroplacque surface lies under nerve terminals (Bourgeois et al., 1972). (Actually, it is possible that even the noninnervated regions of the ventral *Torpedo* electroplacque may be covered with receptors [Rosenbluth, 1975]). Receptor-rich membranes are readily isolated from *Torpedo marmorata* electric tissue after homogenization of the tissue in distilled water (Cohen et al., 1972) or in 0.4 M NaCl-.1 M MgCl$_2$ (Nickel and Potter, 1973) by ultracentrifugation on a sucrose density gradient. Membranes isolated at 38 percent (w/v) sucrose contain 30 percent of the tissue a-neurotoxin binding sites (2-3 μmoles a-toxin site per gram of protein) and 100 times fewer acetylcholinesterase catalytic sites. Based on the number of a-neurotoxin sites, the cholinergic receptor constitutes at least 30 percent of the proteins in these membranes. The bulk of the acetylcholinesterase and Na,K-ATPase is found elsewhere in the gradient, around 28 percent sucrose. A modified isolation procedure utilizes a sucrose cushion (35 percent w/v) in place of the density gradient (Hazelbauer and Changeux, 1974). After ultracentrifugation, most of the membranes are recovered at the supernatant-sucrose interface, and the receptor-rich membranes are recovered as a pellet. The membranes isolated in this manner are not as rich in receptor, but this loss of purity is compensated by the ease with which reasonable quantities of material can be isolated. This procedure has been used for the new studies reported here.

Polyacrylamide gel electrophoresis of the *T. marmorata* receptor-rich membranes in the presence of dodecyl sulfate reveals a simple polypeptide composition (Cohen, Weber, and Changeux, 1974). The dominant polypeptide has an apparent molecular weight of 43,000 daltons, and it is this chain that is labeled by the cholinergic affinity label 4-(N-maleimido)-benzyltrimethylammonium, developed by Karlin and Cowburn (1973). There are several other polypeptides less than 65,000 daltons and another around 105,000 daltons. The four polypeptides of low molecular weight are observed both in the receptor-rich membranes of *T. californica* and in the receptors purified and isolated from that fish by various affinity chromatography resins (Duguid and Raftery, 1973; Weill, McNamee, and Karlin, 1974; Reed et al., 1975; Flanagan, Barondes, and Taylor, 1976). Hence, it appears that the receptor-rich membranes isolated from *Torpedo* electric tissue may contain only the polypeptide chains associated with the isolated purified receptors.

The specialized nature of these membrane fragments is also revealed dramatically by negative-stain electron microscopy (Cartaud et al., 1973; Nickel and Potter, 1973; Raftery et al., 1974). The membrane surfaces are covered at a surface density of 10,000-15,000 particles/μ^2 by rosette-like structures about 9 nm in diameter and consisting of 5-6 subunits that are exactly like those of the isolated receptors purified in detergent solutions from *Electrophorus* (Cartaud et al., 1973) or *Torpedo* (Eldefrawi, Eldefrawi, and Shamoo, 1975) electric tissue. Furthermore, the surface density of particles in these receptor-rich membranes is very similar to the density of a-toxin binding sites characteristic of the postsynaptic, subsynaptic membrane in vivo (about 30,000/μm^2 for *Electrophorus* [Bourgeois et al., 1972] and for skeletal muscle [Fertuck and Salpeter, 1976; Porter and Barnard, 1975]).

Hence, although there is no a priori reason why it should be possible to isolate from all other membranes the specialized portion of the plasma membrane underlying the nerve terminals, this, in fact, appears to be the case. Because of the highly specialized composition of these membranes, electron microscopy and x-ray diffraction (Dupont, Cohen, and Changeux, 1974; Raftery et al., 1975) can be expected to provide considerable information about the manner in which the receptor is integrated in the membranes. However, these receptor-rich membranes are of interest not only for structural studies, but also for functional studies. At least some membranes exist in the form of sealed vesicles, and it is possible to study the effect of cholinergic ligands on the ^{22}Na efflux from these vesicles (Hazelbauer and Changeux, 1974; Popot, Sugiyama, and Changeux, 1974). Basic dose-response relations for agonists and antagonists have been determined, and it has been shown that pharmacological desensitization occurs in vitro as well as in vivo. After a membrane has been exposed to an agonist for several minutes, agonists no longer cause an increased rate of ion efflux. Since the receptor-rich membranes retain functionality in the two senses, ligand binding and permeability control, it should be possible to relate the two functions, that is, to identify receptor conformations and structures involved in permeability control.

Binding of Cholinergic Ligands at Equilibrium

The equilibrium binding of cholinergic agonists and antagonists to the receptor-rich membranes can be determined by many different methods: by equilibrium dialysis (Eldefrawi et al., 1971), by ultracentrifugation (Weber and Changeux, 1974b, c; Raftery et al., 1974), and by the use of fluorescent cholinergic ligands (Cohen and Changeux, 1973; Cohen, Weber, and Changeux, 1974; Raftery et al., 1974). The equilibrium binding properties can also be inferred from the manner in which the reversible ligands modify the rate of binding of a-neurotoxins (Weber and Changeux, 1974a, b, c; Raftery et al., 1976). In this section examples will be presented of the use of these different assays. It will be shown that the same affinities are determined for any one ligand by the different techniques, but there is the surprising result that at equilibrium agonists are bound more tightly than expected; that is, the dissociation constants are smaller than the concentrations necessary to cause a permeability response.

Ultracentrifugation can be used to separate free from bound ligand, and the equilibrium binding of radioactive cholinergic ligands can be measured at specified temperature, pH, and ionic concentration. In any study of this sort, it is necessary to have criteria to identify the specific binding of cholinergic ligands to the receptor: the saturable ligand binding that can be displaced by known nicotinic drugs and a-neurotoxins. Given the high concentration of a-neurotoxin sites in the membranes, it is possible to study ligand binding at micromolar concentrations of receptor binding sites. Since there are residual amounts of acetylcholinesterase present in the membrane suspension, it is necessary to use an appropriate inhibitor before studying the binding of a ligand such as AcCh.

The binding of AcCh to the *Torpedo* membranes, determined several hours after mixing drug and membranes, is of high affinity, with a dissociation constant $K_D \cong 20$ nM. A typical binding curve is shown later (Fig. 5.4). There is a slight variation from one preparation to another, but for 14 different preparations, $K_D = 14 \pm 3$ nM at 4°C. It is also possible to determine whether there is any AcCh bound less tightly, that is, with a dissociation constant approximating the micromolar concentrations necessary for a response in vivo. By determining the binding of AcCh to membrane suspensions micromolar in a-toxin sites, it can be shown that all the bound AcCh that can be displaced by a-neurotoxin is bound with $K_D \cong 20$ nM, and that for free AcCh concentrations up to 10 μM, less than 5 percent of the total amount bound is not displaced by a-neurotoxin (J. B. Cohen, unpublished observations). If there is another receptor conformation binding AcCh with a dissociation constant 5 μM or less, its concentration is less than 10 percent of that of the high-affinity conformation.[1]

The antagonist [14C]-dimethyltubocurarine is bound less tightly to the membrane-bound *Torpedo* receptor than is acetylcholine. Figure 5.1 shows a typical experiment where the observed binding consists of two components, one sensitive to a-neurotoxin and cholinergic ligands and the other not. In the presence of 7 μM dimethyltubocurarine, about 70 percent of the total binding is displaced by a-neurotoxin, and hence is associated with binding to the receptor site. Analysis of concentration dependence of the specific binding component determines that dimethyltubocurarine is bound with about a micromolar dissociation constant, though there is some variation from one preparation to another. For eight different membrane preparations, $K_D = 0.9 \pm 0.2\,\mu M$. By performing parallel experiments determining the number of binding sites for [3H]-AcCh and [14C]dimethyltubocurarine, it is possible to show that there must be a one-to-one site ratio. Binding experiments on membrane preparations from five fish result in a ratio of agonist to antagonist site concentration of $1.1 \pm .1$.

Fluorescence spectroscopy can also be used to determine the equilibrium binding of fluorescent cholinergic ligands to the *Torpedo* receptor-rich membranes. One such ligand is DNS-chol (dimethylaminonaphthalene sulfonamido-ethyl trimethyl ammonium iodide), a quaternary ammonium ligand with emission properties sensitive to local environment (Weber et al., 1971). Studies of the effect of DNS-chol on the membrane potential of the isolated *Electrophorus* electroplacque indicate that the ligand acts both as a cholinergic agonist and as a noncompetitive blocking agent (Cohen, Weber, and Changeux, 1974). This dual pharmacological activity is similar to the activity of another aromatic trimethylammonium compound, benzoylcholine (Bartels, 1965).

The equilibrium interaction of DNS-chol with the *Torpedo* receptor-rich membranes is studied directly by the use of differential fluorescence spectroscopy (Cohen and Changeux, 1973; Cohen, Weber, and Changeux, 1974). DNS-chol interacts with two classes of sites in the membranes: the AcCh binding site itself, and a secondary site apparently related to the site of action of noncompetitive blocking agents. The interaction of DNS-chol with the receptor-rich membranes can be measured without having to physically separate the bound ligand from the free. Although the quantum yield of DNS-chol bound to the receptor is some-

---

[1]These results may appear at variance with earlier reports that the binding of [3H]-AcCh to particulate fractions of *T. marmorata* prepared from lyophilized tissue was characterized by both high- and low-affinity components (Eldefrawi, Britten, and Eldefrawi, 1971). However, recent reports show that the high-affinity binding is a general characteristic of freshly prepared suspensions of membrane-bound AcChR (Eldefrawi, Eldefrawi, and Wilson, 1975) and that the low-affinity component may be the result of aging and storage of the tissue.

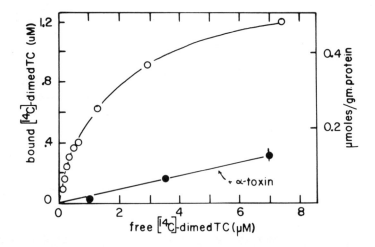

FIGURE 5.1. Binding of [^{14}C] dimethyltubocurarine to *Torpedo* receptor-rich membrane fragments (2.5 g protein/l, 1 μM a-toxin sites) in physiological saline. The free dimethyltubocurarine in equilibrium with the membrane fragments is determined from the supernatant radioactivity after ultracentrifugation, and the bound dimethyltubocurarine is determined from the difference between free and total dimethyltubocurarine. o ———— o represents total binding at equilibrium; •————• represents ligand binding to membranes pretreated with excess a-neurotoxin (10 μM). The specific binding to the receptor site is the difference between the total binding and that in the presence of a-neurotoxin. A double reciprocal plot of the specific binding establishes a site concentration 1.1 μM and a dissociation constant $K_D = 1.1$ μM.

what greater than that of the free ligand in its aqueous environment, a more significant factor is that the DNS-chol bound to the receptor site is excited very efficiently by nonradiative energy transfer from the membrane proteins ($\lambda_{ex} = 287$ nm). In fact, it is only the DNS-chol bound to the receptor site itself and not those molecules bound to the second class of sites that is excited efficiently by energy transfer. The practical result is that when only 2 percent of the total DNS-chol occupies receptor sites, that fraction contributes as much as 50 percent of the total detected fluorescence. Hence, while it would be extremely difficult to use radioactive techniques to measure ligand binding that is only 1 percent of the total concentration, it is possible to measure DNS-chol binding to the receptor when the ratio of bound to free is only 0.001.

The dissociation constant of the membrane-bound AcChR for DNS-chol as determined by the spectroscopic measurements is $K_D \cong 20$ μM. The spectroscopic evidence for the identification of that site as the AcCh binding site is based upon the fact that the presence of reversible cholinergic ligands and a-neurotoxins prevents the interaction of DNS-chol with this site. However, the identification of this site as the AcCh binding site is supported by other lines of evidence (Cohen and Changeux, 1973). When ultracentrifugation techniques are used to separate bound from free DNS-chol, it is found that cholinergic ligands do physically displace DNS-chol from the membrane. DNS-chol displaces [³H]-AcCh from and alters the rate of binding of [³H]-a-neurotoxins to the membranes in a manner consistent with the hypothesis that DNS-chol does bind to the AcChR binding site.

The receptor affinities for nonradioactive and nonfluorescent cholinergic ligands can be determined from competitive binding experiments. The equilibrium interaction of many ligands, both agonists and antagonists, with *Torpedo* receptor-rich membranes has been reported (Cohen and Changeux, 1973; Weber and Changeux, 1974; Raftery et al., 1975a), and in each case the data are consistent with a competitive binding of agonists and antagonists at equilibrium. It is possible, for example, to determine the receptor affinity for the agonist carbamylcholine by its capacity to alter the equilibrium binding of the agonist [³H]-AcCh, the antagonist [¹⁴C]-dimethyl-tubocurarine, or the fluorescent DNS-chol. As is summarized in Table 5.1, carbamylcholine appears to bind competitively with each ligand with a dissociation constant $K_D = 0.5$ μM. Another means to determine the equilibrium affinities of cholinergic ligands is by the manner in which they modify the kinetics of binding of radioactive a-neurotoxins. When membrane fragments are preincubated with cholinergic ligands prior to the addition of a-toxin, the initial rate of a-toxin binding reflects the number of unoccupied receptor sites. Once again, the apparent dissociation constant for carbamylcholine determined in this manner is 0.5 μM, consistent with the hypothesis that under these conditions a-neurotoxins and cholinergic ligands bind in a mutually competitive manner.

Comparison between K_{AP} and K_D

Agonists such as AcCh ($K_D = 20$ nM) or carbamylcholine ($K_D = .5$ μM) are bound with high affinity. This result is pleasing from a practical point of view, namely, that it is easy to study agonist binding. However, it is disturbing because it has been known for some time that higher agonist concentrations are necessary to elicit membrane responses in vivo and in vitro. The apparent dissociation constant (K_{AP}) for the action of AcCh on the isolated *Electrophorus* electroplacque (depolarization response) or on the isolated vesicles (²²Na efflux) is $K_{AP} = 1$ μM, and that of carbamylcholine is $K_{AP} = 50$ μM (Kasai and Changeux, 1971). Estimations of the K_{AP} for carbamylcholine, based on the conductance response, suggest an even weaker interaction ($K_{AP} = 300$ μM) (Lester, Changeux, and Sheridan, 1975).

It may be argued that *Torpedo* electric tissue is fundamentally different from that of *Electrophorus*, but, as has been discussed before (Cohen,

TABLE 5.1. Carbamylcholine binding at equilibrium to membrane-bound receptor of *Torpedo marmorata*.

Method of determination	Dissociation constant (μM)
1. Agonist: ^{3}H-Acetylcholine	0.4 ± .1
2. Antagonist: ^{14}C-Dimethyltubocurarine	0.4 ± .1
3. ^{3}H-a-neurotoxin	0.5 ± .2[a]
4. Fluorescent cholinergic	0.6 ± .3[b]

[a]Weber and Changeux, 1974b.
[b]Cohen and Changeux, 1973.

Weber, and Changeux, 1974), the high-affinity binding of AcCh (K_D = 20 nM) is difficult to reconcile with the duration of the *Torpedo* postsynaptic potential (5 milliseconds) (Fessard, 1958). If the binding of AcCh is the result of a single-step bimolecular association characterized by a diffusion limited association rate ($k_{on} \cong 10^9 M^{-1}$ sec^{-1}), the calculated unimolecular dissociation rate of the AcCh-receptor complex would be about 10 sec^{-1}. In other words, the average time for which a single AcCh molecule could occupy a receptor site would exceed the duration of the conductance event by an order of magnitude. Recently, in fact, Changeux and collaborators have reported apparent dissociation constants for the action of agonists and antagonists in vivo on intact *Torpedo* electroplacques and in vitro on the ^{22}Na efflux from the isolated vesicles (Changeux et al., 1976). Table 5.2 compares the equilibrium dissociation constants (K_D) for the binding of agonists and antagonists to the *Torpedo* receptor-rich membranes with the K_{AP} characterizing the membrane response. For AcCh and carbamylcholine, the drug concentration necessary for membrane response exceeds the K_D by two orders of magnitude. However, unlike the results for the agonists, the antagonists (gallamine, tubocurarine, hexamethonium) are characterized by a K_{AP} that does not differ greatly from the K_D.

Several explanations are possible to account for the discrepancy between the tight equilibrium binding of agonists and the weaker interactions implied by the concentrations necessary for a response:

1. The equilibrium binding bears no relation to the receptor, an unlikely possibility because of the relationship to the binding of agonists, antagonists, and a-neurotoxins.

2. Ninety-nine percent of the receptors must be occupied for a response; again, quite unlikely.

3. The fractionation procedure has altered the binding properties of most receptors, while the ^{22}Na efflux is mediated by a small number of receptors binding with low affinity. If this were true, it might be possible

TABLE 5.2. Interaction of cholinergic ligands with cholinergic receptor of *Torpedo marmorata*.

Ligand	Membranes: binding assays $K_d (\mu M)$[a][b]	Soluble: DNS-chol fluorescence $C_{50} (\mu M)$[c]	Membranes: ^{22}Na efflux $K_{app} (\mu M)$[d]
Acetylcholine (with 10^{-4} DFP)	0.014	3	3
Carbamylcholine	0.5	50	50
Decamethonium	0.8	10	10
Gallamine	10.	20	40
Tubocurarine	0.2	3	3
Hexamethonium	60.	260	200

[a]Weber and Changeux, 1974b.
[b]Cohen and Changeux, 1973.
[c]Cohen, Weber, and Changeux, 1974.
[d]Changeux et al., 1976.

to find preparative conditions yielding membranes binding AcCh with low affinity or to try to separate membrane fragments according to their affinity for agonists.

4. The high-affinity equilibrium binding involves a receptor conformation favored by the prolonged presence of cholinergic ligands in general or agonists specifically. If the permeability control mechanism involves multiple equilibria, it is probable that neither the equilibrium binding functions nor the dose-response relations in vitro or in vivo reflect a single one of these equilibria. This hypothesis can be tested by observing ligand-binding functions at short times after mixing membranes and ligands.

A major portion of the remainder of this report will be concerned with analyses of the origins of the high-affinity equilibrium binding of agonists.

Factors Controlling Receptor Affinity

A first approach is to examine different factors that might alter the equilibrium binding properties of the membrane-bound receptor. The high-affinity equilibrium binding of agonists by the receptor-rich *Torpedo* membranes is not an immutable property of the receptors. Studies of the ligand-binding properties of detergent solutions of receptors reveal that agonist-binding constants can differ significantly from those of membrane-bound receptors, while the antagonist affinities are unchanged. This was first observed by Franklin and Potter (1972) in a comparison of the interactions of solubilized receptors with carbamylcholine and tubo-curarine, and it has since been extended to other agonists and antagonists (Cohen, Weber, and Changeux, 1974; Raftery et al., 1974). Table 5.2 lists ligand affinities of receptors solubilized from the receptor-rich membranes by the anionic detergent sodium cholate. In this case, the agonists AcCh and carbamylcholine are bound two orders of magnitude less tightly than to the membrane-bound receptors, while the antagonist affinities are changed by less than a factor of 3-4. Actually, the binding functions of receptors in cholate solutions must be characterized by multiple affinities (Sugiyama and Changeux, 1975), and the affinities listed in Table 5.2 represent an average value determined by a fluorescence assay. Also, it is now known that the changed binding functions are not solely a result of membrane dissolution. Fresh solutions of receptors in neutral detergents (Triton X-100,, Emulphogene BC-720, Lubrol WX) bind AcCh with high affinities, but there is a conversion to low-affinity binding with time (O'Brien and Gibson, 1975; Sugiyama and Changeux, 1975; Eldefrawi, Eldefrawi, and Wilson, 1975).

Examination of the ligand affinities in Table 5.2 reveals that there is a striking correlation between the agonist affinities of the receptor in cholate solutions and the apparent constants in vivo or in vitro. However, although it is possible that the receptor affinities in cholate are those of a membrane-bound receptor conformation associated with permeability control, it is also likely that the detergent has modified the ligand-binding function and that the correlation is fortuitous.

While the differences between the ligand-binding functions of receptors in membranes and detergent solutions are quite dramatic, the equilibrium binding properties of the membrane-bound receptor itself are also subject to control by drugs and ions. Local anesthetics, drugs that act as noncompetitive blocking agents, do modify the ligand-binding properties of the membrane-bound receptor (Cohen, Weber, and Changeux, 1974). In the presence of physiological concentrations of monovalent and divalent cations, various local anesthetics actually increase by a factor of 2-3 the affinity of the membrane-bound receptor for cholinergic ligands, both agonists and antagonists. For example, in the presence of 3 mM prilocaine, the affinity of the membrane-bound receptor for dimethyltubocurarine is increased from $K_D = 1 \mu M$ to $K_D = 0.3 \mu M$ (Fig. 5.2). Clearly, prilocaine is interacting with the receptor-rich membranes at a site other than the AcCh binding site. For a variety of structurally unrelated local anesthetics, there is a close correlation between the anesthetic concentrations causing a half-maximal increase of receptor affinity and the concentrations active in vivo on the *Electrophorus* electroplacque as noncompetitive-blocking agents. Furthermore, when ligand-binding functions are determined in the presence of 250 mM monovalent cation, the addition of millimolar Ca causes a comparable increase of affinity. This control of affinity by local anesthetics and Ca is no longer observed in detergent solutions.

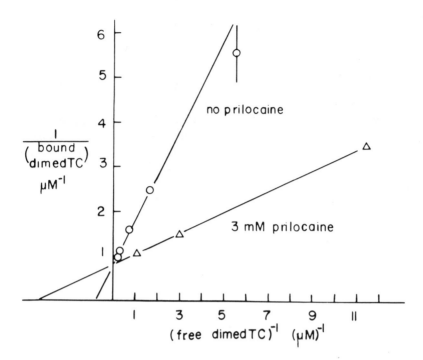

FIGURE 5.2. Double reciprocal plot of the effect of 3 mM prilocaine on the binding of [^{14}C] dimethyltubocurarine to the cholinergic receptor site in receptor-rich membrane fragments. The data are from an experiment similar to that shown in Figure 5.1, except that the membrane suspension contained 1.25 g protein/l and 0.8 μM a-toxin sites. In the absence of prilocaine (o ———— o), the binding is consistent with a site concentration 1.1 μM and a dissociation constant $K_D = 0.9$ μM. In the presence of 3 mM prilocaine (Δ ———— Δ), the equilibrium binding is characterized by the same site concentration, but $K_D = 0.3$ μM.

Another important variable is ionic strength. Raftery and coworkers determined that the affinity of *Torpedo californica* receptors for bis quaternary amine cholinergic ligands increases by a factor of 100 as the salt concentration is decreased from that of normal Ringer's to 20 mM NaCl (Raftery et al., 1975). The affinity of monoquaternary amines such as AcCh is insensitive to ionic strength. This control of affinity is observed for receptors both membrane bound and in detergent solution, and it has proved possible to chemically modify the subsite involved with the binding of the second quaternary amine function (Chao, Vandlen, and Raftery, 1975).

However, although these studies establish that there are drugs and ions that modify the affinity of the membrane-bound receptor, they provide no insight into the significance of the high-affinity binding of AcCh agonists. To date, it has not been possible to identify drugs or ionic conditions causing the membrane-bound receptor to bind AcCh at equilibrium with a dissociation constant larger than 20-30 nM.

Kinetics of Conformational Changes of AcChR

Since it is apparent that the equilibrium binding of AcCh to the *Torpedo* receptor-rich membranes is associated with paradoxically high affinities, it is important to determine the time scale at which the binding occurs. The ultracentrifugation assays involve times on the order of hours, and the fluorescence assays established that high-affinity binding is present within minutes. (The kinetics of equilibrium of DNS-chol with the receptor-rich membranes can in principle be determined by rapid mixing techniques, but this has not yet been done.) Weber, David-Pfeuty, and Changeux (1975) approached the problem by studying the interaction of [³H]-a-neurotoxins with the membrane-bound receptor after the membranes were preexposed to cholinergic ligands for varying lengths of time. They observed that cholinergic agonists decreased the rate of binding of a-neurotoxin only after they had interacted with the membranes for several minutes. For antagonists, no such time dependence occurred. They concluded that the observations were consistent with the hypothesis that, prior to the addition of agonists, the membrane-bound receptors existed in a conformation binding agonists with low affinity, and that, within minutes and in a reversible manner, the agonists stabilized a high-affinity receptor conformation.

It is clearly desirable to study in a direct manner the binding of reversible cholinergic ligands to the membrane-bound receptor on the time scale of seconds or faster. Preliminary experiments indicate that it is feasible to study the binding of [³H]-AcCh on that time scale (Cohen et al., in preparation). The approach is to rapidly mix solutions of AcCh and receptor-rich membrane fragments at the desired concentration and then to separate free AcCh from bound AcCh by ultrafiltration. Mixing can be achieved on the second time scale by injecting a concentrated membrane solution into a dilute AcCh solution, and mixing in about 0.1 second can be achieved by the use of pressure-driven syringes and a mixing chamber.

When nanomolar concentrations of AcCh and membrane-bound receptor are mixed (within seconds) at 4°C, the amount of AcCh bound is not at equilibrium immediately. In Figure 5.3, it is shown that the amount of AcCh bound increases over a period of minutes. After 20 minutes, the equilibrium binding can be further modified, as expected, by the addition of appropriate drugs. Following the addition of the antagonist dimethyltubocurarine, a new equilibrium is achieved within the time resolution of this experiment (20 seconds). The addition of the noncompetitive blocking agent SKF-525a results in a rather slow increase (minute time scale) of the amount of AcCh bound, due to the increase of

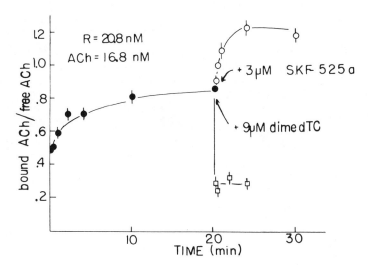

FIGURE 5.3. Time-dependent high-affinity binding of [³H]acetylcholine to *Torpedo* membrane fragments in physiological saline at 4°C. A membrane suspension (2.5 g protein/l, 1 μM a-toxin sites) is incubated with 3.3 mM diisopropyl fluorophosphate (DFP) for 30 minutes, then the suspension is rapidly diluted 50-fold into a solution containing AcCh. The final suspension contains 20 nM a-toxin sites, 16 nM AcCh, and 10^{-4} M DFP. Samples are taken at different times after mixing, and the free AcCh is separated from the membranes by ultrafiltration. Bound AcCh is determined from the difference between total AcCh and free AcCh. The first sample is taken 20 seconds after mixing, and after 10 minutes the reaction mixture is separated into two portions. In one portion (□———□), 20 minutes after mixing AcCh and membranes, the antagonist dimethyltubocurarine is added (final concentration 9 μM), and within 20 seconds the amount of AcCh bound reaches a new time independent value. In the other portion (o ——— o), 20 minutes after mixing AcCh and membranes, the noncompetitive antagonist SKF-525a (3 μM final concentration) is added; the amount of AcCh bound then increases over the next minutes until a new steady state value is obtained.

receptor affinity for AcCh. However, what is the origin of the time-dependent increase of AcCh bound to the membranes in the absence of other ligands? This increase may reflect an increase in the number of high-affinity sites, an increase in the affinity of a fixed number of sites, or a combination of the two. (Because of the low concentration of receptor sites in these experiments, AcCh bound with micromolar affinity would not be measurable.)

The ligand-binding functions have to be determined at a fixed time interval after mixing AcCh and membranes. Fig. 5.4 compares the concentration dependence of the AcCh binding function determined 30 seconds and 40 minutes, respectively, after mixing drug and receptor. Forty minutes after mixing, when the binding has attained equilibrium, the

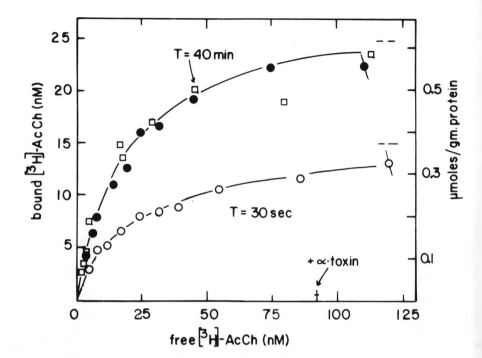

FIGURE 5.4. Binding of [³H] AcCh to *Torpedo* receptor-rich membranes 30 seconds (o ———— o) and 40 minutes (• ———— •, □————□), respectively, after mixing membranes and ligand in physiological salt solution at 4°C. A concentrated membrane suspension (1.6 g protein/l, 0.8 μM a-toxin sites) is incubated with 3.3 mM diisopropyl fluorophosphate for 30 minutes prior to a rapid 40-fold dilution into the appropriate AcCh solutions. Thirty seconds after mixing, the free AcCh is separated from the membrane suspension by ultrafiltration (Millipore HA filters, filtration time less than 2 seconds). Forty minutes after mixing, the bound AcCh is determined by two different methods: ultrafiltration (• ———— •) and ultracentrifugation (□————□). Analysis of the binding data by a double reciprocal plot indicates that at equilibrium the concentration of AcCh sites is 26 nM and the dissociation constant is 17 nM. Thirty seconds after mixing, the binding function can be characterized by a site concentration of 16 nM and an apparent dissociation constant of 22 nM. At equilibrium, over 95 percent of the bound AcCh is displaced by a-neurotoxin (+).

concentration of high-affinity sites is 26 nM and the K_D = 17 nM. These parameters are independent of the assay procedure: ultrafiltration or ultracentrifugation. However, 30 seconds after mixing, only 0.6 of the equilibrium number of high-affinity sites are occupied with a K_D = 22 nM. For AcCh concentrations ranging from 5 nM to 125 nM, the measured binding can be characterized by a single dissociation constant, and hence, over this concentration range, the number of high-affinity sites present at a fixed time is apparently independent of the AcCh concentration with which they have been incubated. This paradoxical result undoubtedly reflects the underlying kinetic processes and it is necessary to extend these studies to higher AcCh concentrations to determine the domain where the kinetic processes are sensitive to ligand concentrations. Comparable data can be collected for other times after mixing: for this preparation at 0.3 second, less than 10 percent of the high-affinity sites are accessible. Hence, the high-affinity sites are appearing within seconds.

These experiments demonstrate in a direct manner that there is not a significant concentration of a receptor conformation that binds AcCh with high affinity prior to the addition of AcCh. AcCh must be bound first at some lower affinity, and this binding is not detected under the experimental conditions. There is then a rather slow process associated with the stabilization of the high-affinity binding.

It is also possible to determine which ligands other than AcCh cause the formation of the receptor conformation that binds AcCh with high affinity. This can be studied qualitatively by preincubating the receptor-rich membranes with another ligand prior to the addition of [^{3}H]-AcCh, and then measuring the rate of attainment of equilibrium for [^{3}H]-AcCh. In Figure 5.5 it is shown that following 20-minute preincubation of a membrane suspension with the antagonist dimethyltubocurarine, the binding of [^{3}H]-AcCh is time independent (occurs within 20 seconds). The AcCh bound at equilibrium is then the same as that measured at equilibrium following the simultaneous addition of AcCh and dimethyltubocurarine to the membrane suspension. These results suggest that the preincubation with dimethyltubocurarine does cause the appearance of the high-affinity receptor conformation, and then when AcCh is added, a competitive reequilibration of bound ligands occurs rapidly. Because of this fact, the amount of AcCh bound shortly after mixing is actually greater when the membranes have been pretreated with the antagonist than when agonist and antagonist are added simultaneously.

In summary, in the presence of low concentrations of AcCh (less than 0.1 μM), a desensitized (high-affinity) receptor conformation of the membrane-bound *Torpedo* receptor appears in seconds at 4°C. The antagonist dimethyltubocurarine also causes the formation of a high-affinity receptor conformation, although the time course of the conformational change in the presence of antagonist is not yet known. That antagonists can act in such a manner has been inferred from in vivo studies of desensitization (Rang and Ritter, 1969), and from membrane structural changes detected by the use of a fluorescent local anesthetic (Grunhagen and Changeux, 1975). There are, therefore, several techniques available to study the dynamics of response of the receptor-rich membranes to the presence of cholinergic ligands.

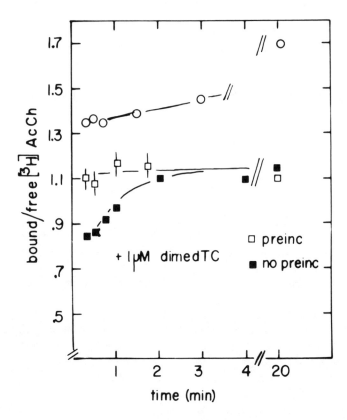

FIGURE 5.5. Change of binding properties of membrane-bound cholinergic receptor of *T. marmorata,* caused by preincubation of the membrane fragments with the antagonist dimethyltubocurarine. In the sample indicated by ■———■, following incubation of a membrane suspension (1.6 g protein/l, 0.8 μM a-toxin sites) with 3.3 mM DFP, the membranes are rapidly diluted 40-fold into a physiological salt solution containing 16 nM [^{3}H] AcCh and 1 μM dimethyltubocurarine. The sample indicated by □———□ is done in the same way, except that the concentrated suspension is preincubated 20 minutes with 1 μM dimethyltubocurarine prior to the 40-fold dilution. In o ——— o, after pretreatment with DFP, the concentrated suspension is diluted directly into the solution containing 16 nM AcCh. Free AcCh is separated from the membranes suspension by ultrafiltration.

Binding of Local Anesthetics to Receptor-Rich Membranes

The preceding sections illustrate that the detailed analysis of the binding of cholinergic agonists and competitive antagonists to the receptor-rich membranes provides considerable information about the cholinergic permeability control mechanism. However, such studies do not provide direct information about the permeability response. Drugs and chemicals acting at the postsynaptic membrane as noncompetitive antagonists must be acting at a site other than the AcCh binding site, and an understanding of their mode of action could shed light on the coupling between receptor occupation and the permeability response. Many surface-active compounds act as noncompetitive blocking agents: for example, detergents, general anesthetics, and local anesthetics (Cohen and Changeux, 1975; Gage, 1976). While many of these chemicals undoubtedly act as nonspecific membrane perturbants, it is also reasonable to try to identify compounds interacting with specific sites in the postsynaptic membrane. For example, is it possible to identify compounds interacting with the site of ion translocation? If the number of such sites is comparable to the number of AcCh binding sites in the receptor-rich membranes, it is possible to have micromolar concentrations of channels in a test tube, and it will be easy to measure ligand binding to that site if it is characterized by a dissociation constant micromolar or lower.

Local anesthetics might interact with specific sites in the postsynaptic membrane. They are aromatic secondary or tertiary amines, cationic at physiological pH, that are defined by the fact that they block reversibly the action potential of nerve and muscle (not the postsynaptic response). Certain local anesthetics do act as noncompetitive blocking agents at the nicotinic synapse (Podleski and Bartels, 1963; Steinbach, 1968; Weber and Changeux, 1974). While many such compounds act only at rather high concentrations (0.1 mM or higher), there are more potent compounds.

Dimethisoquin is a local anesthetic that is one of the most potent noncompetitive blocking agents at the nicotinic synapses, acting in vivo on the *Electrophorus* electroplacque and in vitro on the receptor-rich *Torpedo* membranes at micromolar concentrations (Weber and Changeux, 1974; Cohen, Weber, and Changeux, 1974).

$$O\ CH_2\ CH_2\ N(CH_3)_2$$

Dimethisoquin

Quat-dimethisoquin—(1-β-trimethylaminoethoxy)-3-n-butyl isoquinoline iodide—the monomethyl quaternary derivative of dimethisoquin, is actually an order of magnitude more potent than the parent compound. It blocks the *Electrophorus* response in vivo at 0.1 μM, and it acts in vitro at similar concentrations to increase the affinity of the membrane-bound *Torpedo* receptor for AcCh (Fig. 5.6). Hence, it is of interest to study the

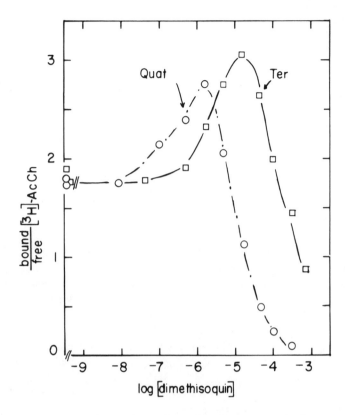

FIGURE 5.6. Effect of the local anesthetic dimethisoquin and of a quaternized derivative on the binding of [³H] AcCh to receptor-rich membrane fragments. A membrane suspension (4 g protein/l, 1.5 µM a-toxin sites) is incubated 30 minutes with 3.3 mM DFP, then diluted 50-fold into physiological saline solution containing 7 nM total [³H] AcCh and the indicated concentrations of local anesthetics: □————□, dimethisoquin; o ————— o, quaternized dimethisoquin (1-β-trimethylaminoethoxy)-3-n-butylisoquinoline iodide). Binding was determined by an ultracentrifugation assay.

binding of this compound to the receptor-rich membranes (Cohen, in preparation).

The equilibrium binding of [¹⁴C]-quat dimethisoquin to the receptor-rich membranes in normal physiological salt solution is determined by an ultracentrifugation assay (Fig. 5.7). In the absence of cholinergic ligands there is a nonsaturable binding occurring for ligand concentrations ranging from .03 µM to 3 µM. However, even at 1 µM drug, the number of anesthetic molecules bound is no greater than the number of AcCh binding sites in the membranes. If the receptor sites are occupied with a-neurotoxin, there is no change in that binding. However, in the presence of the agonist carbamylcholine, there is an additional binding of quat-

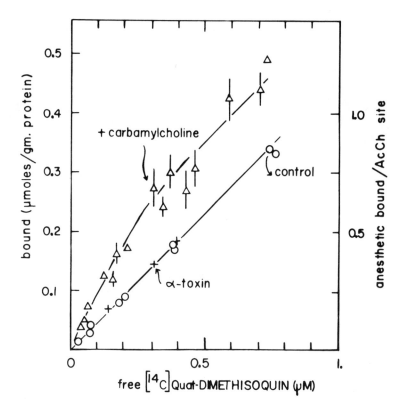

FIGURE 5.7. Binding at equilibrium of the noncompetitive antagonist [^{14}C] quat-dimethisoquin to the receptor-rich membranes of *T. marmorata*. The free quat-dimethisoquin in equilibrium with the membrane fragments (0.3 g protein/l, 0.1 μM a-toxin sites) in physiological salt solution is determined from the supernatant radioactivity after ultracentrifugation, and the bound ligand is determined from the difference between the free and total ligand. Equilibrium binding is shown in the absence of any cholinergic ligand (o ———— o), in the presence of 1 μM a-toxin (+ ———— +), and in the presence of 26 μM carbamylcholine (Δ ———— Δ).

dimethisoquin that is saturable. Although the data lack the precision of the experimental data obtained for the binding of cholinergic ligands, the additional anesthetic binding observed in the presence of carbamylcholine can be characterized by a K_D of about 0.2 μM, and the number of molecules bound appears about one-quarter the number of AcCh binding sites. Because of the conditions of the experiment, the additional anesthetic bound in the presence of carbamylcholine must be bound at a site other than the acetylcholine binding site: (1) over the range of anesthetic concentrations studied, the anesthetic only causes an increased binding of AcCh to the membranes; (2) since the concentration of carbamyl-

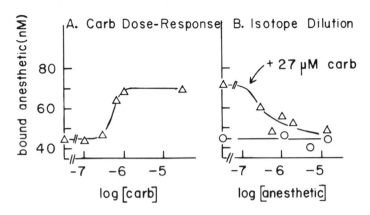

FIGURE 5.8. Binding at equilibrium of [^{14}C] quat-dimethisoquin (0.4 μM) to the receptor-rich membranes of *T. marmorata* in the presence of various concentrations of the cholinergic agonist carbamylcholine or of nonradioactive quat-dimethisoquin. A membrane suspension (0.3 g protein/l, 0.1 μM α-toxin sites) is equilibrated with quat-dimethisoquin (0.4 μM). (*A*) Carb dose-response: the amount bound in the presence of the indicated concentrations of carbamylcholine. (*B*) Isotope dilution: the amount bound in the presence of the indicated concentrations of non-radioactive quat-dimethisoquin, either in the presence of 27 μM carbamylcholine (Δ ——— Δ) or in the absence of any added cholinergic ligand (o ——— o).

choline (27 μM) is over an order of magnitude greater than the equilibrium dissociation constant for its binding to the receptor site, essentially all receptor sites are occupied by carbamylcholine.

Further evidence is available strengthening the interpretation that there is a specific binding site for this quaternary local anesthetic in the membranes and that its occupation is related to the presence of carbamylcholine:

1. If receptor sites are occupied by α-neurotoxin, the presence of carbamylcholine results in no increased binding of the local anesthetic.

2. The concentration of carbamylcholine necessary to increase the amount of bound anesthetic is characterized by $K_{AP} = 0.5$ μM, that is, the K_D for the carbamylcholine-receptor interaction (Fig. 5.8A).

3. The increased binding of quat-dimethisoquin associated with the presence of carbamylcholine can be displaced by nonradioactive quat-dimethisoquin, as is necessary if the binding is to a finite number of sites (Fig. 5.8B).

Hence, although there is a nonspecific partitioning of quat-dimethisoquin into the membrane, there is also binding to a specific site in the receptor-rich membranes. However, since this specific equilibrium binding is only seen in the presence of carbamylcholine, quat-dimethisoquin is actually binding to the receptor-rich membranes when the receptor is in the conformation binding agonists with high affinity, the "desensitized" receptor. Further studies are necessary to identify the nature of this binding site and its relationship to the mechanism of ion translocation.

Receptor Conformation and Receptor Function: Speculations

Although it is not yet possible to identify in a direct manner the rates and equilibria involved, it is useful to try to speculate how the ligand-binding properties of the receptor-rich membranes might be related to the different functional forms of the receptor: R_c, the receptor conformation existing in the absence of ligands; R_o, the receptor associated with an open channel; and R_d, the desensitized receptor.

The following mechanism is a modification of the cyclic mechanism proposed by Katz and Thesleff (1957) to account for the kinetics of receptor desensitization:

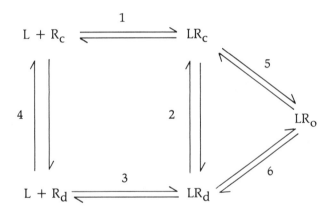

Such a mechanism implies (arbitrarily) that there is no active receptor unoccupied by agonist, but that there can be direct ligand binding to R_d. Since the antagonist dimethyltubocurarine can cause the appearance of high affinity AcCh binding, it seems wise to suggest that agonists may desensitize either via (1)-(2) or (1)-(5)-(6), while antagonists are involved in (1)-(2).

The binding studies presented here set limits on some of the equilibria. The apparent equilibrium dissociation constant for AcCh is about 10^{-8}M. This necessitates that $K_1 K_2 = K_3 K_4 = 10^{-8}$M, that is, that none of the individual conformations need bind AcCh with that dissociation constant. It is reasonable to assume that at least the bimolecular association K_1 occurs rapidly relative to the conformational changes (2) and (4). Since the high-affinity binding develops seconds after the mixing of AcCh and membranes, there is little high-affinity receptor (R_D) in the absence of AcCh. Hence, $K_4 = R_c/R_d > 10$ and $K_3 < 10^{-9}$M. In this case it is quite possible that even if K_3 involves a diffusion-limited bimolecular association, the dissociation of LR_d may take seconds, or longer, to occur. In any event, the kinetics of appearance of LR_d must reflect rate constants both for $LR_c \rightarrow LR_d$ and $LR_c \rightarrow LR_o \rightarrow LR_d$. There is, unfortunately, no data for K_1 or K_2. If the equilibrium binding measured in cholate solutions is relevant, perhaps $K_1 \sim 10^{-6}$M. In that case it will be possible to determine K_1 by an appropriate kinetic experiment.

The fact that the binding of AcCh involves multiple equilibria and that some of the rates are quite slow may permit a reinterpretation of previous binding experiments thought to be performed at equilibrium. For certain preparations of receptor-rich membranes, AcCh binding is sigmoidal in nature, with the first molecules binding with an affinity of about 30 nM and the last with an affinity of 6 nM (Weber and Changeux, 1974; Cohen, Weber, and Changeux, 1974). Such binding functions may reflect some of the slow kinetic processes reported here (Fig. 5.3). A second feature is the manner by which local anesthetics modify the equilibrium binding functions. Prilocaine converts a sigmoidal AcCh binding function into a hyperbolic function characterized by $K_d = 6$ nM, and a comparable increase of affinity is seen for a cholinergic antagonist such as dimethyltubocurarine (Fig. 5.2) that binds at equilibrium with a single dissociation constant. In the present study two further aspects are established: one such anesthetic, quat-dimethisoquin, actually interacts preferentially with a high-affinity receptor conformation (Fig. 5.6), and even twenty minutes after mixing AcCh and membranes the increased affinity for AcCh associated with the presence of SKF-525a (Fig. 5.3) still takes minutes to develop. Further studies are necessary to determine whether local anesthetics modify the binding properties of only the high-affinity receptor conformation.

The studies do not yet provide the means of identifying the receptor conformation associated with active receptors. However, if the cyclic mechanism is reasonable, it is interesting to note that the drug concentrations necessary for a response, that is, the formation of LR_o, are not going to be related in a simple manner to any of the individual equilibria. If the formation of desensitized receptors is clearly the rate-limiting step, then the apparent constant from the dose response relation will approach K_1 if $K_5 = LR_c/LR_o \gg 1$. Indirect evidence led Katz and Miledi (1973) to suggest that less than 10 percent of the receptors occupied by AcCh were associated with open channels ($K_5 > 10$). However, this reasoning only suggests the type of data that must be obtained and analyzed in order to understand the mechanism of permeability control by cholinergic receptors.

Conclusion

When a cell surface containing nicotinic, cholinergic receptors is exposed to acetylcholine (AcCh) or other cholinergic agonists, there is a resultant increase in the membrane permeability to Na and K. During normal synaptic functioning, AcCh is only present for milliseconds, and for that length of time the ion channels remain open. However, when a cholinergic agonist is present for extended periods of time (seconds or minutes) at a constant concentration, there is not a steady-state permeability response. Instead, there is a gradual loss of tissue responsiveness to the agonist; desensitization occurs. In order to understand the molecular mechanism by which the ligand binding is coupled to the control of permeability, it is necessary to utilize a variety of experimental approaches and to study the problem at levels of organization ranging from that of intact cells to isolated molecules.

The receptor-rich membranes isolated from *Torpedo* electric tissue provide a unique material for the study of the molecular mechanisms involved in the control of permeability by AcCh. The dense clustering of cholinergic receptors characteristic of the nicotinic postsynaptic membrane in vivo is also seen in these isolated membranes. Individual receptor oligomers of diameter 9 nm are directly visualized at a surface density of $12,000/\mu^2$ by electron microscopy. Analysis of the peptide composition of the isolated membranes reveals that the major polypeptides, if not the only ones, are those found in the purified isolated cholinergic receptor. This high degree of structural specialization permits detailed studies of the molecular architecture of the postsynaptic membrane.

The fact that isolated membranes contain such a high concentration of receptors and that these receptors remain functional permits the use of various biophysical techniques to study the mechanisms of permeability control. This report has emphasized one approach, the analysis of the ligand-binding properties of these membranes. It is possible to study not only the binding at equilibrium of AcCh and cholinergic ligands to the membrane-bound receptor, but also the binding of noncompetitive blocking agents to these membranes. Furthermore, it is possible to begin a kinetic analysis of the AcCh binding function. The membrane-bound receptor exists in multiple conformations. A conformation binding AcCh with high affinity is stabilized by the presence of cholinergic ligands, and its rate of appearance is about that expected for receptor desensitization. Further data are necessary to determine the individual rates and equilibria involved.

ACKNOWLEDGMENTS

I gratefully acknowledge the technical assistance of Dan Medynski, and I thank Drs. Menez, Morgat, Fromageot, and Boquet for a generous gift of [³H]-*a*-neurotoxin of *N. nigricollis*, Dr. C. Cazaux at the Institut de Biologie Marine, Arcachon, for supplying *Torpedo*, and L. Garabaldi and J. Dayton for maintaining the *Torpedo*. This work was supported by U.S. Public Health Service research grant NS 12408 from the National Institutes of Health, by a grant from the

Pharmaceutical Manufacturers Association, and by an NIH Research Career Development Award (NS 00155).

REFERENCES

Anderson, C. R., and C. F. Stevens. 1973. Voltage clamp analysis of acetylcholine produced end-plate current fluctuations at frog neuromuscular junction. *J. Physiol.* 235:655-691.

Bartels, E. 1965. Relationship between acetylcholine and local anesthetics. *Biochim. Biophys. Acta* 109:194-203.

Bartels, E., and T. L. Rosenberry. 1972. Correlation between electrical activity and splitting of phospholipids by snake venom in the single electroplax. *J. Neurochem.* 19:1251-1265.

Bourgeois, J. P., A. Ryter, A. Menez, P. Fromageot, P. Bouquet, and J.-P. Changeux. 1972. Localization of the cholinergic récepteur protein in *Electrophorus* electroplax by high resolution autoradiography. *FEBS Letters* 25: 127-133.

Brisson, A., P. Devaux, and J.-P. Changeux. 1975. Effet anesthesique local de plusieurs composes liposolubles sur la reponse de l'electroplacque de gymnote a la carbamylcholine et sur la liaison de l'acetylcholine au récepteur cholinergique de Torpille. *C. R. Acad. Sci.* D280: 2153-2156.

Cartaud, J., L. Benedetti, J. B. Cohen, J. C. Meunier, and J.-P. Changeux. 1973. Presence of a lattice structure in membrane fragments rich in nicotinic receptor protein from the electric organ of *Torpedo marmorata. FEBS Letters* 33:109-113.

Changeux, J.-P. 1975. The cholinergic receptor protein from fish electric organ. In L. L. Iversen et al., eds., *Handbook of psychopharmacology.* Vol. 6, pp. 235-301. Plenum Press, New York.

Changeux, J.-P., L. Benedetti, J. P. Bourgeois, A. Brisson, J. Cartaud, P. Devaux, H. Grunhagen, M. Moreau, J. L. Popot, A. Sobel, and M. Weber. 1976. Some structural properties of the cholinergic receptor protein in its membrane environment relevant to its function as a pharmacological receptor. *Cold Spring Harbor Sym. Quant. Biol.* 40: 211-230.

Chao, Y., R. L. Vandlen, and M. A. Raftery. 1975. Preferential chemical modification of a binding subsite on the acetylcholine receptor. *Biochem. Biophys. Res. Commun.* 63:300-307.

Cohen, J. B., and J.-P. Changeux. 1973. Interaction of a fluorescent ligand with membrane bound cholinergic receptor from *Torpedo marmorata. Biochemistry* 12:4855-4864.

Cohen, J. B., and J.-P. Changeux. 1975. The cholinergic receptor protein in its membrane environment. *Annu. Rev. Pharmacol.* 15:83-103.

Cohen, J. B., M. Weber, and J.-P. Changeux. 1974. Effects of local anesthetics and calcium on the interaction of cholinergic ligands with the nicotinic receptor protein from *Torpedo marmorata. Mol. Pharmacol.* 10:904-932.

Cohen, J. B., M. Weber, M. Huchet, and J.-P. Changeux. 1972. Purification from *Torpedo marmorata* electric tissue of membrane fragments particularly rich in cholinergic receptor. *FEBS Letters* 26:43-47.

delCastillo, J., and B. Katz. 1954. The membrane change produced by the neuromuscular transmitter. *J. Physiol.* 125:546-565.

Duguid, J. R., and M. A. Raftery. 1973. Fractionation and partial characterization of membrane particles from *Torpedo californica* electroplax. *Biochemistry* 12:3593-3597.

Dupont, Y., J. B. Cohen, and J.-P. Changeux. 1974. X-ray diffraction study of membrane fragments rich in acetylcholine receptor protein prepared from the electric organ of *Torpedo marmorata*. *FEBS Letters* 41: 130-133.

Eldefrawi, M. E., A. Britten, and A. T. Eldefrawi. 1971. Acetylcholine binding to *Torpedo* electroplax: relationship to acetylcholine receptors. *Science* 173: 338-340.

Eldefrawi, M. E., A. T. Eldefrawi, L. P. Gilmour, and R. D. O'Brien. 1971. Multiple affinities for binding of cholinergic ligands to a particulate fraction of *Torpedo* electroplax. *Mol. Pharmacol.* 7:420-428.

Eldefrawi, M. E., A. T. Eldefrawi, and A. E. Shamoo. 1975. Molecular and functional properties of the acetylcholine receptor. *Ann. N.Y. Acad. Sci.* 264: 183-202.

Eldefrawi, M. E., A. T. Eldefrawi, and D. B. Wilson. 1975. Tryptophan and cysteine residues of the acetylcholine receptors of *Torpedo* species: relationship to binding of cholinergic ligands. *Biochemistry* 14: 4304-4310.

Fatt, P., and B. Katz, 1951. An analysis of the end-plate potential recorded with an intracellular electrode. *J. Physiol.* 115:320-370.

Fertuck, H. C., and M. M. Salpeter. 1976. Quantitation of junctional and extrajunctional acetylcholine receptors by electron microscope autoradiography after [125]I a-bungarotoxin binding at mouse neuromuscular junctions. *J. Cell Biol.* 69:144-158.

Fessard, A. 1958. Les organes electriques. In P. P. Grasse, ed., *Traité de zoologie.* Vol. 13A, pp. 1143-1238. Masson, Paris.

Flanagan, S. D., S. H. Barondes, and P. Taylor. 1976. Affinity partitioning of membranes. *J. Biol. Chem.* 251:858-865.

Franklin, G. I., and L. T. Potter. 1972. Studies of the binding of a-bungarotoxin to membrane bound and detergent-dispersed acetylcholine receptors from *Torpedo* electric tissue. *FEBS Letters* 28: 101-106.

Gage, P. W. 1976. Generation of end-plate potentials. *Physiol. Rev.* 56: 177-247.

Grunhagen, H., and J.-P. Changeux. 1975. Transitions structurales du récepteur cholinergique de Torpille dans son état membranaire mises on évidence par l'aide d'un anesthetique local fluorescent: la quinacrine. *C. R. Acad. Sci.* D 281:1047-1050.

Haydon, D. A., and S. B. Hladky. 1972. Ion transport across thin lipid membranes: a critical discussion of mechanisms in selected systems. *Q. Rev. Biophys.* 5: 187-282.

Hazelbauer, G. L., and J.-P. Changeux. 1974. Reconstitution of a chemically excitable membrane. *Proc. Natl. Acad. Sci. USA* 71:1479-1483.

Hess, G. P., J. P. Andrews, G. E. Struve, and S. E. Coombs. 1975. Acetylcholine-receptor-mediated ion flux in electroplax membrane preparations. *Proc. Natl. Acad. Sci. USA* 72:4371-4375.

Israel, M., J. Gautron, and B. Lesbats. 1970. Subcellular fractionation of the electric organ of *Torpedo marmorata*. *J. Neurochem.* 17:1441-1450.

Karlin, A. 1974. The acetylcholine receptor: a progress report. *Life Sci.* 14: 1385-1415.

Karlin, A., and D. Cowburn. 1973. The affinity-labeling of partially purified acetylcholine receptor from electric tissue of *Electrophorus. Proc. Natl. Acad. Sci. USA* 70:3636-3640.

Karlin, A., C. L. Weill, M. B. McNamee, and R. Valderrama. 1976. Facets of the structures of acetylcholine receptors from *Electrophorus* and *Torpedo. Cold Spring Harbor Symp. Quant. Biol.* 40: 203-210.

Kasai, M., and J.-P. Changeux. 1971. *In vitro* excitation of purified membrane fragments by cholinergic agonists. *J. Membrane Biol.* 6:1-80.

Katz, B., and R. Miledi. 1972. The statistical nature of the acetylcholine potential and its molecular components. *J. Physiol.* 224:665-699.

Katz, B., and R. Miledi. 1973. The binding of acetylcholine to receptors and its removal from the synaptic cleft. *J. Physiol.* 231:549-574.

Katz, B., and S. Thesleff. 1957. A study of the desensitization produced by acetylcholine at the motor end-plate. *J. Physiol.* 138:63-80.

Lee, C. Y. 1972. Chemistry and pharmacology of polypeptide toxins in snake venoms. *Ann. Rev. Pharmacol.* 12:265-281.

Lee, C. Y., and C. C. Chang. 1966. Modes of actions of purified toxins from elapid venoms on neuromuscular transmission. *Mem. Inst. Butantan Symp. Internat.* 33:555-572.

Lester, H. A., J.-P. Changeux, and R. E. Sheridan. 1975. Conductance increases produced by bath application of cholinergic agonists to *Electrophorus* electroplacques. *J. Gen. Physiol.* 65:797-816.

Michaelson, D. M., and M. A. Raftery. 1974. Purified acetylcholine receptor: its reconstitution to a chemically excitable membrane. *Proc. Natl. Acad. Sci. USA* 71:4768-4772.

Nachmansohn, D. 1959. *Chemical and molecular basis of nerve activity.* Academic Press, New York.

Neher, E., and B. Sakman. 1976. Single channel currents recorded from membrane of denervated frog muscle fibers. *Nature* 260:799-801.

Nickel, E., and L. T. Potter. 1973. Ultrastructure of isolated membranes of *Torpedo* electric tissue. *Brain Res.* 57:508-517.

O'Brien, R. P., and Gibson, R. E. 1975. Conversion of high affinity acetylcholine receptor from *T. californica* electroplax to altered form. *Arch. Biochem. Biophys.* 169:458-463.

Podleski, T. R., and E. Bartels. 1963. Difference between tetracaine and *d*-tubocurarine in the competition with carbamylcholine. *Biochim. Biophys. Acta* 75:387-396.

Popot, J. L., H. Sugiyama, and J.-P. Changeux. 1974. Demonstration de la desensibilisation pharmácologique du récepteur de l'acétylcholine *in vitro* avec des fragments de membrane excitable de Torpille. *C. R. Acad. Sci.* D 279: 1721.

Porter, C. W., and E. A. Barnard. 1975. Density of cholinergic receptors at the endplate postsynaptic membrane: ultrastructural studies in two mammalian species. *J. Membrane Biol.* 20:31-49.

Potter, L. T. 1974. *a*-Bungarotoxin (and similar *a*-neurotoxins) and nicotinic acetylcholine receptors. *Methods Enzymol.* 32B: 309-323.

Raftery, M. A., J. Bode, R. Vandlen. Y. Chao, J. Deutsch, J. R. Duguid, K. Reed, and T. Moody. 1974. Characterization of an acetylcholine receptor. In L. Jaenicke, ed., *Biochemistry of sensory functions,* pp. 541-564. Springer-Verlag, New York.

Raftery, M. A., J. Bode, R. Vandlen, D. Michaelson, J. Deutsch, T. Moody, M. J. Ross, and R. M. Stroud. 1975. Structural and functional studies of an acetylcholine receptor. In H. Sund, ed., *Protein-ligand interactions*, pp. 328-355. Walter de Gruyter, Berlin.

Raftery, M. A., R. L. Vandlen, K. L. Reed, and T. Lee. 1976. Characterization of *Torpedo californica* acetylcholine receptors: its subunit composition and ligand binding properties. *Cold Spring Harbor Sym. Quant. Biol.* 40: 193-202.

Rang, H. P., 1975. Acetylcholine receptors. *Q. Rev. Biophys.* 7:283-399.

Rang, H. P., and J. M. Ritter. 1969. A new kind of drug antagonism: evidence that agonists cause a molecular change in acetylcholine receptors. *Mol. Pharmacol.* 5:394-411.

Reed, K., R. Vandlen, J. Bode, J. Duguid, and M. A. Raftery. 1975. Characterization of acetylcholine receptor-rich and acetylcholinesterase-rich membrane particles from *Torpedo californica* electroplax. *Arch. Biochem. Biophys.* 167:138-144.

Rosenbluth, J. 1975. Synaptic membrane structure in *Torpedo* electric organ. *J. Neurocytol.* 4:697-712.

Sheridan, M. N. 1965. The fine structure of the electric organ of *Torpedo marmorata. J. Cell Biol.* 24: 129-141.

Steinbach, A. B. 1968. A kinetic model for the action of xylocaine on receptors for acetylcholine. *J. Gen. Physiol.* 52:162-180.

Sugiyama, H., and J.-P. Changeux. 1975. Interconversion between different states of affinity for acetylcholine of the cholinergic receptor protein from *Torpedo marmorata. Eur. J. Biochem* 55: 505-515.

Takeuchi, A., and N. Takeuchi. 1960. On the permeability of end-plate membrane during the action of transmitters. *J. Physiol.* 154:52-67.

Weber, G., D. Borris, E. DeRobertis, F. Barrantes, J. LaTorre, and M. Carlin. 1971. The use of cholinergic fluorescent probe for the study of the receptor proteolipid. *Mol. Pharmacol.* 7:530-537.

Weber, M., and J.-P. Changeux. 1974a. Binding of *Naja nigricollis* [^{3}H] a-toxin to membrane fragments from *Electrophorus* and *Torpedo* electric organs. I. Binding of the tritiated a-neurotoxin in the absence of effects. *Mol. Pharmacol.* 10:1-14.

Weber, M., and J.-P. Changeux. 1974b. Binding of *Naja nigricollis* [^{3}H] a-toxin to membrane fragments from *Electrophorus* and *Torpedo* electric organs. II. Effect of the cholinergic agonists and antagonists on the binding of the tritiated a-neurotoxins. *Mol. Pharmacol.* 10:15-34.

Weber, M., and J.-P. Changeux. 1974c. Binding of *Naja nigricollis* [^{3}H] a-toxin to membrane fragments from *Electrophorus* and *Torpedo* electric organs. III. Effect of local anesthetics on the binding of the tritiated a-neurotoxins. *Mol. Pharmacol.* 10:35-40.

Weber, M., T. David-Pfeuty, and J.-P. Changeux. 1975. Regulation of the binding properties of the nicotinic receptor protein by cholinergic ligands in membrane fragments from *Torpedo marmorata. Proc. Natl. Acad. Sci. USA* 72:3443.

Weill, C. L., M. G. McNamee, and A. Karlin. 1974. Affinity-labeling of purified acetylcholine receptors from *Torpedo californica. Biochem. Biophys. Res. Commun.* 61:997-1003.

6 The Functional Roles of Band 3 Protein of the Red Blood Cell

ASER ROTHSTEIN

Band 3 of the red blood cell membrane is defined as a particular, heavily stained band of protein of about 90,000 daltons observed after sodium dodecyl sulfate (SDS)[1] acrylamide gel electrophoresis of dissolved red-cell membranes (Fairbanks, Steck, and Wallach, 1971). Although it was reported as an entity only five years ago, almost a hundred papers describing its properties and functions have appeared in that time. The intense interest in band 3 has arisen for several reasons: (1) it is one of the major "intrinsic," hydrophobic proteins of the membrane; (2) it is accessible from both the outside and the inside of the cell, and so appears to "span" the membrane; (3) it appears to be involved in a number of transport functions including anion, sugar, cation, and water permeation, and cation active transport; and (4) it is reported to contain acetylcholinesterase and to be a substrate for phosphorylation by endogenous protein kinase. The apparent association of transport functions with transmembrane proteins is particularly intriguing because it suggests that the protein itself may provide a pathway through which the transfers across the membrane may occur.

This paper will be primarily concerned with possible roles of band 3 protein in transport. The chemical and physical properties of band 3, its arrangement in the membrane, and its association with other membrane

[1]Abbreviations: DASA, diazosulfonic acid; DIDS, 4,4'-diisothiocyano-2,2'-stilbene disulfonic acid; FDNB, 1-fluoro-2,4-dinitrobenzene; FMMP, formyl methionyl methyl phosphate; IBSA, 1-isothiocyano-4-benzene sulfonic acid; IMP, intramembranous particles; NAP-taurine, N-(4-azido-2-nitrophenyl)-2-aminoethyl sulfonic acid; PDP, pyridoxal phosphate; SDS, sodium dodecyl sulfate; TNBS, 2,4,6 trinitrobenzene sulfonic acid.

components have been extensively reported; they will be briefly outlined to provide the context for the transport studies. More detailed information can be found in a number of reviews (Bretscher, 1973; Carraway, 1975; Guidotti, 1972; Juliano, 1972; Singer, 1974; Singer and Nicolson, 1972; Steck, 1974; Wallach, 1972).

Chemical and Physical Properties of Band 3 Protein

The molecular weight of band 3 has been determined as about 90,000 daltons by SDS-acrylamide gel electrophoresis (Fairbanks, Steck, and Wallach, 1971), 87,000 by sedimentation under centrifugal force (Clarke, 1975), and 80,000 by gel filtration in SDS (Clarke, 1975; Ho and Guidotti, 1975). Band 3 is, however, an unusually wide band on acrylamide gels (marked by an arrow in the gel at the top of Fig. 6.1), so that the apparent molecular weight covers the range of perhaps 85,000 to 110,000. The band spread has been attributed to the heterogeneity of the carbohydrates (Steck, 1974; Steck and Dawson, 1974), which constitute a small fraction of the total mass of the protein. Heterogeneity of carbohydrates has also been demonstrated by the use of lectins. Thus, concanavalin A binds to band 3 with a high degree of specificity, as compared to the other membrane glycoproteins (Furthmayr, Kahane, and Marchesi, 1976; Tanner and Anstee, 1976), but only a fraction (as little as 15 percent) of the band 3 interacts (Adair and Kornfeld, 1974; and Findlay, 1974).

The carbohydrate content of band 3 is estimated at 4-9 percent, based on various reports (Furthmayr, Kahane, and Marchesi, 1976; Ho and Guidotti, 1975; Steck, 1974; Tanner and Boxer, 1972), with galactose, mannose, and N-acetyl glucosamine as the principal sugars, but with little sialic acid (Furthmayr, Kahane, and Marchesi, 1976). The amino acid composition is unusual in terms of the high proportion (almost 40 percent) of nonpolar residues (Furthmayr, Kahane, and Marchesi, 1976; Guidotti, 1972; Ho and Guidotti, 1975; Tanner and Boxer, 1972; Yu and Steck, 1975). This finding is consistent with the solubility characteristics of band 3. It is a highly hydrophobic protein that cannot be extracted by high or low ionic strength solutions, or by protein perturbants (Juliano, 1972; Juliano and Rothstein, 1971; Steck, 1972a; Steck, 1974; Steck and Yu, 1973; Tanner and Boxer, 1972). It can, however, be dissolved in ionic or nonionic detergents (Fairbanks, Steck, and Wallach, 1971; Juliano, 1972; Steck, 1974; Tanner and Boxer, 1972; Yu, Fischman, and Steck, 1973). The high degree of hydrophobicity is indicated by the observation that solubilization by Triton X-100 is associated with the binding of .85 grams of detergent per gram of protein (Clarke, 1975).

Although band 3 is an abundant protein in the membrane, the exact amount present is not precisely known. Estimates of 25 percent to 30 percent of the total membrane protein are often cited, based on densito-

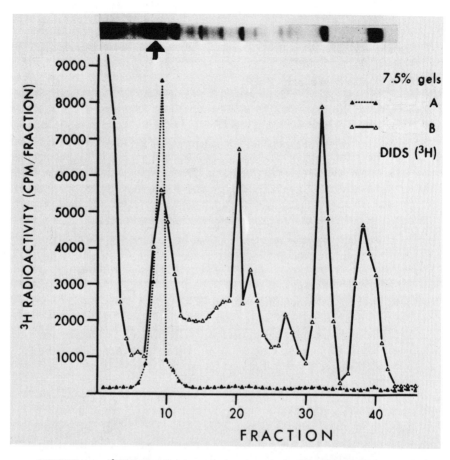

FIGURE 6.1. [³H] DIDS binding to membrane proteins of intact cells (A) and leaky ghosts (B) as determined by SDS-acrylamide gel electrophoresis. The gel at the top is stained for protein with Coomassie blue. (From Cabantchik and Rothstein, 1974a.)

metric tracings from SDS-acrylamide gels stained with Coomassie Blue (Fairbanks, Steck, and Wallach, 1971; Steck, 1974). This value is, however, dependent on the procedure used for preparing ghosts, and on the assumption that the staining with the dye is proportional to the amount of each protein. Another estimate can be made by extracting all of the "extrinsic" proteins, which can be done by using variations in pH and ionic strength and by the addition of urea. The "intrinsic" (hydrophobic) proteins (largely sialoglycoproteins and band 3) remaining in the vesicles constitute about 30 percent of the total based on chemical analysis (Juliano and Rothstein, 1971). The sialoproteins constitute about 6-7 percent of the total membrane protein (Steck, 1974), so that band 3 would be about 24 percent, in reasonable agreement with the staining procedure.

The number of band 3 monomers per cell has been calculated to be about a million (Fairbanks, Steck, and Wallach, 1971; Steck, 1974). The calculation is, however, only approximate. It is based on a molecular weight of 89,000, on the assumption that band 3 represents 30 percent of total membrane protein, and on a particular value for the amount of protein per ghost.

Functional studies suggest that the polypeptide portion of band 3 may be heterogeneous. It has been reported, for example, that band 3 is involved in anion transport (Cabantchik and Rothstein, 1972, 1974a; Ho and Guidotti, 1975; Lepke et al., 1976; Rothstein, Cabantchik, and Knauf, 1976; Zaki et al., 1975), sugar transport (Batt, Abbott, and Schachter, 1976; Kahlenberg, 1976; Kasahara and Hinkle, 1976a; Lin and Spudich, 1974; Taverna and Langdon, 1973). Na,K-ATPase activity (Avruch and Fairbanks, 1972; Knauf, Proverbio, and Hoffman, 1974; Williams, 1972), water permeation (Brown, Feinstein, and Sha'afi, 1975), and cholinesterase activity (Bellhorn, Blumenfeld, and Gallor, 1970). Each function presumably involves a unique polypeptide. Some of the functions may, however, involve amounts of protein too small to be distinguished as separate species by present fractionation techniques. The Na,K-ATPase, for example, involves only about 300 copies per cell (Sachs, Knauf, and Dunham, 1974).

Attempts to demonstrate chemical heterogeneity by end-group analysis have led to some confusion. Earlier reports of multiple N-terminal amino acids (Knupferman, Bhakdi, and Wallach, 1975; Langdon, 1974) have not been confirmed and have been attributed to impurities. In fact, the N-terminal amino acids seem to be completely blocked (Furthmayr, Kahane, and Marchesi, 1976; Ho and Guidotti, 1975; Yu and Steck, 1975). Other evidence does, however, indicate heterogeneity. For example, many investigators have reported that treatment of intact cells with the proteolytic enzyme pronase results in partial digestion of most of band 3 so that it no longer is found in its normal location in SDS acrylamide gel electrophoresis. At the original location of band 3 in gels, three distinct bands are revealed that are not susceptible to the proteolytic attack (see gels at the top of Fig. 6.2). They represent components that were masked by the large amount of pronase-sensitive band 3 component. The phosphorylated intermediate of the Na,K-ATPase is also not susceptible to pronase in the intact membrane (Knauf, Proverbio, and Hoffman, 1974), but it is not visible by staining because it is present in such small amounts. Recently it has been demonstrated that band 3 can be physically separated into a major and four or five minor components by two-dimensional gel electrophoresis (Ansellstetter and Horstman, 1975; Conrad and Penniston, 1976), or by isoelectric focusing (M. Mor-

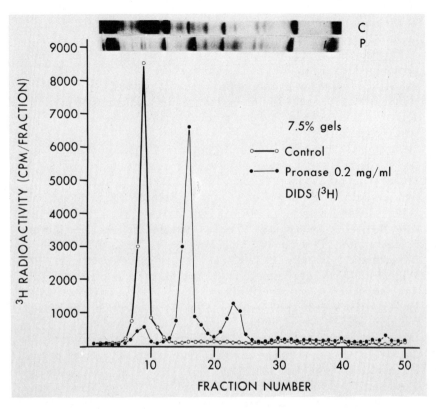

FIGURE 6.2. The effect of pronase treatment of the intact cell on DIDS binding to membrane proteins as determined by SDS-acrylamide gel electrophoresis. The gels are stained for protein with Coomassie blue. The upper (C) is the control and the lower (P) is from pronase-treated cells. (From Cabantchik and Rothstein, 1974b.)

rison, personal communication; Y. Shami and A. Rothstein, unpublished observations). Thus band 3 seems to include a major, pronase-sensitive component (probably heterogeneous) and a number of minor components.

In earlier studies, small amounts of band 3 were obtained for chemical study by cutting out the appropriate segment of the acrylamide gels and eluting the protein. More recently, procedures have been published for isolating more substantial amounts of the protein using a variety of procedures (Findlay, 1974; Furthmayr, Kahane, and Marchesi, 1976; Knuferman, Bhakdi, and Wallach, 1975; Yu and Steck, 1975). Although the latest preparations represent a considerable degree of enrichment and achieve good separation from sialated proteins and other distinct bands, absolute criteria for purity are difficult to establish. The total recovery of band 3 is less than 20-30 percent so the recovered proteins may represent largely a subspecies of band 3 protein.

Arrangement of Band 3 in the Membrane

A portion of band 3 is exposed on the external surface of the membrane. This conclusion is based on several criteria: (1) interactions of band 3 with a number of "nonpenetrating" covalent probes (Carraway, 1975) such as formyl methionyl methyl phosphate (FMMP) (Bretscher, 1971a, b, 1973), 2,4,6 trinitrobenzene sulfonic acid (TNBS) (Arrotti and Garvin, 1971, 1972), diazosulfanilic acid (DASA) (Berg, 1969; Juliano and Rothstein, 1971), 1-isothiocyano-4-benzene sulfonic acid (IBSA) (Ho and Guidotti, 1975), isethionyl acetimidate (Whitely and Berg, 1974), 4,4'-diisothiocyano-2,2'-stilbene disulfonic acid (DIDS) (Cabantchik and Rothstein, 1972, 1974a; Lepke et al., 1976), pyridoxal phosphate (PDP) (Cabantchik et al., 1975a, b) and N-(4-azido-2-nitrophenyl)-2-amino ethyl sulfonic acid (NAP-taurine) (Staros and Richards, 1974); (2) on the splitting of band 3 by proteolytic enzymes from the outside of intact membranes (Bender, Garen, and Berg, 1971; Bretscher, 1971; Cabantchik and Rothstein, 1974b; Hubbard and Cohn, 1972; Jenkins and Tanner, 1975; Passow et al., 1976; Phillips and Morrison, 1971b; Steck, 1972b, 1974; Steck, Fairbanks, and Wallach, 1971; Steck, Ramos, and Strapazon, 1976; Triplett and Carraway, 1972; Wallach, 1972); (3) on its iodination by lactoperoxidase from the outside of intact membranes (Boxer, Jenkins, and Tanner, 1974; Hubbard and Cohn, 1972; Mueller and Morrison, 1974; Phillips and Morrison, 1970, 1971a, b, c, 1973; Reichstein and Blostein, 1973, 1975; Shin and Carraway, 1974); and (4) on its interaction with concanavalin A coupled to ferritin in the intact cell (Pinto da Silva and Nicolson, 1974). Part of band 3 is also accessible from the inside of the membrane, and, therefore, band 3 must span the membrane. The earlier experiments concerning inside and outside accessibility were based on comparisons of intact cells with leaky ghosts. For example, trypsin does not cleave band 3 from the outside of intact membranes, but it does cleave band 3 in leaky ghosts (Carraway, Kobylka, and Triplett, 1971; Steck, 1972b); pronase cleaves band 3 into a 65,000-dalton segment in the intact cell, but extensively degrades it in leaky ghosts (Bender, Garen, and Berg, 1971); DIDS reacts with additional sites in band 3 in leaky ghosts that are not accessible in cells (Cabantchik and Rothstein, 1974); the nonpenetrating agent FMMP labels certain peptides derived from band 3 in intact cells and additional peptides, also derived from band 3, in leaky ghosts (Bretscher, 1971a, 1973). In these experiments, it was assumed that new sites in band 3 made accessible to the enzymes or probes in the leaky ghosts were located at the inner face of the membrane. This assumption was subject to the criticism that exposed sites in leaky ghosts are not necessarily located on the inner face of the membrane, but may result from disorganization of the membrane during hemolysis. This objection has been overcome to some extent by the use of resealed ghosts or by comparison of inside-out and right-side-out vesicles derived from ghosts, on the basis that the protein organization in these preparations seems to be essentially the same as in cells (Boxer, Jenkins, and Tanner, 1974; Cabantchik et al., 1975; Reichstein and Blostein, 1975; Triplett and Carraway, 1972), with some minor differences noted by Staros and Richards (1974). The probes used have included proteolytic enzymes (Passow et al., 1975, 1976; Steck, Fairbanks, and Wallach, 1971) and iodination by lactoperoxidase (Boxer, Jenkins, and Tanner, 1974; Mueller and Morrison, 1974; Reichstein and Blostein, 1973, 1975; Shin and Carraway, 1974). In one study, it was demonstrated that the same molecules of band 3 that were segmented by proteolytic enzymes from the outside were iodinated from the inside (Reichstein and Blostein, 1975), and, in another study, that the peptide segments labeled from the inside were different from those labeled from the outside (Boxer, Jenkins, and Tanner, 1974). The latter studies provide unequivocal evidence that certain band 3 peptide chains span the membrane in ghosts or vesicles. Unfortunately, the iodination procedure only labels a small fraction of band 3 protein, as few as 30,000 monomers per cell (Reichstein and Blostein, 1975) out of a total of 1,000,000, so it must be assumed that the small iodinated sample is representative of the bulk of band 3. In the intact cell, it is not so easy to directly demonstrate that band 3 spans the membrane, but several studies with probes that can be used to label from outside or inside offer supporting evidence (Cabantchik et al., 1975a, b; Whitely and Berg, 1974).

Taken together, the studies cited above provide quite convincing evidence that at least a substantial fraction of band 3 protein spans the membrane so that certain peptide segments are exposed to the outside and others to the inside. Implicit in this conclusion is the corollary that the arrangement of band 3 in the membrane is asymmetric and permanent. Sugar residues are only exposed to the outside, as indicated by surface labeling of carbohydrates by concanavalin A coupled to ferritin (Pinto da Silva and Nicolson, 1974) and by reactions involving galactose oxidase (Gahmberg and Hakamori, 1973; Itaya, Gahmberg, and Hakamori, 1975; Steck and Dawson, 1974). Most of the sugar seems to be associated with the 35,000-dalton segment derived from band 3 by proteolysis of the intact cell (Steck, 1974). A more detailed analysis of the arrangement of band 3 is based on its susceptibility to proteolytic attack from the inside as well as the outside (Jenkins and Tanner, 1975; Passow et al., 1975, 1976; and Steck, Ramos, and Strapazon, 1976). It has been suggested that a loop of the protein is exposed at the outside that can be cleaved to produce the 35,000- and 65,000-dalton segments. Both segments are assumed to be inserted into the lipid bilayer because they are not water soluble, but the 35,000-dalton piece can be further digested to give some water-soluble segments (Jenkins and Tanner, 1975; Passow et al., 1976). The 65,000-dalton segment spans the membrane so that it is exposed at the inner surface, subject to proteolytic attack at that side. The spanning segment is 17,000 to 20,000 daltons (Jenkins and Tanner,

1975; Steck, Ramos, and Strapazon, 1976). The remaining 40,000-dalton piece, after proteolysis from the inside, also seems to penetrate the membrane, so that it is exposed in part at the outer face. Thus it can be iodinated from the outside by lactoperoxidase (Jenkins and Tanner, 1975) or react from the outside with nonpenetrating agents such as DIDS (Passow et al., 1975, 1976). On this basis, it has been suggested that band 3 is S-shaped in the membrane, with three segments inserted into lipid, and with at least two of them spanning the bilayer (Jenkins and Tanner, 1975). On the other hand, the 40,000 dalton segment is largely released as a soluble peptide after proteolysis of inside out vesicles, suggesting that it is entirely located on the inner face and not inserted through the bilayer (Steck, Ramos, and Strapazon, 1976; Grinstein, Ship, and Rothstein, 1977). On this basis, band 3 might only possess a single membrane spanning segment (the 17,000 dalton segment).

Band 3 probably exists as a dimer in the membrane. For example, after gentle oxidation of membranes to produce disulfide bonds, its molecular weight (based on SDS-acrylamide gel electrophoresis) is doubled (Steck, 1972c). The disulfide group is located on the portion of band 3 exposed to the inside (Steck, Ramos, and Strapazon, 1976). Cross-linking can also be produced by treating ghosts with bifunctional chemical agents (Wang and Richards, 1974) or by the action of transglutaminase (Dutton and Singer, 1975). The dimer form also persists (without cross-linking) after extraction by Triton X-100 (Clarke, 1975; Yu and Steck, 1973, 1975b).

High-molecular-weight, highly hydrophobic, membrane-spanning proteins, such as band 3, would be expected to be visible as intramembrane particles (IMP) by the technique of freeze-fracture electron microscopy. Early studies of the IMP established their connection with the sialoglycoproteins of the red-cell membrane. It was evident, however, that the hydrophobic portion of the glycoproteins could only account for a small fraction of the total mass of all of the observed particles (Bretscher, 1973; Guidotti, 1972; Pinto da Silva, Moss, and Fudenberg, 1973; Steck, 1974), and on this basis it was suggested that band 3, the most abundant hydrophobic protein, might account for the remainder. More direct evidence for association of band 3 with IMP has recently been presented (Furthmayr, Kahane, and Marchesi, 1976; Pinto da Silva and Nicolson, 1974; Rothstein and Cabantchik, 1974; Rothstein, Cabantchik, and Knauf, 1976). The total number of intramembrane particles is about half a million (Pinto da Silva, Moss, and Fudenberg, 1973), and of dimers of band 3, about the same (Steck, 1974). Thus the dimers of band 3 can account for the number of particles. The number of sialoglycoprotein monomers is also about half a million, suggesting that each particle might contain one band 3 dimer and one monomer of sialoglycoprotein.

The intramembrane particles seem to represent, at least in part, a visualization of the lipid-spanning elements of band 3. The particles are too bulky (about 80Å in diameter) in terms of the diameter of a single peptide chain, but they may be of about the correct magnitude for a dimer of 180,000 daltons, given reasonable assumptions about packing. It has been proposed (see above) that each monomer may have three segments inserted into the lipid, with two of them spanning the membrane (Jenkins and Tanner, 1975). Thus, for each dimer there may be six peptide chains within the lipid, probably sufficient to account for the diameter of 80Å. The segments are not extensively folded, for about 50 percent of the protein is in a helical form (Yu and Steck, 1975a). Within the bilayer, it would be expected that the penetrating segments would be arranged so that the exposure of hydrophobic residues toward the lipid components of the bilayer would be maximized and the exposure of the hydrophilic residues minimized. This condition could be met if the lipid-spanning elements were essentially hydrophobic. It could also be met if the six inserted segments of the dimer were arranged so that they collectively presented a hydrophobic face toward the outside (toward the lipid components of the bilayer) and presented hydrophilic residues toward an inside core. Such an arrangement would constitute an essentially hydrophilic (aqueous) channel through the membrane. It could provide aqueous pathways across the lipid bilayer through which transport or permeation of certain ions and hydrophilic molecules might occur, a possibility that has been proposed recently with increasing frequency (Bretscher, 1973; Ho and Guidotti, 1975; Pinto da Silva and Nicolson, 1974; Rothstein, Cabantchik, and Knauf, 1976; Singer, 1974; Steck, 1974). It is consistent with the observation that the intramembrane particles, largely band 3, behave as though substantially hydrated (Pinto da Silva, 1973).

Associations of Band 3 with Other Membrane Proteins

In the previous section, evidence that both the sialoglycoproteins and band 3 protein are associated with IMP was summarized. The finding that the number of IMP (about half a million) is about the same as the number of dimers of band 3 and the number of monomers of sialoglycoprotein suggests that each particle might contain two monomers of band 3 and one monomer of sialoprotein. Other evidence also supports an association. Thus, digestion of the sialoglycopeptide by the action of trypsin on the intact cell exposes additional sites on band 3 for interaction by DIDS (Cabantchik and Rothstein, 1974b) or lactoperoxidase (Phillips and Morrison, 1973). Thus, at the least, part of band 3 seems to be "covered" by the glycoprotein. On the other hand, band 3 and sialoglycoproteins extracted by Triton X-100 can be readily separated (Yu and Steck, 1975a, 1975b). Based on indirect evidence, band 3 also seems to be associated with spectrin, a high-molecular-weight, fibrous protein located on the inner face of the cell (Pinto da Silva and Nicolson, 1974), but no direct evidence of binding was found between spectrin and Triton

X-100 extracted band 3 (Yu and Steck, 1975b). Bands 4.2 and 6, on the other hand, remain associated with band 3 after extraction with nonionic detergent. The former can be dissociated by protein perturbants such as PCMBS (Yu and Steck, 1975a), while the latter, which is the enzyme glyceraldehyde-3-phosphate dehydrogenase, is dissociated at high ionic strength (Yu and Steck, 1975b; Billak et al., 1976). The binding shows one-to-one stoichiometry, and the site is located in a tryptic fragment of 22,000 daltons that is exposed at the inside face of the membrane (Kant and Steck, 1973; Yu and Steck, 1975b). The enzyme aldolase is also associated with band 3 at the inner face of the membrane (Strapazon and Steck, 1976). Indeed, a number of glycolytic enzymes may be associated with band 3 at the inner face of the membrane in an organized array of functional enzyme complexes (Fossel and Solomon, 1977).

The possible arrangement of band 3 in the bilayer and its associations with other proteins, as outlined in the preceding sections, is summarized in the highly schematic models of Figures 6.3 and 6.4. In the first, the S-shaped arrangement proposed by Jenkins and Tanner (1975) is modified and embellished by the data of others: Passow et al. (1975, 1976) on the location of the DIDS-binding sites, and Steck and his collaborators on the existence of the dimer that can be cross-linked by disulfide bond formation, on the location of the carbohydrate moieties, and on the location of the glycolytic enzyme binding site. In Figure 6.4, the concept of band 3 as a transmembrane aqueous channel is presented. This model will be discussed in more detail in relation to the mechanisms of anion transport.

Functions of Band 3

Band 3 has been identified with a variety of functions, including anion transport (Cabantchik and Rothstein, 1972, 1974a, b; Ho and Guidotti, 1975; Lepke et al., 1976; Passow et al., 1976; Rothstein and Cabantchik, 1974; Rothstein, Cabantchik, and Knauf, 1976; Rothstein et al., 1975; Zaki et al., 1975), sugar transport (Batt, Abbott, and Schachter, 1976; Kahlenberg, 1976; Kasahara and Hinkle, 1976a; Lin and Spudich, 1974; Taverna and Langdon, 1973), Na-K transport (Avruch and Fairbanks, 1972; Knauf, Proverbio, and Hoffman, 1974; Williams, 1972), water permeation (Brown, Feinstein, and Sha'afi, 1975), acetyl cholinesterase activity (Bellhorn, Blumenfeld, and Gallor, 1970), and protein phosphorylation (Roses and Appel, 1975). The functional associations have been established by three different criteria:

1. Band 3 has been demonstrated to be the product of enzymic phosphorylation by endogenous protein kinase and Na,K-ATPase, with ATP as the phosphate donor. The ATPase activity is assumed to be associated with active transport of Na and K.

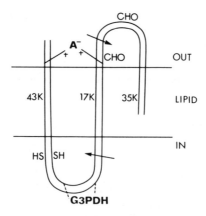

FIGURE 6.3. A diagram of the arrangement of band 3. Arrows indicate the sites of proteolysis in intact membrane. (Based on Jenkins and Tanner, 1975; Passow et al., 1976; Steck, Ramos, and Strapazon, 1976.)

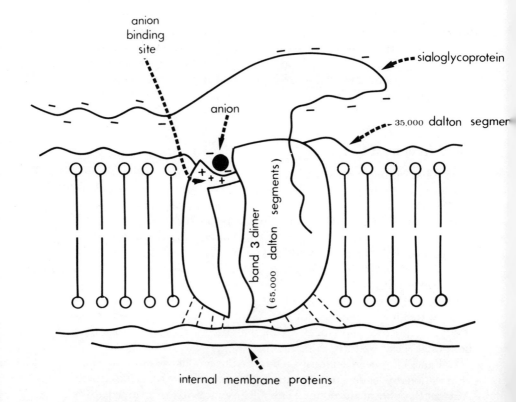

FIGURE 6.4. A schematic diagram of band 3 as a protein channel for anion transfer through the membrane. (From Rothstein, Cabantchik, and Knauf, 1976.)

2. Band 3 is the primary site of action of covalently reacting inhibitors of anion transport, water transfer, sugar transport, and acetylcholinesterase activity.

3. Band 3 enriched preparations derived from red-cell membranes, or reconstituted from red-cell proteins and phospholipids, display sugar transport (Kahlenberg, 1976; Kasahara and Hinkle, 1976) or anion transport (Rothstein et al., 1975) activities.

The usefulness of each technique is dependent on the specificity of the measured effect, and on the certainty with which the effect can be attributed to band 3. Enzymic activity is a highly specific marker, because band 3 is the substrate. With inhibitors, the specificity depends on a number of factors. Diisopropylfluorophosphate is, for example, a highly site-specific inhibitor of the enzyme acetyl cholinesterase (Bellhorn, Blumenfeld, and Gallor, 1970). The probes used for inhibition of anion transport, on the other hand, although they are anionic, are not highly specific in a chemical sense. They depend for covalent reaction on groups such as isothiocyano, diazo, aldehyde, or nitrene, which have varying degrees of ligand specificity (for example, for sulfhydryl, guanidino, or amino groups). Such ligands are, however, present in every protein, so that the protein-specificity might be very low. Nevertheless, some of the agents are relatively specific when used with the intact cell because of accessibility factors. For example, DIDS, the most potent inhibitor, can react with all of the membrane proteins in a leaky ghost, but reacts largely with band 3 in the intact cell (Fig. 6.1). The specificity can be further improved by treating the cells with pronase. The small amount of DIDS (less than 10 percent) bound to sialoglycoprotein is solubilized, with the glycopeptide cleaved by the enzyme. Most of the DIDS remaining in the membrane is found in the larger of two segments (65,000 and 35,000 daltons) derived from the splitting of most of band 3 (Fig. 6.2). Accessibility factors may also account for the differences between the binding of PDP and DIDS. The former agent is highly specific for amino groups because of formation of a Schiff base, whereas the latter can react with amino, guanidino, and sulfhydryl groups. Nevertheless, in the intact membrane, PDP is far less specific in its interactions, binding substantially to three glycoprotein components in addition to band 3 (Fig. 6.5).

Specificity can be improved by use of transport substrates as protective agents. For example, the substrate for the anion system, Cl^-, can protect against the binding of the probe NAP-taurine to band 3 (Rothstein, Cabantchik, and Knauf, 1976). Glucose or cytochalasin B (a highly specific and potent inhibitor of sugar transport) can protect band 3 against reaction with sulfhydryl agents (Batt, Abbott, and Schachter, 1976).

Despite relatively high specificities of binding of irreversible inhibitors in some cases, it is difficult to unequivocally exclude the involvement of minor components that may be visible or that may be obscured by the

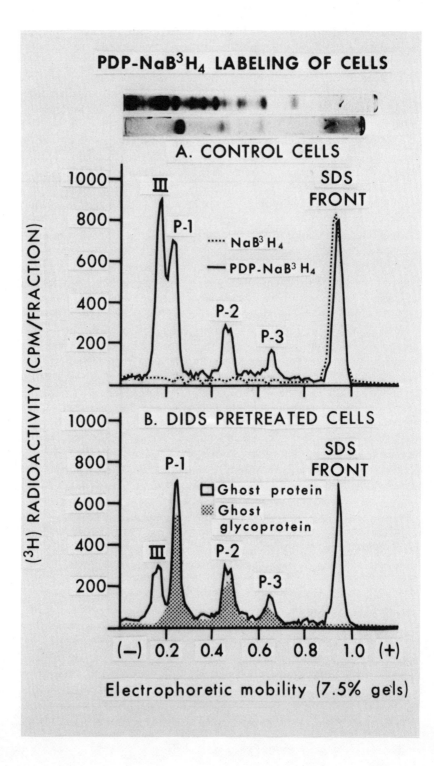

PDP-NaB³H₄ LABELING OF CELLS

A. CONTROL CELLS

SDS FRONT

III

P-1

······ NaB³H₄
——— PDP-NaB³H₄

P-2

P-3

B. DIDS PRETREATED CELLS

SDS FRONT

P-1

☐ Ghost protein
▨ Ghost glycoprotein

III

P-2

P-3

(³H) RADIOACTIVITY (CPM/FRACTION)

Electrophoretic mobility (7.5% gels)

(−) 0.2 0.4 0.6 0.8 1.0 (+)

general background of radioactivity in the acrylamide gels used to achieve protein separations (Passow et al., 1976).

In the studies with vesicles derived from red cells (Kahlenberg, 1976), or in reconstitution experiments (Kasahara and Hinkle, 1976a; Rothstein et al., 1975), the attribution of transport function to band 3 depends on the "purity" of the protein components. Although the systems are enriched in band 3 in each case, they are not pure fractions, so it is also possible that minor components may be responsible for the functional activities. For example, recent reconstitution experiments (Kasahara and Hinkle, 1976b) suggest that a minor component in the detergent extract, band 4.5, may be responsible for the sugar transport activity.

Despite the fact that it is difficult to "prove" that band 3 is responsible for various membrane activities, the cumulative weight of evidence favors such a view (with the exception of sugar transport, as noted above). With each improvement in the resolution of the procedures, band 3 continues to be the most likely candidate.

It has already been pointed out that because band 3 is implicated in many different functions, it must be heterogeneous. Although some sorting out of components has been observed by two-dimensional gel electrophoresis (Anselstetter and Horstmann, 1975; Conrad and Penniston, 1976), isoelectric focusing (M. Morrison, personal communication; Y. Shami and A. Rothstein, unpublished results) and lectin-binding (Adair and Kornfeld, 1974; Findlay, 1974), none of the subfractions have as yet been associated with particular activities. Cholinesterase activity can be largely segregated from a purified fraction of band 3 on an ion exchange column (Yu and Steck, 1975a). The Na,K-ATPase intermediate can be distinguished from the bulk of band 3 after treatment with pronase. The enzyme remains in the acrylamide gels at the original band 3 location (Knauf, Proverbio, and Hoffman, 1974), whereas band 3 is largely split into segments of 65,000 and 35,000 daltons (Fig. 6.2).

The anion transport system has certain advantages as a model system for study of the role of band 3 protein. The flux activity is exceptionally high, with a half time in milliseconds (Sachs, Knauf, and Dunham, 1974). Even allowing for an exceptionally high turnover rate, it must be a very abundant system. The studies with specific inhibitors certainly suggest that a substantial fraction of the protein is involved (Cabantchik and Rothstein, 1974a; Ho and Guidotti, 1975; Lepke et al., 1976; Rothstein,

FIGURE 6.5. PDP binding to membrane proteins from normal cells and DIDS-treated cells, as determined by SDS-polyacrylamide gel electrophoresis. The stippled portion of the lower segment represents the PDP binding to glycoproteins isolated from the membranes. The upper gel is stained for protein with Coomassie blue and the lower gel for carbohydrate with periodic acid Schiff. (From Cabantchik et al., 1975b.)

Cabantchik, and Knauf, 1976). If so, the various properties attributed to the bulk of band 3, such as its arrangement in the membrane, its composition, its accessibility on the two sides of the membrane, and its susceptibility to proteolytic attack, can also be attributed to the anion transport system. Furthermore, anion transport has been intensively investigated and its properties are well defined. The remainder of this paper will, therefore, be largely concerned with the role of band 3 in anion transport, with some attempt to generalize the findings to other kinds of transport. For more details, the individual papers (Cabantchik et al., 1975a, b; Cabantchik and Rothstein, 1972, 1974a, b; Ho and Guidotti, 1975; Lepke et al., 1976; Passow et al., 1975, 1976; Rothstein, Cabantchik, and Knauf, 1976; Sachs, Knauf, and Dunham, 1974; Zaki et al., 1975) or reviews (Rothstein, Cabantchik, and Knauf, 1976; Sachs, Knauf, and Dunham, 1974) can be consulted.

The Anion Transport System

The red cell carries CO_2 from tissues to lungs by a mechanism that is facilitated by the exchange of Cl for HCO_3. The process is unusually rapid, proceeding by a pathway that has the properties of a carrier system, involving an interaction of anions with a specific transport site in the membrane (Dalmark, 1976; Deuticke, 1970; Gunn et al., 1973; Passow, 1969; Sachs, Knauf, and Dunham, 1974). Techniques for the possible identification of the components of the anion transport system are limited. No enzyme activity is involved and the binding of Cl to the transport site is of such low affinity ($K_m > 75mM$) (Dalmark, 1976; Gunn et al., 1973) that it cannot be used for identification purposes. The primary technique has, therefore, involved the use of irreversible inhibitors as markers of membrane proteins. A number of agents have been used, including DASA (P. A. Knauf, personal communication, for inhibition; Berg, 1969, and Juliano and Rothstein, 1971, for binding), IBSA (Ho and Guidotti, 1975), FDNB (Zaki et al., 1975), TNBS (Knauf and Rothstein, 1971, for inhibition; Arrotti and Garvin, 1972, for binding), NAP-taurine (Cabantchik et al., 1976), PDP (Cabantchik et al., 1975b), and DIDS (Cabantchik and Rothstein, 1974a, b; Lepke et al., 1976; Passow et al., 1976). DIDS is the most potent and the most specific for band 3. DIDS blocks more than 99 percent of anion transport at concentrations lower than 10 μM. It has no reported effects on other transport or permeation processes, such as those for sugar or for cations. Of the few identifiable DIDS-binding peaks on acrylamide gels, band 3 is by far the most substantial (Fig. 6.1, control A, and Fig. 6.2, control C). About 5 percent is found in the sialoglycoprotein (PAS 1), but that component can be removed by trypsin treatment of the cells without any effect on band 3, on the transport, or the inhibition of transport by DIDS (Cabantchik and Rothstein, 1974b). A small "satellite" band has also been reported that moves ahead of band 3 during electrophoresis (Lepke et al.,

1976; Passow et al., 1976). It contains about 5 percent of the total DIDS. No other discrete peaks are observed.

It is difficult to prove that some minor component other than band 3 is not the anion transport protein. A case can be made, however, that the turnover of Cl transport is very rapid (10^5 per second), even for a protein as abundant as band 3 (Ho and Guidotti, 1975; Passow et al., 1976; Sachs, Knauf, and Dunham, 1974) and that for a minor component, an excessively high rate of turnover would be required. One recent finding that points directly to band 3 is based on the effects of cross-linking agents. Mild oxidation by copper and orthophenanthrolene cross-links band 3 to form dimers, with no observed effect on any other proteins (Steck, 1972c). The characteristics of the anion permeability of resealed ghosts are altered by the procedure (Rice and Steck, 1976).

The DIDS associated with band 3 can be subdivided into three components by treating cells with pronase (Fig. 6.2). Band 3 is largely split into segments of 65,000 and 35,000 daltons. Most of the DIDS (80 percent) is associated with the larger segment, with lesser amounts in the smaller segment or in the pronase-resistant fraction. The possibility that the 65,000-dalton segment is the site of inhibition is reinforced by the finding that Triton X-100 extracts enriched either in band 3 or in the 65,000-dalton segment can markedly increase the anion transport in lecithin vesicles (Rothstein et al., 1975).

The studies with inhibiting covalent probes lead to the conclusion that band 3 probably contains an inhibitory site. The arrangement of ligands within the site can be crudely mapped by comparing the inhibitory potency of the various compounds, including a series of analogs of DIDS (Cabantchik and Rothstein, 1972; Zaki et al., 1975). Monofunctional anionic compounds are less potent than bifunctional, suggesting that the high affinity of the disulfonic stilbenes is due to a chelation of the sulfonic acid groups by two positively charged sites. That this form of reversible ionic binding is sufficient for inhibition is indicated by the fact that compounds lacking a covalent binding group can be potent inhibitors. Even with compounds containing covalent binding groups, a reversible binding precedes the irreversible reaction (Cabantchik and Rothstein, 1972; Lepke et al., 1976) and the inhibition is not altered as the irreversible reaction proceeds (Lepke et al., 1976). Thus, the site must have, in addition to two charged ligands, a group such as guanidino or amino capable of reacting with the isothiocyano group of DIDS (Fig. 6.6). The precise identity of the three ligands is not known, but one is presumably an amino group of lysine, it is suggested on the basis of studies with PDP (Cabantchik et al., 1975a). This probe is an inhibitor of anion transport that reacts with sites in band 3 that are common to DIDS (Fig. 6.5). It forms a highly specific Schiff base with amino groups, and, in confirmation, a lysine-PDP conjugate has been isolated from membranes prepared from PDP-treated cells.

The inhibitory site identified by use of the probes might be the sub-

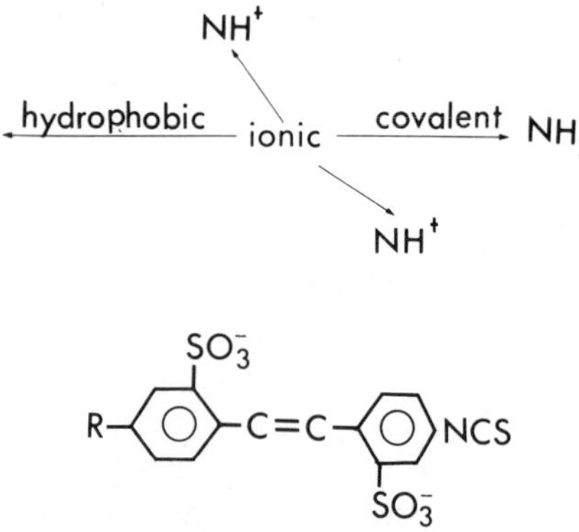

FIGURE 6.6. A schematic map of the inhibitory site of anion transport based on the potency of stilbene disulfonate analogs and other organic anions. (From Cabantchik and Rothstein, 1972.)

strate binding site of the transport system. This conclusion would be consistent with the fact that most of the probes are themselves anions, and that in the case of one, IBSA, the inhibitory effect is reduced by high concentrations of a transported anion, phosphate (Ho and Guidotti, 1975). On the other hand, many nonionic compounds can inhibit anion transport (Deuticke, 1970). In the absence of rigorous data demonstrating that the probes inhibit by competing for the transport site, conclusions concerning the inhibitory site and its relationship to the transport system must, therefore, be considered tentative. Inhibition of membrane functions might be indirect, due to interactions of the probes at some distance from the transport site, or even with an adjacent component.

The only kind of probe that would unequivocally interact with the transport site would itself have to be a transport substrate. But to mark the site, it would also have to possess the capacity to react covalently. Although these two requirements seem to be contradictory, they have been met in two cases. The first is PDP (Cabantchik et al., 1975b); under normal circumstances, this anion permeates the red-cell membrane slowly, but on addition of $NaBH_4$ it is fixed in situ in an irreversible bond. Its aldehyde group forms a relatively stable but reversible bond with amino groups (Schiff base), which when reduced by $NaBH_4$ becomes irreversible due to insertion of hydrogen (or tritium if tritiated borohydride is used).

The uptake of PDP involves a fast and a slow component (Fig. 6.7). If tritiated $NaBH_4$ is added at various times to fix the PDP in an irreversible bond, almost all of the agent is found in the membrane in the first ten minutes, but after longer periods of time increasing amounts are found associated with hemoglobin. The membrane-bound and hemoglobin-bound components of PDP distribution correspond almost exactly to the

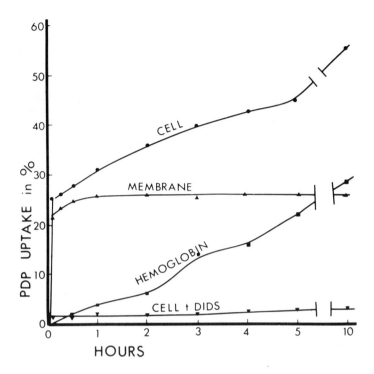

FIGURE 6.7. The effects of DIDS on the membrane binding (fast component) and the uptake of PDP (slow component). (From Cabantchik et al., 1975b.)

fast and slow components of uptake. Thus the fast component represents association of PDP with membrane sites, and the slow component represents permeation into the cell. Both components are almost completely blocked by pretreatment of the cells with DIDS. This finding suggests that the binding reaction (fast component) may be an essential step in the permeation and that the pathway may be the normal anion system (which is also specifically blocked by DIDS).

PDP is an effective inhibitor of the permeation of inorganic anions (Cl and SO$_4$). The inhibition is normally reversible but after fixation by NaBH$_4$ it becomes irreversible (Cabantchik et al., 1975b). The extent of the inhibition is not, however, altered by the addition of NaBH$_4$, suggesting that PDP becomes fixed to the same inhibitory sites with which it was reversibly associated. Thus the location of the PDP should reflect its reversible associations with the transport system.

At low concentrations (within the inhibitory range), PDP is bound primarily to band 3, but as its concentration is raised, it is also found in three glycoprotein bands as well. In the presence of DIDS, however, the binding to band 3 is markedly and specifically reduced, whereas that to the glycoproteins is not (Fig. 6.5). Furthermore, in cells pretreated with

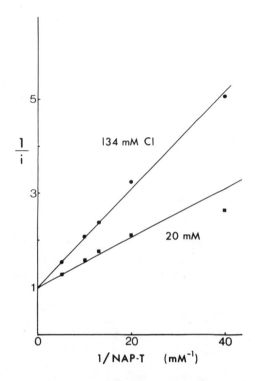

FIGURE 6.8. The kinetics of NAP-taurine inhibition of sulfate fluxes in the presence of low and high Cl (from Knauf et al., 1976). The reciprocal of the fractional inhibition $(1/i)$ is plotted against the reciprocal of the inhibitor concentration. Straight lines of increasing slope (with increasing Cl) are consistent with competition of Cl and NAP-taurine for the transport site.

PDP, the binding of DIDS to band 3 is also markedly reduced. Thus the majority of binding sites in band 3 are common to PDP and DIDS. They presumably include those that are responsible for the inhibition of inorganic anion transport as well as those involved in the transport of PDP.

The most useful probe for studying transport sites has been NAP-taurine (Cabantchik et al., 1976; Knauf et al., 1976; Rothstein, Cabantchik, and Knauf, 1976). In the dark, NAP-taurine is a reversible inhibitor of inorganic anion transport. Its effects are reversed by inorganic anions such as Cl and HCO_3, with the latter being more effective. This finding is consistent with the finding that the K_m for transport is lower for HCO_3 than for Cl (Dalmark, 1976). The kinetic behavior of the transport in the presence of NAP-taurine (Fig. 6.8) is consistent with the conclusion that it competes for the transport site (Knauf et al., 1976).

NAP-taurine in the dark permeates into the cell except at low temperatures (Staros and Richards, 1974; Staros, Haley, and Richards, 1976). Its flux is inhibited (Cabantchik et al., 1976) by two chemically distinct inhibitors of inorganic anion transport, DIDS (Cabantchik and Rothstein, 1972, 1974a) and dipyridamole (Deuticke, 1970). Furthermore, NAP-taurine (in the dark) can prevent DIDS binding to band 3 and protects the cells against its inhibitory effect (Table 6.1) (Rothstein, Cabantchik, and Knauf, 1976). These findings are all consistent with the supposition that NAP-taurine is a substrate of the normal anion transport system and that, as such, it is a competitive inhibitor of inorganic anion transport and that it protects the transport sites from interaction with DIDS.

On exposure to light, NAP-taurine is irreversibly fixed to the membrane (Staros and Richards, 1974). From the outside of the cell, it is largely localized in band 3 (Staros and Richards, 1974), but from the inside it reacts with many additional proteins (Staros, Haley, and Richards, 1976). In order to establish that some of the binding sites are those connected with anion transport, it is necessary to demonstrate that the irreversible binding of the agent after exposure to light almost exactly reflects its reversible binding as an anion in the dark. An agent that meets

TABLE 6.1. The protective effect of reversibly bound NAP-taurine on the inhibition of sulfate efflux by DIDS. (From Cabantchik et al., 1976.)

	Inhibition of control (%)	
	8μM DIDS 50μM NAP-taurine	4μM DIDS 100μM NAP-taurine
DIDS only	88, 92	72, 83
NAP-T (dark) + DIDS, then washed	63, 58	22, 25
NAP-T (dark) then washed + DIDS	95, —	66, 78

Note: Cells (10 percent hematocrit) preloaded with (^{35}S) sulfate (5mM) were exposed to DIDS for 5 min at 10°C in the presence or absence of NAP-taurine. They were then washed twice with albumin-containing buffer (0.5 percent albumin) and twice with buffer. In some aliquots of cells, NAP-taurine, 50 or 100μM, was present during DIDS treatment; in others it was added and then washed out prior to DIDS treatment. Control samples of blood cells were subjected to the same washing procedures. Fluxes were measured at 37°C at 5 percent hematocrit after the washing procedure. All procedures were carried out in semidarkness to avoid photolysis of the NAP-taurine.

this condition would be classified as a photoaffinity probe. NAP-taurine meets the necessary criteria (Knauf et al., 1976):

1. Its effects in the dark are reversible, but become irreversible after exposure to light.
2. The extent of the irreversible inhibition parallels that of the reversible inhibition at any concentration of the agent.
3. Prephotolysed NAP-taurine does not cause irreversible inhibition.
4. The presence of "scavengers" in the medium does not reduce the inhibitory effect.

NAP-taurine binding to band 3 is markedly reduced by pretreatment with DIDS and vice versa, so that most of the sites in that protein are common to both agents (Fig. 6.9, upper panel). Furthermore, like DIDS (Cabantchik and Rothstein, 1974b), most of the NAP-taurine is found in the 65,000-dalton segment derived from band 3 (Fig. 6.9, lower panel) as a result of pronase treatment of intact cells. As already mentioned, in the dark NAP-taurine protects against DIDS inhibition (Table 6.1). The sites in band 3 or in the 65,000-dalton segment can also be protected against NAP-taurine interaction by high concentrations of Cl (Fig. 6.10), in parallel with the protective effect of Cl against the NAP-taurine inhibition of transport. These findings suggest that some fraction of the sites revealed by NAP-taurine in band 3 are transport sites and that the inhibitory actions of the agents NAP-taurine, DIDS, and PDP are due at least in part to binding to those sites.

The agents may bind sites in band 3 (or other proteins) in addition to those involved in transport. Thus, the number of binding sites for the agents will represent a maximal estimate of the number of transport sites. The first estimates were made with tritiated DIDS (Cabantchik and Rothstein, 1974a). About 300,000 sites per cell were related to inhibition in a linear fashion. Later studies, however, indicated that this was an underestimate (Lepke et al., 1976; Passow et al., 1976). The tritiated DIDS was in a reduced form (dihydro-DIDS) that reacts more slowly than DIDS. Because the tritiated sample contained DIDS as well as reduced tritiated DIDS, the binding was underestimated. The total number of binding sites in band 3 necessary for complete inhibition with dihydro-DIDS has been reported to be 1,700,000 per cell, and the number of binding sites for DIDS has been estimated indirectly to be about the same (Lepke et al., 1976). Recently, using tritiated DIDS (not reduced) an estimate of 1,200,000 sites per cell was obtained, with about 90 percent in band 3 (Shipp, Shami, and Rothstein, 1977). This agrees with an estimate of 1,200,000 sites per cell based on the amount of DIDS added to achieve complete inhibition of transport (Halestrap, 1976). It also agrees with a value of 1,400,000 sites per cell based on NAP-taurine binding (Knauf et

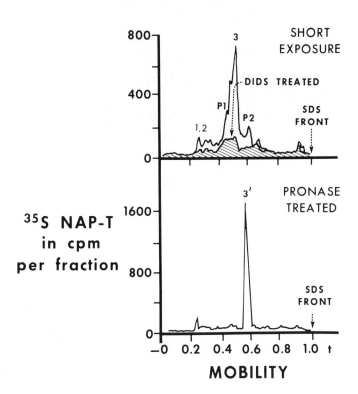

FIGURE 6.9. NAP-taurine binding to membrane proteins from normal, DIDS-treated, and pronase-treated cells as indicated by SDS-acrylamide gel electrophoresis. (From Cabantchik et al., 1976.)

al., 1976). Only about 80 percent of the membrane binding of NAP-taurine is attributable to band 3 and only about half of that is displaced by Cl when inhibition is reversed (Fig. 6.10), suggesting that the inhibition-related binding might be as low as 500,000 sites per cell. It is not clear how many molecules of agent bind per monomer of band 3, but in view of the fact that about 1,000,000 monomers are present in each cell (Fairbanks, Steck, and Wallach, 1971), it seems likely that a substantial fraction of band 3 monomers are involved in anion transport.

Any proposed mechanism of anion transport must take into account the following information:

1. The kinetics of transport are of the saturation type, with competition between pairs of anions, indicating that binding to a site is required (Dalmark, 1976; Gunn et al., 1973; Passow, 1969; Sachs, Knauf, and Dunham, 1974).

2. The kinetics are consistent with a mobile carrier model (Gunn et al., 1973) or a fixed-charge model (Passow, 1969).

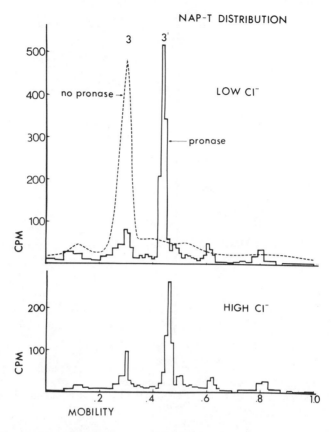

FIGURE 6.10. The effect of high Cl concentrations on NAP-taurine binding to membrane proteins from pronase treated cells as indicated by SDS-acrylamide gel electrophoresis (Knauf et al., 1976). The dotted line represents the distribution of NAP-taurine in proteins from normal cells. The inhibition of SO_4 efflux was 46 percent in low Cl (25mM) and zero in high Cl (135mM).

3. Most of the anion fluxes involve a nonconductive, obligatory, one-for-one exchange (Sachs, Knauf, and Dunham, 1974), suggesting that the anions cross the membrane as an uncharged complex.

4. An inhibitory site is exposed to nonpenetrating agents from the outside (Cabantchik and Rothstein, 1972, 1974a; Ho and Guidotti, 1975; Lepke et al., 1976), but in the case of the disulfonic stilbenes, the site is demonstrated to be inaccessible from the inside (Zaki et al., 1975).

5. The site involves a cluster of positively charged groups (Cabantchik and Rothstein, 1972; Passow, 1969; Rothstein, Cabantchik, and Knauf, 1976), including a lysine amino group (Cabantchik et al., 1975b).

6. The site appears to be located in a large protein (band 3) of 90,000 daltons (Cabantchik and Rothstein, 1972, 1974a; Ho and Guidotti, 1975;

Lepke et al., 1976; Zaki et al., 1975). The protein is oriented in the membrane with its carbohydrate groups exposed to the outside (Itaya, Gahmberg, and Hakimori, 1975; Pinto da Silva and Nicolson, 1974; Steck and Dawson, 1974), and with particular peptide segments exposed to the outside and inside (Jenkins and Tanner, 1975; Steck, Ramos, and Strapazon, 1976). It may span the membrane twice, with a third segment also inserted into the lipid. In view of the fact that band 3 probably exists as a dimer in the membrane (Clarke, 1975; Steck, 1972c, 1974; Yu and Steck, 1975b), as many as six polypeptide segments may be inserted into the lipid. They are probably arranged in the membrane lipid with most of the hydrophobic groups facing outward toward the lipid and with the hydrophilic groups and associated water facing toward the core, forming a largely aqueous channel through the bilayer (Bretscher, 1973; Guidotti, 1972; Pinto da Silva and Nicolson, 1974; Rothstein, Cabantchik, and Knauf, 1976; Singer, 1974; Steck, 1974).

A model that is consistent with the above information has been proposed (Fig. 6.4). It involves a site in band 3 capable of binding transportable anions or of reacting with irreversible inhibitors such as DIDS (Cabantchik and Rothstein, 1972, 1974a). Because some of these inhibitors are "nonpenetrating," and can only reach the site from the outside (Zaki et al., 1975), the site must be operationally located near the outside. Furthermore, even in the process of transport, the site cannot become accessible to the disulfonic stilbene from the inside. This restriction would preclude any mechanism in which the site moved from one side of the membrane to the other, either by rotation of the whole band 3 protein or by movement of a site-containing protein segment through the lipid bilayer. Rotation of the whole protein or movement of a large segment through the lipid bilayer is also highly unlikely, both in terms of the extremely high rate of turnover of the anion transport (Ho and Guidotti, 1975; Passow et al., 1976; Sachs, Knauf, and Dunham, 1974), and in terms of the binding energies of lipids and protein that would have to be overcome (Singer, 1974). Furthermore, the protein is arranged asymmetrically in the membrane, with sugars and certain peptides permanently facing outward and other peptides facing inward. It is proposed, therefore, that band 3 provides a proteinaceous, largely aqueous channel through which the anions can pass. The channel cannot be continuous but must be blocked by a diffusion barrier near the outside that largely prevents conductive diffusion but that allows a rapid, nonconductive, one-for-one exchange of those anions that become associated with the binding site. The diffusion barrier would involve only a small distance within the channel that could be traversed by a small movement of the anion-site complex. A local conformational change in which the site oscillates between two positions within the channel, one facing outward and the other inward, would be an appropriate mechanism, with the obligatory exchange accounted for by allowing the conformational

TABLE 6.2. PDP present on the inside of the membrane reduces the (^{3}H) DIDS binding to the outside. (From Rothstein, Cabantchik, and Knauf, 1976.)

| | (^{3}H) DIDS binding to cells | |
Location of PDP	Dipyridamole present	Dipyridamole absent
None	100	100
Outside	12	11
Outside then washed	85	83
Inside	55 ─────────────→ 89	

Note: For "PDP outside," cells (30 percent Hct) were exposed to PDP (10 mM) for 10 min before addition of DIDS. For "outside then washed," the exposure to PDP was followed by two washes with buffered saline containing albumin (0.5 percent) and two with buffered saline. Exposure to (^{3}H)DIDS was for 10 min. For "inside," the cells were exposed to PDP for 20 hours at 37°C at pH 7.0 with the medium replaced every 5 hours. The washing and treatment with (^{3}H)DIDS were the same as above. All data are averages for three experiments, given as percent binding relative to controls with no PDP. Statistical analysis indicates that the value for "inside," in the presence of dipyridamole (55 percent), is significantly lower than the control with no dipyridamole (89 percent) at a level of 0.01.

change to occur only if the site were loaded with an anion. Such a mechanism would obey mobile carrier kinetics, but it differs from the classical model inasmuch as the mobile element is a protein site that moves a small distance within the core of the protein, rather than a lipid-soluble complex that moves through the lipid bilayer.

Some preliminary evidence to support the concept of a conformational change in anion transport has already been published (Rothstein, Cabantchik, and Knauf, 1976). If cells are loaded with PDP, an anion with a high affinity for the binding site, and if it is washed away from the outside in the presence of dipyridamole to inhibit the operation of the transport system (Deuticke, 1970), then the binding of DIDS to the site from the outside is substantially reduced because of the presence of PDP inside the cell (Table 6.2). If the dipyridamole is removed to allow an efflux of PDP to occur, the DIDS binding sites can once more be titrated. The trans effect of inside PDP on the binding of DIDS from the outside suggests that the transport sites can be recruited to face inward and that when the blocking agent is removed, they can return to face outward.

How applicable is a model involving a protein channel traversing the bilayer to other transport systems? It has already been noted that water (Brown, Feinstein, and Sha'afi, 1975), Na-K (Avruch and Fairbanks, 1972; Knauf, Proverbio, and Hoffman, 1974; Williams, 1972), and sugar transfer (Batt, Abbot, and Schachter, 1976; Kahlenberg, 1976; Kasahara and Hinkle, 1976; Lin and Spudich, 1974; Taverna and Langdon, 1973) through the red-cell membrane also involve band 3. One limitation in generalizing the anion model is that if the number of sites involved in a particular transport system is small, the properties attributable to the bulk of band 3 would not necessarily apply to a small fraction of the protein. In the case of Na-K transport, for example, the number is very small, about three hundred per cell. A transmembrane protein is obviously involved, however, because the K and ouabain binding sites are located on the outside of the membrane, whereas the Mg, Na, and ATP binding sites are on the inside (Sachs, Knauf, and Dunham, 1974). Furthermore, the transport seems to involve a conformational change (Bodeman and Hoffman, 1976). The number of sites for sugar transport and for water permeation is not precisely known. It seems reasonable, nevertheless, that protein channels may be generally involved in the transport of hydrophilic, biologically important substances through the membrane. If so, it is an intriguing question, why a number of different activities seem to be associated with hydrophobic proteins (band 3) of about the same molecular weight.

ACKNOWLEDGMENTS

This study was supported in part by a grant from the Medical Research Council (Canada), MT-4665. The advice of Dr. P. A. Knauf is gratefully acknowledged.

REFERENCES

Adair, W. L., and S. Kornfeld. 1974. Isolation of receptors for wheat germ agglutinin and the *Ricinus communis* lectins from human erythrocytes using affinity chromatography. *J. Biol. Chem.* 249: 4676-4704.

Anselstetter, V., and J. H. Horstmann. 1975. Two-dimensional polyacrylamide gel electrophoresis of the proteins and glycoproteins of the human erythrocyte membrane. *Eur. J. Biochem.* 56: 259-269.

Arrotti, J. J., and J. E. Garvin. 1972. Selective labelling of human erythrocyte membrane components with tritiated trinitrobenzene sulfonic acid and picryl chloride. *Biochem. Biophys. Res. Commun.* 49:205-211.

Avruch, J., and C. Fairbanks. 1972. Demonstration of a phosphopeptide intermediate in the Mg^{++} dependent Na^+- and K^+- stimulated adenosine triphosphatase reaction of the erythrocyte membrane. *Proc. Natl. Acad. Sci. USA* 69:1261-2120.

Batt, E. R., R. E. Abbott, and D. Schachter. 1976. An exofacial component of glucose transport mechanism of human erythrocyte. *Fed. Proc.* 5:606a.

Bellhorn, M. B., O. O. Blumenfeld, and P. M. Gallop. 1970. Acetyl cholinesterase of the human erythrocyte membrane label with tritiated diisopropylfluorophosphate. *Biochem. Biophys. Res. Commun.* 39:267-273.

Bender, W. W., H. Garen, and H. C. Berg. 1971. Proteins of the human erythrocyte membrane as modified by pronase. *J. Mol. Biol.* 58: 783-797.

Berg, H. C. 1969. Sulfanilic acid diazonium salt: a label for the outside of the human erythrocyte membrane. *Biochim. Biophys. Acta* 183: 65-78.

Billak, M. M., J. B. Finean, R. Coleman, and R. H. Michell. 1976. Preparation of erythrocyte ghosts by a glycol-induced osmotic lysis under isionic conditions. *Biochim. Biophys. Acta* 433:54-62.

Bodeman, H. H., and J. F. Hoffman. 1976. Side dependent effects of internal versus external Na and K on ouabain binding to reconstituted human red blood cell ghosts. *J. Gen. Physiol.* 67:497-526.

Boxer, D. H., R. E. Jenkins, and M. J. A. Tanner. 1974. The organization of the major protein of the human erythrocyte membrane. *Biochem. J.* 137:531-534.

Bretscher, M. S. 1971a. A major protein which spans the human erythrocyte membrane. *J. Mol. Biol.* 59: 351-357.

Bretscher, M. S. 1971b. Human erythrocyte membranes: specific labelling of surface proteins. *J. Mol. Biol.* 58: 775-781.

Bretscher, M. S. 1973. Membrane structure: some general principles. *Science* 181: 622-629.

Brown, P. A., M. B. Feinstein, and R. I. Sha'afi. 1975. Membrane proteins related to water transport in human erythrocytes. *Nature* 254:523-525.

Cabantchik, Z. I., M. Balshin, W. Breuer, H. Markus, and A. Rothstein. 1975a. A comparison of intact human red blood cells and resealed and leaky ghosts with respect to their interactions with surface labelling agents and proteolytic enzymes. *Biochim. Biophys. Acta* 382: 621-633.

Cabantchik, Z. I., M. Balshin, W. Breuer, and A. Rothstein. 1975b. Pyridoxal phosphate: an anionic probe for protein amino groups exposed on the outer and inner surfaces of intact human red blood cells. *J. Biol. Chem.* 250:5130-5136.

Cabantchik, Z. I., P. A. Knauf, T. Ostwald, H. Markus, L. Davidson, W. Breuer, and A. Rothstein. 1976. The interaction of an anionic photoreactive probe with the anion transport system of the human red blood cell. Unpublished.

Cabantchik, Z. I., and A. Rothstein. 1972. The nature of the membrane sites controlling anion permeability of human red blood cells as determined by studies with disulfonic stilbene derivatives. *J. Membrane Biol.* 10: 311-330.

Cabantchik, Z. I., and A. Rothstein. 1974a. Membrane proteins related to anion permeability of human red blood cells. I. Localization of disulfonic stilbene binding sites in proteins involved in permeation. *J. Membrane Biol.* 15:207-226.

Cabantchik, Z. I., and A. Rothstein. 1974b. Membrane proteins related to anion permeability of human red blood cells. II. Effects of proteolytic enzymes on disulfonic stilbene sites of surface proteins. *J. Membrane Biol.* 15: 227-248.

Carraway, K. L. 1975. Covalent labelling of membranes. *Biochim. Biophys. Acta* 415:379-410.

Carraway, K. L., D. Kobylka, and R. B. Triplett. 1971. Surface proteins of erythrocyte membranes. *Biochim. Biophys. Acta* 241:934-940.

Clarke, S. 1975. The size and detergent binding of membrane proteins. *J. Biol. Chem.* 250:5459-5469.

Conrad, M. J., and J. T. Penniston. 1976. The resolution of erythrocyte membrane proteins by two dimensional electrophoresis. *J. Biol. Chem.* 251: 253-255.

Dalmark, M. 1976. Effects of halide and bicarbonate on chloride transport in human red blood cells. *J. Gen. Physiol.* 67:223-234.

Deuticke, B. 1970. Anion permeability of the red blood cell. *Naturwiss.* 57:172-279.

Dutton, A., and S. J. Singer. 1975. Crosslinking and labelling of membrane proteins by transglutaminase-catalysed reactions. *Proc. Natl. Acad. Sci. USA* 72:2568-2571.

Fairbanks, G., T. L. Steck, and D. F. H. Wallach. 1971. Electrophoretic analysis of the major polypeptides of the erythrocyte membrane. *Biochemistry* 10:2606-2617.

Findlay, A. 1974. The receptor proteins for concanavalin A and lens culinaris phyto-hemaglutinin in the membrane of the human erythrocyte. *J. Biol. Chem.* 249:4378-4403.

Fossel, E. T., and A. K. Solomon. 1977. Membrane mediated link between ion transport and metabolism in human red cells. *Biochim. Biophys. Acta* 464:82-92.

Furthmayr, H., I. Kahane, and V. T. Marchesi. 1976. Isolation of the major intrinsic transmembrane protein of the human erythrocyte. *J. Membrane Biol.* 26: 173-187.

Gahmberg, C. G., and S. Hakamori. 1973. External labelling of cell surface galactose and galactose amine in glycolipid and glycoprotein of human erythrocytes. *J. Biol. Chem.* 348:4311-4317.

Grinstein, S., S. Ship, and A. Rothstein. 1977. Anion transport in relation to proteolytic dissection of band 3. *Biochim. Biophys. Acta* (in press).

Guidotti, G. 1972. Membrane proteins. *Annu. Rev. Biochem.* 41:731-752.

Gunn, R. B., M. Dalmark, D. C. Tosteson, and J. O. Wieth. 1973. Characteristics of chloride transport in human red blood cells. *J. Gen. Physiol.* 61:185-206.

Halestrap, A. P. 1976. Transport of pyruvate and lactate into human erythrocytes. *Biochem. J.* 156:193-207.

Ho, M. K., and G. Guidotti. 1975. A membrane protein from human erythrocytes involved in anion exchange. *J. Biol. Chem.* 250:675-683.

Hubbard, A. L., and Z. A. Cohn. 1972. The enzymatic iodination of the red cell membrane. *J. Cell Biol.* 55:390-405.

Itaya, K., C. G. Gahmberg, and S. I. Hakamori. 1975. Cell surface labelling of erythrocyte glycoproteins by galactose oxidase and Mn^{++} = catalyzed coupling reaction with methionine hydrazide. *Biochem. Biophys. Res. Commun.* 64:1028-1035.

Jenkins, R. E., and M. J. A. Tanner. 1975. The major human erythrocyte membrane protein: evidence for an S-shaped structure which traverses the membrane twice and has duplicated sets of sites. *Biochem. J.* 147:393-399.

Juliano, R. L. 1972. The solubilization and fractionation of human erythrocyte membrane proteins. *Biochim. Biophys. Acta* 266:301-306.

Juliano, R. L. 1973. The proteins of the erythrocyte membrane. *Biochim. Biophys. Acta* 300:341-378.

Juliano, R. L., and A. Rothstein. 1971. Properties of an erythrocyte membrane lipoprotein fraction. *Biochim. Biophys. Acta* 249:227-235.

Kahlenberg, A. 1976. Partial purification of a membrane protein from human erythrocytes involved in glucose transport. *J. Biol. Chem.*, in press.

Kant, J. A., and R. L. Steck. 1973. Specificity in the association of glyceraldehyde 3-phosphate dihydrogenase with isolated human erythrocyte membrane. *J. Biol. Chem.* 248:8437-8464.

Kasahara, M., and P. C. Hinkle. 1976a. Reconstitution of D-glucose transport catalysed by a protein fraction from human erythrocytes in sonicated liposomes. *Proc. Natl. Acad. Sci. USA* 73:396-400.

Kasahara, M., and P. C. Hinkle. 1976b. D-glucose transport in vesicles formed from an erythrocyte protein fraction and phospholipids. In G. Semenza and E. Carafoli, eds. *FEBS Symposium on Biochemistry of Membrane Transport,* in press.

Knauf, P. A., L. Davidson, W. Breuer, T. Ostwald, and A. Rothstein. 1976. NAP-taurine as a probe for the anion transport site of the human erythrocyte membrane. In preparation.

Knauf, P. A., F. Proverbio, and J. F. Hoffman. 1974. Chemical characterization and pronase susceptibility of the Na:K pump-associated phosphoprotein of human red blood cells. *J. Gen. Physiol.* 63:305-323.

Knauf, P. A., and A. Rothstein. 1971. Chemical modification of membranes. I. Effects of sulfhydryl and amino reactive reagents on anion and cation permeability of the human red blood cell. *J. Gen. Physiol.* 58:190-210.

Knuferman, H., S. Bhakdi, and D. F. H. Wallach. 1975. Rapid preparative isolation of major erythrocyte membrane proteins using polyacrylamide gel electrophoresis in sodium dodecyl sulfate. *Biochim. Biophys. Acta* 387:464-476.

Langdon, R. G. 1974. Serum lipoprotein apoprotein as major protein constituents of the human erythrocyte membrane. *Biochim. Biophys. Acta* 342:213-228.

Lepke, S., H. Fasold, M. Pring, and H. Passow. 1976. A study of the relationship between inhibition of anion exchange and binding to the red blood cell membrane of 4,4'-diisothiocyano stilbene-2,2'-disulfonic acid (DIDS) and its dihydro derivative (H_2DIDS). *J. Membrane Biol.,* in press.

Lin, S., and J. A. Spudich. 1974. Binding of cytochalasin B to a red cell membrane protein. *Biochem. Biophys. Res. Commun.* 61:1471-1476.

Mueller, T. J., and M. Morrison. 1974. The transmembrane proteins in the plasma membrane of normal human erythrocytes. *J. Biol. Chem.* 249:7568-7573.

Passow, H. 1969. Passive ion permeability of the erythrocyte membrane. In J. A. V. Butler and D. Noble, eds. *Progress in biophysics.* Vol. 19, pp. 425-467. Pergamon, New York.

Passow, H., H. Fasold, S. Lepke, M. Pring, and B. Schuhmann. 1976. Chemical and enzymic modification of membrane proteins and anion transport in human red blood cells. In M. Miller and A. E. Shamoo, eds., *Membrane toxicity.* Plenum Press, New York, in press.

Passow, H., H. Fasold, R. Zaki, L. Schumann, and S. Lepke. 1975. Membrane proteins and anion exchange in human erythrocytes. In G. Gardos and I. Szasz, eds., *Biomembranes, structure and function. FEBS Symp. Ser.* 35: 197-214.

Phillips, D. R., and M. Morrison. 1970. The arrangement of proteins in the human erythrocyte membrane. *Biochem. Biophys. Res. Commun.* 40:284-289.

Phillips, D. R., and M. Morrison. 1971a. Exposed protein on the intact human erythrocyte. *Biochemistry* 10:1766-1771.

Phillips, D. R., and M. Morrison. 1971b. Exterior proteins on the human erythrocyte membrane. *Biochem. Biophys. Res. Commun.* 45:1103-1108.

Phillips, D. R., and M. Morrison. 1971c. Position of glycoprotein peptide chain in the human erythrocyte membrane. *FEBS Letters* 18:95-97.

Phillips, D. R., and M. Morrison. 1973. Changes in accessibility of plasma membrane protein as the result of tryptic hydrolysis. *Nature New Biol.* 242:213-215.

Pinto da Silva, P. 1973. Membrane intercalated particles in human erythrocyte ghosts: sites of preferred passage of water molecules at low temperature. *Proc. Natl. Acad. Sci. USA* 70:1339-1343.

Pinto da Silva, P., P. S. Moss, and H. H. Fudenberg. 1973a. Anionic sites on the membrane intercalated particles of human erythrocyte ghost membranes: freeze-etch localization. *Exp. Cell Res.* 81:127-138.

Pinto da Silva, P., and G. L. Nicolson. 1974. Freeze etch localization of concanavalin A receptors to the membrane intercalated particles of human erythrocyte ghost membranes. *Biochim. Biophys. Acta* 363:311-319.

Reichstein, E., and R. Blostein. 1973. Assymetric iodination of the human erythrocyte membrane. *Biochem. Biophys. Res. Commun.* 54:494-500.

Reichstein, E., and R. Blostein. 1975. Arrangement of human erythrocyte membrane proteins. *J. Biol. Chem.* 25: 6256-6263.

Rice, W. R., and T. L. Steck. 1976. Pyruvate flux into resealed ghosts from human erythrocytes. *Biochim. Biophys. Acta* 433:39-53.

Roses, A. D., and S. M. Appel. 1975. Phosphorylation of component and of the human erythrocyte membrane in myotonic muscular dystrophy. *J. Membrane Biol.* 20:51-58.

Rothstein, A., and Z. I. Cabantchik. 1974. Protein structures involved in the anion permeability of the red blood cell membrane. In K. Bloch, L. Bolis, and S. E. Luria, eds., *Comparative biochemistry and physiology of transport*, pp. 354-362. North Holland Publishing Co., Amsterdam.

Rothstein, A., Z. I. Cabantchik, M. Balshin, and R. Juliano. 1975. Enhancement of anion permeability in lecithin vesicles by hydrophobic proteins extracted from red blood cell membranes. *Biochem. Biophys. Res. Commun.* 64:144-150.

Rothstein, A., Z. I. Cabantchik, and P. Knauf. 1976. Mechanism of anion transport in red blood cells: role of membrane proteins. *Fed. Proc.* 35:3-10.

Sachs, J. R., P. A. Knauf, and P. B. Dunham. 1974. Transport through red cell membranes. In D. M. Surgenor, ed., *The red blood cell*, 2nd ed. Vol. 2, pp. 613-705. Academic Press, New York.

Schmidt-Ulrich, R., H. Knufermann, and D. F. H. Wallach. 1973. The reaction of dimethylaminonaphthalene-5-sulfonyl chloride (Dansyl) with erythrocyte membranes: a new look at "vectorial" membrane probes. *Biochim. Biophys. Acta* 307:353-365.

Shin, B. C., and K. L. Carraway. 1974. Lactoperoxidase labelling of erythrocyte membranes from the inside and outside. *Biochim. Biophys. Acta* 345: 141-153.

Ship, S., Y. Shami, and A. Rothstein. 1977. Synthesis of tritiated 4,4'-diisothiocyano-2,2'-stilbene disulfonic acid ([^{3}H] DIDS) and its covalent reaction with sites related to anion transport in human red blood cells. *J. Membrane Biol.* 33:311-324

Singer, S. J. 1974. The molecular oganization of membranes. *Annual. Rev.*

Biochem. 43:805-833.

Singer, S. J., and G. L. Nicolson. 1972. The fluid mosaic model of the structure of cell membranes. *Science* 175:720-731.

Staros, J. V., B. E. Haley, and F. M. Richards. 1976. Photochemical labelling of the cytoplasmic surface of the membranes of intact human erythrocytes. *J. Biol. Chem.*, in press.

Staros, J. W., and F. M. Richards. 1974. Photochemical labelling of the surface proteins of human erythrocytes. *Biochemistry* 13:2720-2726.

Steck, T. L. 1972a. Selective solubilization of red blood cell membrane proteins with guanidine hydrochloride. *Biochim. Biophys. Acta* 255: 553-556.

Steck, T. L. 1972b. In C. F. Fox, ed., *Membrane research*, p. 71. Academic Press, New York.

Steck, T. L. 1972c. Cross-linking the major proteins of the isolated erythrocyte membrane. *J. Mol. Biol.* 66:295-305.

Steck, T. L. 1974. The organization of proteins in the human red cell membrane. *J. Cell Biol.* 62:1-19.

Steck, T. L., and G. Dawson. 1974. Topographical distribution of complex carbohydrates in the erythrocyte membrane. *J. Biol. Chem.* 249:2135-2142.

Steck, T. L., G. Fairbanks, and D. F. H. Wallach. 1971. Disposition of the major proteins in the isolated erythrocyte membrane proteolytic dissection. *Biochemistry* 10:2617-2624.

Steck, T. L., B. Ramos, and E. Strapazon. 1976. Proteolytic dissection of band 3 the predominant transmembrane polypeptide of the human erythrocyte membrane. *Biochemistry* 15:1154-1161.

Steck, T. L., and J. Yu. 1973. Selective solubilization of proteins from red blood cell membranes by protein perturbants. *J. Supramol. Struct.* 1:220-232.

Strapazon, E., and T. L. Steck. 1976. Binding of rabbit muscle aldolase to band 3, the predominant polypeptide of the human erythrocyte membrane. *Biochemistry* 15:1421-1424.

Tanner, M. J. A., and D. J. Anstee. 1976. A method for the direct demonstration of the lectin binding component of the human erythrocyte membrane. *Biochem. J.* 153:265-270.

Tanner, M. J. A., and D. H. Boxer. 1972. Separation and some properties of the major proteins of the human erythrocyte membrane. *Biochem. J.* 129: 333-347.

Taverna, R. D., and R. G. Langdon. 1973. D-glucosyl-isothiocyanate an affinity label for the glucose transport proteins of the human erythrocyte membrane. *Biochem. Biophys. Res. Commun.* 54: 593-599.

Triplett, R. B., and K. L. Carraway. 1972. Proteolytic digestion of erythrocytes, resealed ghosts and isolated membranes. *Biochemistry* 11:2897-2903.

Wallach, D. F. H. 1972. The disposition of proteins in the plasma membranes of animal cells: analytical approaches using controlled peptidolysis and protein labels. *Biochim. Biophys. Acta* 265:61-83.

Wang, K., and F. M. Richards. 1974. Reaction of dimethyl-3,3'-dithiobisproprionimidate with intact human erythrocytes. *J. Biol. Chem.* 249:8005-8018.

Whiteley, H. M., and H. C. Berg. 1974. Amidation of the outer and inner surfaces of the human erythrocyte membrane. *J. Mol. Biol.* 87:541-561.

Williams, R. O. 1972. The phosphorylation and isolation of two erythrocyte membrane proteins in vitro. *Biochem. Biophys. Res. Commun.* 47:671-678.

Yu, J., D. A. Fischman, and T. L. Steck. 1973. Selective solubilization of proteins and phospholipids from red blood cell membranes by nonionic detergents. *J. Supramol. Struct.* 1:233-248.

Yu, J., and T. L. Steck. 1973. Selective solubilization, isolation, and characterization of a predominant polypeptide from human erythrocyte membranes. *J. Cell Biol.* 59: 374a.

Yu, J., and T. L. Steck. 1975a. Isolation and characterization of band 3, the predominant polypeptide of the human erythrocyte membrane. *J. Biol. Chem.* 250:9170-9175.

Yu, J., and T. L. Steck. 1975b. Associations of band 3, the predominant polypeptide of the human erythrocyte membrane. *J. Biol. Chem.* 250: 9176-9184.

Zaki, L., H. Fasold, B. Schuhmann, and H. Passow. 1975. Chemical modification of membrane proteins in relation to inhibition of anion exchange in human red blood cells. *J. Cell. Physiol.* 86:471-494.

7 Energetics and Molecular Biology of Membrane Transport

H. R. KABACK, S. RAMOS, D. E. ROBERTSON, P. STROOBANT, AND H. TOKUDA

It has become increasingly apparent over the past ten years or more that membrane vesicles isolated from bacteria as well as from eukaryotes provide a useful system for the study of active transport (Fogarty International Symposium, 1976). Vesicles are devoid of cytoplasmic constituents and their metabolic activities are restricted to those provided by the membrane itself, which constitute considerable advantages over intact cells. Moreover, transport by membrane vesicles per se is practically nil. Thus, the driving force for uptake of a particular solute can be determined by studying which compounds or experimental manipulations result in accumulation. In addition, the accumulated solutes and the energy source undergo minimal biochemical alteration, allowing clear definition of the reactions involved. Finally, removal or disruption of the cell wall allows the use of certain probes, inhibitors, and ionophores that cannot be used in the intact cell because the plasma membrane is inaccessible.

Since many aspects of the bacterial membrane vesicle system have been reviewed (Kaback, 1970, 1971, 1972, 1974a, 1976), this contribution will be more of a progress report than a review. Although various aspects of the work are summarized, emphasis is placed upon more recent observations dealing with respiration-linked active transport. It should be clear, though, that studies with this system are relevant not only to active transport, but to the general area of energy transduction and other membrane-related phenomena (Weissbach, Thomas, and Kaback, 1971; Thomas, Weissbach, and Kaback, 1972, 1973; Cox, Kaback, and Weissbach, 1974; Cox et al., 1973; Grollman et al., 1977). In mitochondria and chloroplasts, respiratory energy and light are converted to another form of chemical energy (ATP). In the experimental system, on the other hand, respiratory energy is converted primarily into work in the form of solute accumulation against an electrochemical or osmotic gradient.

General Aspects of the Vesicle System

Transport activity in isolated membrane vesicles is assayed in various ways (Kaback, 1971, 1974a). In the most widely used assay, the vesicles are incubated with a radioactive solute in the presence of an appropriate energy source and, at a given time, the reaction mixtures are diluted and the vesicles immediately separated from the medium by rapid membrane filtration. Although the method is rapid and convenient, it suffers from the disadvantage that the reaction mixtures must be diluted prior to filtration and the vesicles must be washed on the filters to diminish background radioactivity. In many instances, these operations result in significant losses of certain solutes (Ramos, Schuldiner, and Kaback, 1976). The vesicles can also be separated from the reaction mixtures by centrifugation, but in this case samples cannot be assayed rapidly, corrections must be made for radioactive solute trapped in the pellet, and the reaction mixtures become anaerobic even during rapid centrifugation (Ramos, Schuldiner, and Kaback, 1976). These problems have been resolved recently by utilizing a technique called flow dialysis, which allows the rapid, continuous determination of changes in the concentration of solutes in the external medium under aerobic conditions without manipulation of the experimental system (Ramos, Schuldiner, and Kaback, 1976; Ramos and Kaback, 1977a, b, c; Tokuda and Kaback, 1977; Ramos, Schuldiner, and Kaback, 1977). The technique was introduced originally by Colowick and Womack (1969) to determine binding of small molecular weight ligands to enzymes, and the method is based upon the measurement of the rate of dialysis of ligand. Under steady-state conditions, the concentration of ligand in the dialysate is proportional to the concentration of unbound ligand on the other side of a dialysis membrane, and changes in binding or uptake are easily ascertained by assaying the dialysate. Flow dialysis is readily adaptable to transport studies and has provided a uniquely important technique for the quantification of pH gradients and solute accumulation in isolated bacterial membrane vesicles (Ramos, Schuldiner, and Kaback, 1976; Ramos and Kaback, 1977a, b, c; Tokuda and Kaback, 1977).

As early as 1960 (Kaback, 1960), evidence was presented that suggested that bacterial membrane vesicles might provide a useful model system for studies of active transport. Subsequently, it was demonstrated that the vesicles catalyze the transport of a plethora of metabolites in the presence of appropriate energy sources (Kaback, 1974b). Moreover, initial rates of transport and steady-state levels of accumulation of many of these metabolites are comparable to those found in the intact cell (Lombardi and Kaback, 1972; Short, White, and Kaback, 1972a). Initial progress with the system was slow, primarily because of preconceived notions regarding the physical nature of the vesicles and the energetics of active transport. It was generally argued that the vesicles could not be osmotically intact, since their cytoplasm had been removed. Moreover, considerable time was expended in an effort to document the supposition that active transport should be driven by ATP or other high-energy phosphate intermediates. It is now clear that the vesicles are osmotically intact and that the energy source for transport in the vesicles varies with the organism and with the substance transported, and that ATP and other high-energy phosphate intermediates do not drive transport in these vesicles (see Kaback 1974b, for a review of the evidence).

Generally, the vesicular transport systems are of three types: (1) group translocation systems, in which a covalent change is exerted upon the transported substrate (Kaback, 1970, 1974b); (2) active transport systems, in which solute is accumulated against an electrochemical or osmotic gradient without covalent modification; and (3) passive diffusion of certain weak acids and lipophilic ions, followed by equilibration with the pH gradient and the electrical potential, respectively, across the membrane.

Transport of many sugars, amino acids, organic acids, and ions by *Escherichia coli* and *Salmonella typhimurium* membrane vesicles occurs by active transport. These transport systems are driven most effectively by the oxidation of D-lactate to pyruvate, catalyzed by a flavin-adenine-dinucleotide-linked, membrane-bound D-lactate dehydrogenase (D-LDH) which has been purified to homogeneity (Kohn and Kaback, 1973; Futai, 1973). Electrons extracted from D-lactate are passed to oxygen via a membrane-bound respiratory chain, and during this series of reactions, respiratory energy is converted into work in the form of solute accumulation (Lombardi and Kaback, 1972; Barnes and Kaback, 1971; Kaback and Barnes, 1971; Kaback, 1972; Stroobant and Kaback, 1975). Although other oxidizable substrates stimulate transport to some extent, they are not nearly as effective as D-lactate unless ubiquinone-1 (CoQ_1) is added to the vesicles, and this property of the system is elucidated most dramatically with NADH (Stroobant and Kaback, 1975; and see below). It should be emphasized, however, that generation of NADH from inside the vesicles stimulates transport in the absence of exogenous CoQ_1 (Futai, 1974a) and that D-lactate is not an electron donor for active transport in all bacterial membrane systems (Kaback, 1974b). Active transport in the vesicles is driven most effectively by the nonphysiological electron donors reduced phenazine methosulfate (PMS) (Konings, Barnes, and Kaback, 1971) or pyocyanine (Stroobant and Kaback, 1975), both of which donate electrons to the respiratory chain at a site prior to cytochrome b_1, the first cytochrome in the *E. coli* respiratory chain. The use of nonphysiological electron donors has allowed the generalization of the vesicle system to many bacteria (Kaback, 1974b; Konings, Barnes, and Kaback, 1971).

E. coli membrane vesicles also catalyze active transport under anaerobic conditions when the appropriate electron transfer systems are present (Konings and Kaback, 1973; Boonstra et al., 1975; Konings and Boonstra, 1977a, b). Lactose and amino acid transport under anaerobic conditions can be coupled to the oxidation of a-glycerol-P with fumarate as an electron acceptor, or to the oxidation of formate-utilizing nitrate as an electron acceptor. Both of these anaerobic electron transfer systems are derepressed by growth of the organism under appropriate conditions, and components of both systems are relatively loosely bound to the membrane.

Finally, active transport can be driven by artificially induced ion diffusion gradients of appropriate polarity (Hirata, Altendorf, and Harold, 1973, 1974). When vesicles containing potassium are diluted into solutions lacking this cation in the presence of the ionophore valinomycin, the diffusion of potassium out of the vesicles creates an electrical potential across the vesicle membrane (interior negative), and uptake of certain solutes is observed. Although the extent of transport observed under these conditions is considerably less than that observed with D-lactate or reduced PMS, this finding has important implications with respect to the energetics of respiration-linked active transport (see below).

Sidedness of Membrane Vesicles and Its Relationship to the Specificity of D-Lactate as an Electron Donor for Active Transport

One of the most controversial aspects of the vesicle system is the degree of specificity of the physiologic electron donors that energize active transport. In E. coli vesicles, of a large number of potential energy sources tested, very few replace D-lactate to any extent whatsoever, and none is as effective, although many are oxidized at least as rapidly (Lombardi and Kaback, 1972; Barnes and Kaback, 1971; Kaback and Barnes, 1971; Kaback and Milner, 1970; Barnes and Kaback, 1970). It should also be emphasized, moreover, that the vesicles convert radioactive D-lactate, L-lactate, succinate and a-glycerol-P stoichiometrically to pyruvate, fumarate, and dihydroxyacetone-P, respectively (Kaback and Milner, 1970; Barnes and Kaback, 1970; Short, White, and Kaback, 1972b). Thus, in each case, the ability of these compounds to drive transport is related to a relatively clearly defined enzymatic reaction.

Since each electron donor reduces the same membrane-bound cytochromes, qualitatively and quantitatively (Barnes and Kaback, 1971; Short, White, and Kaback, 1972b), it was suggested that the energy-coupling site for active transport in E. coli and Staphylococcus aureus vesicles is located in a relatively specific segment of the respiratory chain between D-LDH and cytochrome b_1, the first cytochrome in the common portion of the respiratory chain. Some critics have argued persistently, however, that a significant number of vesicles are inverted, and that these inverted vesicles oxidize NADH and other electron donors and hydrolyze ATP but do not catalyze active transport (Mitchell, 1973; Harold, 1972; Hare, Olden, and Kennedy, 1974). This is potentially a very important consideration, but it should have been clear from even the earlier experiments (Barnes and Kaback, 1971; Short, White, and Kaback, 1972b) that this possibility is highly unlikely. Moreover, there is now a large body of evidence that demonstrates virtually unequivocally that the membrane of each vesicle retains the same orientation as the membrane in the intact cell. Another explanation that has been proposed for the inability of certain electron donors to drive transport is dislocation of dehydrogenases from the inner to the outer surface of the vesicle membrane during preparation (Altendorf and Staehelin, 1974; Weiner, 1974; Futai, 1974b). In the one instance in which this consideration has been studied extensively (D-LDH, see below), it is clearly not the case. On the other hand, it does seem clear that the calcium magnesium-stimulated ATPase complex can be dislocated from the inner to the outer surface of the membrane under certain conditions (Short, Kaback, and Kohn, 1975). Some of the evidence supporting the contention that the vesicles retain the same polarity as the membrane in the intact cell is as follows:

(1) Initial rates of transport in the vesicles are, in many cases, similar to those observed in whole cells (Lombardi and Kaback, 1972; Short, White, and Kaback, 1972a). Moreover, steady-state levels of accumulation of transport substrates are usually comparable to those observed with intact cells, and the electrochemical proton gradient generated by E. coli and S. typhimurium vesicles (see below) is at least as great as that of the intact cells (compare the data presented in Ramos, Schuldiner, and Kaback, 1976; Ramos and Kaback, 1977a, b; Tokuda and Kaback, 1977; with that in Padan, Zilberstein, and Rottenberg, 1976).

(2) Freeze-fracture studies of membrane vesicles in at least three different laboratories (Kaback, 1971, 1974b; Altendorf and Staehelin, 1974; Konings et al., 1973) demonstrate that the "texture" of the convex surface of the vesicles is distinct from that of the concave surface and that the vesicles are homogeneous in this respect. Moreover, the texture observed on the respective surfaces is exactly the same as that observed in the intact cell.

(3) As mentioned previously, all electron donors that are oxidized by the vesicles reduce the same cytochromes, qualitatively and quantitatively (Barnes and Kaback, 1971; Short, White, and Kaback, 1972b). If a percentage of the vesicles is inverted and only the inverted vesicles oxidize NADH, it is difficult to understand how NADH can reduce all of the cytochromes in the preparations.

(4) Although NADH is generally a poor electron donor for transport in

E. coli vesicles, it is the best physiological electron donor for transport in *B. subtilis* vesicles that are prepared in a similar manner (Konings and Freese, 1972). Moreover, addition of ubiquinone (CoQ_1) to *E. coli* ML 308-225 vesicles in the presence of NADH results in rates and extents of lactose and amino acid transport that are comparable to those observed with D-lactate (Stroobant & Kaback, 1975). Since this effect of CoQ_1 is not observed in the presence of NADPH nor in vesicles lacking NADH dehydrogenase activity, it seems apparent that CoQ_1 is able to shunt electrons from NADH dehydrogenase to an energy-coupling site that is not located between NADH dehydrogenase and the cytochromes. As such, the observations provide direct evidence for the argument that the energy-coupling site is located in a specific segment of the respiratory chain.

(5) Reeves, Lombardi, and Kaback (1972) demonstrated that fluorescence of 1-anilino-8-naphthalene sulfonate (ANS) is dramatically quenched during D-lactate oxidation, an observation similar to that observed in energized mitochondria and in intact *E. coli* treated with ethylenediaminotetraacetic acid. In chloroplasts and submitochondrial particles where the polarity of the membrane is inverted, ANS fluorescence is enhanced upon energization. It follows that any inverted membrane vesicles in the preparations would exhibit enhanced ANS fluorescence during D-lactate oxidation. Thus, if 50 percent of the vesicles were inverted, no net change in ANS fluorescence should have been observed by Reeves, Lombardi, and Kaback (1972) because half of the vesicles would exhibit quenching and half would exhibit enhancement. Similarly, Rosen and McClees (1974) have shown that inverted membrane preparations catalyze calcium accumulation but no D-lactate-dependent proline transport. In contrast, vesicles prepared by osmotic lysis (Kaback, 1971) do not exhibit calcium transport but accumulate proline effectively during D-lactate oxidation.

(6) Although D-LDH mutants exhibit normal transport, and vesicles prepared from these mutants do not exhibit D-lactate-dependent transport, addition of succinate to these vesicles drives transport to the same extent as D-lactate in wild-type vesicles (Hong and Kaback, 1972). Since succinate oxidation by wild-type and mutant vesicles is similar, it seems apparent that the coupling between succinate oxidation and transport is increased in the mutant vesicles. In vesicles prepared from double mutants defective in both D-LDH and succinate dehydrogenase, the coupling between L-lactate oxidation and transport is increased, and L-lactate is the best physiological electron donor for transport (F. Grau, J.-s. Hong, and H. R. Kaback, unpublished information). Moreover, in vesicles prepared from a triple mutant defective in D-LDH, succinate dehydrogenase, and L-lactate dehydrogenase, the coupling between NADH oxidation and transport is markedly increased, and NADH drives transport as well as D-lactate in wild-type vesicles. In addition, it is note-

worthy that vesicles prepared from galactose-grown *E. coli* exhibit high rates and extents of lactose transport during NADH oxidation. These observations indicate that the coupling between a particular dehydrogenase and the energy-coupling site for transport is subject to stringent regulation, and that it may be difficult, if not impossible, to demonstrate specificity of energy-coupling in the intact cell. In some bacteria, however, evidence in favor of the hypothesis has been presented for intact cells. In *Arthrobacter pyridinolis* (Wolfson and Krulwich, 1974), hexose transport in both intact cells and membrane vesicles is coupled to malate oxidation; and in a marine pseudomonad (Thompson and MacLoed, 1974), it has been shown that amino acid transport in whole cells and membrane vesicles is coupled to alcohol dehydrogenase.

(7) As will be discussed below (see section on "reconstitution"), studies with antibodies against D-LDH demonstrate that this membrane-bound enzyme is present exclusively on the inner surface of the vesicle membrane (Short et al., 1975; Short, Kaback, and Kohn, 1975; Futai, 1975). Moreover, D-lactate oxidation drives transport normally in *dld⁻* membrane vesicles reconstituted with D-LDH, and in this system the enzyme is located exclusively on the outer surface of the vesicle membrane. Thus, none of the wild-type vesicles can be inverted or sufficiently damaged to allow access of antibody to the inner surface of the membrane, and D-LDH can drive transport normally even when it is bound to the wrong side of the membrane.

(8) 2-Hydroxy-3-butenoic acid (vinylglycolate) is an analogue of lactate that is actively transported by the lactate transport system and oxidized by D-LDH and L-LDH. As opposed to normal substrates, however, oxidation of this compound yields a reactive electrophile (2-keto-3-butenoate), which is attacked by sulfhydryl-containing proteins in the membrane. There is considerable evidence supporting these conclusions (Short et al., 1974; Walsh and Kaback, 1973, 1974; Shaw et al., 1975), but only two points are critical for the present discussion: (a) Vinylglycolate transport is the limiting step for labeling the membrane proteins; and (b) essentially all of the vinylglycolate taken up is covalently bound to the vesicles. In experimental terms, the rate of covalent binding of vinylglycolate is stimulated at least ten-fold by reduced PMS; stimulation is completely abolished by proton conductors or phospholipase treatment, neither of which affect vinylglycolate oxidation.

Using extremely high specific activity [³H]vinylglycolate, vesicles have been labeled for an appropriate time in the presence of reduced PMS and examined by radioautography in the electron microscope (Short et al., 1974). Each vesicle that has transported vinylglycolate is overlaid with exposed silver grains, and examination of the preparations reveals that 85-90 percent of the vesicles are labeled. It should be emphasized that this is a minimal estimate. Virtually all of the large vesicles are labeled, while the size of the smaller vesicles is such that their proximity to individual silver grains in the emulsion may be limiting. Moreover, identical radioautographic results are obtained with [³H]acetic anhydride, a reagent that reacts nonspecifically with the vesicles. Thus, the great preponderance of the vesicles in the preparations catalyze active transport, making it extremely unlikely that a significant number can be inverted.

Energetics of Active Transport

An initial model proposed by Kaback and Barnes (1971) depicted the carriers as electron transfer intermediates in which a change from the oxidized to the reduced state results in translocation of the carrier-substrate complex to the inner surface of the membrane and a concomitant decrease in the affinity of the carrier for substrate. The model was posed as a tentative working hypothesis that could provide a role for sulfhydryl groups in translocation and at the same time account for the observation that only certain electron transfer inhibitors cause efflux of accumulated solutes. A very different hypothesis, one that emphasizes indirect coupling with an electrochemical gradient of protons as the driving force for active transport, was proposed by Peter Mitchell early in the past decade (Mitchell, 1973; Harold, 1972; Mitchell, 1966, 1967, 1968, 1970a, b; Greville, 1969), and over the past few years, it has become overwhelmingly apparent that Mitchell's so-called "chemiosmotic hypothesis" provides the best explanation for respiration-linked active transport in bacterial membrane vesicles.

As visualized by Mitchell, oxidation of electron donors via the membrane-bound respiratory chain or the hydrolysis of ATP catalyzed by the membraneous Ca^{2+}-, Mg^{2+}-stimulated ATPase complex is accompanied by the expulsion of protons into the external medium, leading to an electrochemical gradient of protons ($\Delta \bar{\mu}_H+$), which is composed of an electrical and a chemical parameter according to the following relationship:

$$\Delta \bar{\mu}_H+ = \Delta \Psi - \frac{2.3\,RT}{F} \Delta pH \tag{1}$$

where $\Delta \Psi$ represents the electrical potential across the membrane, and ΔpH is the chemical difference in proton concentrations across the membrane ($2.3RT/F$ is equal to 58.8 mV at room temperature). According to this hypothesis, it is the electrochemical gradient of protons or one of the gradient's components that provides the immediate driving force for the inward movement of various solutes. Accumulation of organic acids is postulated to be dependent upon the pH gradient (ΔpH) (that is, the undissociated acid is transported through the membrane and is presumed to accumulate in the ionized form due to the relative alkalinity of the internal milieu); while the transport of positively charged compounds such as lysine or potassium is purportedly driven by the electrical component ($\Delta\Psi$); and the uptake of neutral substrates such as lactose or proline is thought to be coupled to $\Delta\bar{\mu}_H+$ and to occur via symport with protons.

Using lipophilic cations and rubidium (in the presence of valinomycin), it has been demonstrated that *E. coli* membrane vesicles generate a $\Delta\Psi$ (interior negative) of approximately -75 mV in the presence of reduced PMS or D-lactate (Hirata, Altendorf, and Harold, 1973; Altendorf, Hirata, and Harold, 1975; Schuldiner and Kaback, 1975). Furthermore, the potential causes the appearance of high affinity binding sites for dansylgalactosides, azidophenylgalactosides, and *p*-nitrophenyl-*a*-D-galactopyranoside on the surface of the vesicle membrane (Schuldiner et al., 1976; Rudnick, Schuldiner, and Kaback, 1976; Schuldiner and Kaback, 1977) and is partially dissipated as a result of lactose accumulation (Schuldiner and Kaback, 1975). Although these findings lend strong support to the chemiosmotic hypothesis, it is apparent that $\Delta\Psi$ in itself is insufficient to account for the magnitude of solute accumulation by the vesicles if it is assumed that the stoichiometry between protons and solute is one-to-one (Mitchell, 1973). This deficiency, in addition to the apparent absence of a transmembrane pH gradient, has left reasonable doubt as to the quantitative relationship between $\Delta\bar{\mu}_H+$ and solute accumulation (Schuldiner and Kaback, 1975). However, recent experiments (Ramos et al., 1976; Kaback, 1976; Ramos and Kaback, 1977a, b, c; Tokuda and Kaback, 1977) have resolved this problem to a large extent, extended the chemiosmotic hypothesis in certain important respects, and provided explanations for a number of earlier observations that appeared to preclude the importance of chemiosmotic phenomena in active transport in this system.

Utilizing flow dialysis, a technique uniquely suited to the measurement of ΔpH across the vesicle membrane, it has been shown that membrane vesicles isolated from *E. coli* generate a transmembrane pH gradient of about two units (interior alkaline) under appropriate conditions. Using the distribution of weak acids (acetate, propionate, butyrate, and 5,5-dimethyloxazolidine-2,4-dione) to measure ΔpH and the distribution of the lipophilic cation triphenylmethylphosphonium to measure $\Delta\Psi$, it is apparent that the vesicles develop a $\Delta\bar{\mu}_H+$ of almost -200 mV (interior negative and alkaline) at pH 5.5 in the presence of reduced PMS or D-lactate and that the major component is a ΔpH of about -120 mV (Fig. 7.1.). As external pH is increased, ΔpH decreases, reaching 0 at about pH 7.5 and above, while $\Delta\Psi$ remains at about -75 mV and internal pH remains at pH 7.5-7.8. To some extent, these variations in ΔpH are probably caused by changes in the oxidation of reduced PMS (Fig. 7.1) or D-lactate (Ramos and Kaback, 1977a), both of which vary with external pH in a manner similar to that described for ΔpH. However, it should also be mentioned that recent experiments (Eisenbach et al., 1977; S. Ramos, H. Rottenberg, and H. R. Kaback, unpublished information)

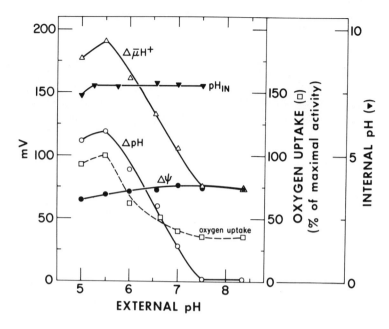

FIGURE 7.1. Effect of external pH on internal pH, ΔpH, $\Delta\Psi$, $\Delta\bar{\mu}_H$+, and oxidation of reduced PMS. The experiments were carried out as described by Ramos and Kaback (1977a) using membrane vesicles prepared from *E. coli* ML 308-225 grown on succinate. (From Kaback, 1976.)

suggest the operation of a mechanism that catalyzes the exchange of external protons for intravesicular cations at relatively alkaline pH. Finally, and importantly, ΔpH and $\Delta\Psi$ can be varied reciprocally in the presence of valinomycin and nigericin with little or no change in $\Delta\bar{\mu}_H$+ (Fig. 7.2) and no apparent change in respiratory activity (Ramos and Kaback, 1977a). In addition to providing direct support for some of the general predictions of the chemiosmotic hypothesis, these results provide a powerful experimental framework within which to test the relationship between $\Delta\bar{\mu}_H$+, ΔpH, and $\Delta\Psi$ and the accumulation of specific transport substrates.

Addition of solutes such as lactose or glucose-6-P, which are accumulated in relatively large amounts by the appropriate vesicles, causes partial collapse of $\Delta\Psi$ (Schuldiner and Kaback, 1975), of ΔpH (Ramos and Kaback, 1977b), or both, demonstrating that respiratory energy can drive active transport via the $\Delta\bar{\mu}_H$+ across the vesicle membrane. Moreover, it should be apparent from the experimental data presented in Figure 7.2 that it is a relatively straightforward matter to investigate the coupling between individual transport systems and $\Delta\bar{\mu}_H$+ or one of its components by means of valinomycin and nigericin titrations at pH 5.5 and

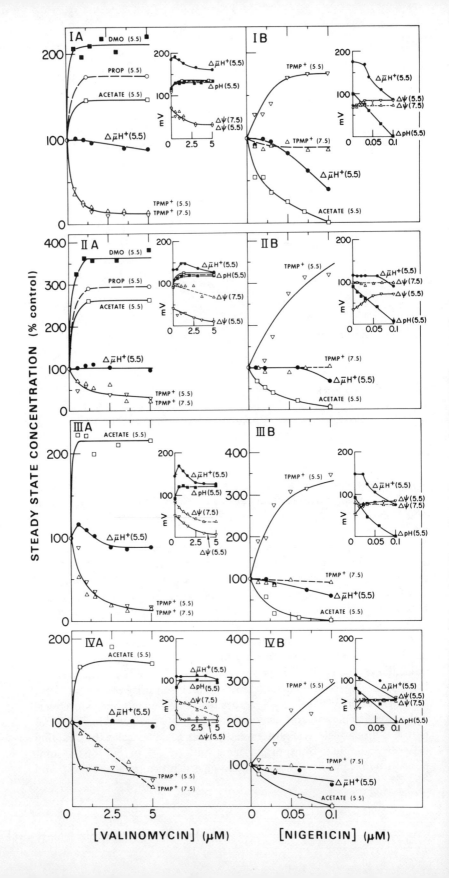

pH 7.5. If, for instance, accumulation of a particular solute is stimulated by valinomycin at pH 5.5., it is clear that transport of this solute is driven primarily by ΔpH, since this component of $\Delta\tilde{\mu}_H+$ is enhanced by valinomycin at this external pH (Fig. 7.2, A panels). On the other hand, stimulation of accumulation by nigericin at pH 5.5 would indicate that transport is driven primarily by $\Delta\Psi$ (Fig. 7.2, B panels), while lack of stimulation or mild inhibition by both ionophores would indicate that accumulation is driven by $\Delta\tilde{\mu}_H+$ (Fig. 7.2, A and B). Such studies have been carried out with 14 solutes that are translocated by different carriers or porters (Ramos and Kaback, 1977b), and the results demonstrate that at pH 5.5, there are two general classes of transport systems: Those that are coupled primarily to $\Delta\tilde{\mu}_H+$ (lactose, proline, serine, glycine, tyrosine, glutamate, leucine, lysine, cysteine, and succinate); and those that are coupled primarily to ΔpH (glucose-6-P, lactate, glucuronate, and gluconate). Strikingly, however, it is eminently clear that at pH 7.5, all of the transport systems, including those that are coupled to ΔpH at pH 5.5., are driven by $\Delta\Psi$, which comprises the only component of $\Delta\tilde{\mu}_H+$ at this external pH. In addition, when the effect of external pH on the steady-state level of accumulation of various transport substrates is examined, none of the pH profiles corresponds to those observed for $\Delta\tilde{\mu}_H+$, $\Delta\Psi$, or ΔpH, and at external pH values exceeding 6.0-6.5, $\Delta\tilde{\mu}_H+$ is insufficient to account for the concentration gradients observed for most of the substrates if it is assumed that the proton:substrate stoichiometries remain constant at 1:1 (Ramos and Kaback, 1977b). This finding and the observation that the accumulation of organic acids is coupled to $\Delta\Psi$ at relatively high external pH values indicates that the stoichiometry between protons and transport substrates may vary as a function of external pH, exhibiting a value of 1:1 at relatively low external pH and increasing to 2:1 or more as external pH is increased.

Evidence supporting the contention that proton:substrate stoichiometries vary as a function of external pH has been obtained recently (Ramos and Kaback, 1977c). The proton electrochemical gradient in *E. coli* membrane vesicles was measured under a variety of conditions (in the presence of valinomycin or nigericin, alone and in combination, or carbonylcyanide-*m*-chlorophenylhydrazone) and compared with steady-state levels of accumulation of lactose, proline, D-lactate, and glucose-6-P measured under identical conditions. Accumulation of lactose and proline is proportional to the magnitude of $\Delta\tilde{\mu}_H+$ at pH 5.5, where ΔpH and $\Delta\Psi$ both contribute to $\Delta\tilde{\mu}_H+$, and at pH 7.5, where $\Delta\Psi$

FIGURE 7.2. (*Facing page*) Effect of valinomycin (*A* panels) and nigericin (*B* panels) on ΔpH, $\Delta\Psi$, and $\Delta\tilde{\mu}_H+$ in membrane vesicles prepared from *E. coli* ML 308-225 grown on succinate (*I*), glucuronate (*III*), or gluconate (*IV*) and *E. coli* GN-2 grown on glucose-6-P (*II*). The experiments were carried out as described by Ramos and Kaback (1977a).

represents the only component of $\Delta\tilde{\mu}_H+$ (Fig. 7.1). Moreover, proportionality constants between $\Delta\tilde{\mu}_H+$ and lactose or proline accumulation indicate that the proton:substrate stoichiometries are 1:1 at pH 5.5 and 2:1 at pH 7.5. In addition, the functional group responsible for the increase in proton:proline stoichiometry has a pK of about 6.8. In contrast to lactose and proline, accumulation of D-lactate and glucose-6-P is directly related to the magnitude of ΔpH at pH 5.5, and stoichiometry values of 1.0 and approximately 1.7 are obtained for D-lactate and glucose-6-P, respectively, at this pH. At pH 7.5, on the other hand, accumulation of each organic acid bears a linear relationship to $\Delta\Psi$, and proton:substrate stoichiometries of 1:1 are observed in both instances.

One attractive conceptual aspect of the chemiosmotic hypothesis for bacterial active transport is its analogy to the mechanism suggested for sugar and amino acid transport in eukaryotic cells (Crane, 1977). In eukaryotic systems an electrochemical gradient of sodium rather than protons is generated through the action of the membraneous sodium, potassium-dependent ATPase, and accumulation of sugars and amino acids occurs via coupled movements with sodium (this process is referred to traditionally as cotransport rather than symport).

Although it is almost certain that many bacterial transport systems catalyze proton:substrate symport, several instances have been reported in which the transport of specific solutes is dependent upon sodium or lithium ion (see Tokuda and Kaback, 1977, for a review). Moreover, some of these studies, in particular those of Stock and Roseman (1971) and Lanyi, Renthal, and MacDonald (1976), indicate that symport or cotransport mechanisms are operative. Since the basic energy-yielding process in bacteria is thought to be proton extrusion and since bacteria apparently do not possess a sodium, potassium-dependent ATPase or a primary sodium pump, such transport systems present certain obvious conceptual problems, among which are: (1) the relationship between the proton electrochemical gradient and these transport systems; (2) the mechanism by which the internal sodium concentration is maintained at a low level.

Tokuda and Kaback (1977) have shown that membrane vesicles isolated from *S. typhimurium* G-30 grown in the presence of melibiose catalyzed methyl 1-thio-β-D-galactopyranoside (TMG) transport in the presence of sodium or lithium as shown initially with intact cells (Stock and Roseman, 1971). TMG-dependent sodium uptake is also observed, but only when a potassium diffusion potential (interior negative) is induced across the vesicle membrane. Cation-dependent TMG accumulation varies with the electrochemical gradient of protons generated as a result of D-lactate oxidation, and the vesicles catalyze D-lactate-dependent sodium efflux in a manner that is consistent with the operation of a proton/sodium exchange mechanism. Although the stoichiometry between sodium and TMG appears to be 1:1 when transport is induced by a potassium diffusion potential, the relationship may exceed 1:1 at relatively alkaline pH. The results are consistent with a model (Fig. 7.3) in which TMG/sodium (lithium) symport is driven by an electrochemical gradient of protons, which functions to maintain low intravesicular sodium or lithium concentrations through proton/sodium (lithium) antiport. A similar mechanism has been suggested for light-dependent glutamate transport in vesicles from *Halobacterium halobium* (Lanyi, Renthal, and MacDonald, 1976).

Reconstitution of D-Lactate-Dehydrogenase-Dependent Functions in Mutants Defective in D-Lactate Dehydrogenase

The membrane-bound D-LDH from *E. coli* has a molecular weight of 75,000 ± 7 percent, contains approximately one mole of flavin adenine dinucleotide per mole of enzyme, and exhibits low activity towards L-lactate (Kohn and Kaback, 1973; Futai, 1973). Oxidized diphosphopyridine nucleotide (NAD) has no effect on the conversion of D-lactate to pyruvate, and recent work carried out in collaboration with Drs. John Salerno and Tomoko Ohnishi of the Johnson Foundation of the University of Pennsylvania indicates that the enzyme probably does not contain a non-heme iron center (P. Stroobant, J. Salerno, H. R. Kaback, and T. Ohnishi, unpublished information).

While much of this work was in progress, Reeves, Hong, and Kaback (1973) demonstrated that guanidine HCl extracts from wild-type membrane vesicles containing D-LDH activity are able to reconstitute D-lactate oxidation and D-lactate-dependent active transport in membrane vesicles from *E. coli* and *S. typhimurium* mutants defective in D-LDH (*dld⁻*). These studies have been extended by Short, Kaback, and Kohn (1974) using the homogeneous preparation of D-LDH described above, and Futai (1974) has independently confirmed some of the observations.

Reconstituted *dld⁻* vesicles carry out D-lactate oxidation and catalyze the transport of many substrates when supplied with D-lactate. D-lactate is not oxidized, and will not support transport in unreconstituted *dld⁻* membranes. Binding of D-LDH to wild-type vesicles produces an increase in D-lactate oxidation but has little or no effect on active transport. Reconstitution of *dld⁻* membranes with increasing amounts of D-LDH produces a corresponding increase in D-lactate oxidation, and transport approaches an upper limit that is similar to the specific transport activity of wild-type vesicles.

Binding of 2-(N-dansyl)aminoethyl-β-D-thiogalactoside (Dns²-Gal) to membrane vesicles containing the *lac* transport system is dependent upon D-lactate oxidation, and this fluorescent probe can be utilized to quantitate the number of functional *lac* carrier protein molecules in the membrane vesicles (see subsequent discussion). When *dld⁻* membrane vesicles are reconstituted with increasing amounts of D-LDH, there is a corre-

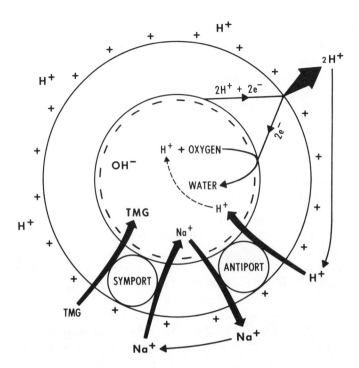

FIGURE 7.3. Schematic representation of sodium-dependent methyl-1-thio-β-D-galactopyranoside (TMG) transport in *S. typhimurium* membrane vesicles. (From Tokuda and Kaback, 1977.)

sponding increase in Dns^2-Gal binding. Assuming that each *lac* carrier protein molecule binds one molecule of Dns^2-Gal, it can be estimated that there is at least a seven- to eight-fold excess of *lac* carrier protein relative to D-LDH in reconstituted *dld*$^-$ vesicles. A similar determination can be made for wild-type vesicles. These vesicles contain approximately 0.07 nmole of D-LDH per mg membrane protein (based on the specific activity of the homogeneous enzyme preparation) and about 1.1 nmoles of *lac* carrier protein per mg membrane protein, yielding a ratio of about 15:1 for *lac* carrier protein relative to D-LDH.

Although the rate and extent of transport increases dramatically on reconstitution, the rate and extent of labeling of *dld*$^-$ vesicles with vinylglycolate remains constant. As discussed above, this compound is transported via the lactate transport system and oxidized to a reactive product by LDH on the inner surface of the vesicle membrane. The observation that reconstituted *dld*$^-$ membranes do not exhibit enhanced labeling by vinylglycolate suggests that bound D-LDH is present on the outer surface of the vesicles.

The suggestion that D-LDH is located on the outer surface of reconstituted *dld*$^-$ membrane vesicles, as opposed to the inner surface of native

ML 308-225 vesicles, has received strong support from experiments with antibody prepared against homogeneous D-LDH (Short et al., 1975; Short, Kaback, and Kohn, 1975; Futai, 1975). Incubation of wild-type membrane vesicles with anti-D-LDH does not inhibit D-LDH activity (assayed by tetrazolium dye reduction, oxygen uptake, D-lactate-dependent transport, or all three) unless the vesicles are disrupted physically or spheroplasts are lysed in the presence of antibody. In contrast, treatment of reconstituted dld^- vesicles with anti-D-LDH causes marked inhibition of activity. The titration curves obtained with reconstituted dld^- membrane vesicles are almost identical quantitatively to that obtained with homogeneous preparations of D-LDH. The conclusion that the enzyme can drive transport from the outer surface of the membrane is also consistent with experiments of Konings (1975) and Short et al. (1975) demonstrating that reduced 5-N-methylphenazonium-3-sulfonate, an impermeable electron carrier, drives transport as well as reduced PMS, its lipophilic analogue. In addition to providing information about the localization of D-LDH in native and reconstituted vesicles, the results with the native vesicles are consistent with many other experiments that demonstrate that essentially none of the vesicles is inverted or sufficiently damaged to allow access of antibody to D-LDH. Given these conclusions and the suggested mechanism for generation of an electrochemical proton gradient (Mitchell, 1973; Harold, 1972), it is amazing that oxidation of D-lactate by reconstituted dld^- vesicles leads to an electrochemical proton gradient that is indistinguishable in polarity and magnitude from that observed in wild-type vesicles where the enzyme is on the inner surface of the vesicle membrane (S. Ramos, S. Schuldiner, and H. R. Kaback, unpublished information).

The flavin moiety of the holoenzyme appears to be critically involved in binding D-LDH to the membrane (Short, Kaback, and Kohn, 1974). Treatment with [1-^{14}C]hydroxybutynoate leads to inactivation of D-LDH by reacting specifically with the flavin adenine dinucleotide coenzyme bound to the enzyme. Enzyme labeled in this manner does not bind to dld^- membrane vesicles. Thus, the flavin coenzyme itself may mediate binding or, alternatively, inactivation of the flavin may result in a conformational change that does not favor binding. It is tempting to speculate on the relevance of this finding to the synthesis of membrane-bound dehydrogenase in the intact cell. Possibly, the apoprotein moiety of D-LDH is synthesized on cytoplasmic ribosomes, but is not inserted into the membrane until coenzyme is bound. If this is so, D-LDH mutants that are defective in the flavin binding site should exhibit soluble material that cross-reacts immunologically with native D-LDH.

Carrier Function at the Molecular Level

It should be apparent from the foregoing discussion that a unifying concept has emerged that can account for the energetics of respiration-

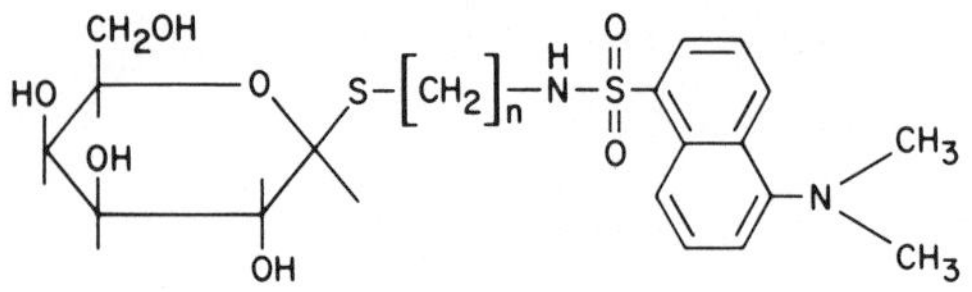

R=	
	DG$_0$
S(CH$_2$)$_2$	DG$_2$
S(CH$_2$)$_3$	DG$_3$
S(CH$_2$)$_4$	DG$_4$
S(CH$_2$)$_5$	DG$_5$
S(CH$_2$)$_6$	DG$_6$
O—CH$_2$CH$_2$	OXY DG$_2$

FIGURE 7.4. Structural formulae of various dansylgalactosides.

dependent active transport in bacterial systems. However, it should be emphasized that this is merely the beginning; ultimately, carrier molecules must be solubilized, purified, and reconstituted so that these concepts can be tested on a molecular level. Preliminary results in this direction have already been reported (Hirata et al., 1976; Altendorf, Müller, and Sandermann, 1977). In any case, within the past few years, some important insights into *lac* carrier function at a more refined level of resolution have been achieved in situ using physical techniques (Schuldiner et al., 1976; Schuldiner and Kaback, 1977, for reviews). (These studies have involved a collaborative effort between our laboratory and that of Dr. Rodolf Weil of Sandoz Forschungsinstitut in Vienna, Austria.)

Substituted Galactopyranosides

The structures of the compounds that we have found to be particularly useful are shown in Figures 7.4 and 7.5. They are of two general varieties: (N-dansyl)aminoalkyl-β-D-galactopyranosides [dansylgalactosides] (Reeves et al., 1973; Schuldiner et al., 1977) and (2-nitro-4-azidophenyl)-β-D-galactopyranosides [azidophenylgalactosides] (Rudnick, Weil, and Kaback, 1975a, b). In both cases, a β-D-galactopyranoside (usually 1-thio-β-D-galactopyranoside) is linked directly or through an alkyl chain of varying length to the appropriate probe. In one instance, a fluorescent dansyl group is attached to the galactopyranoside (Fig. 7.4). The dansyl fluorophore is sensitive to solvent polarity, exhibiting an increase in quantum yield and a blue shift in the emission spectrum as the solvent becomes less polar. Thus, the dansyl moiety is useful as a reporter group,

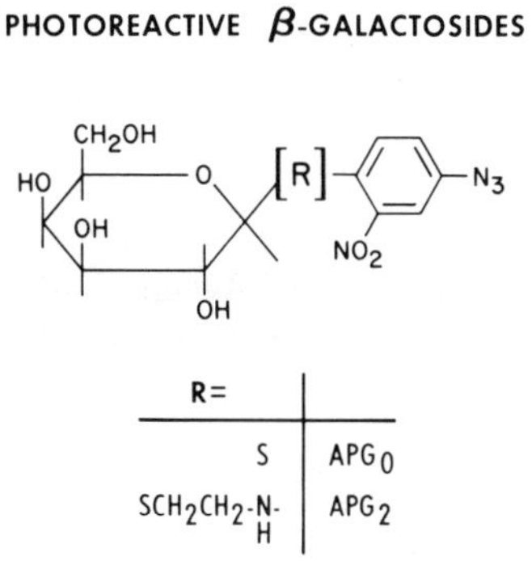

FIGURE 7.5. Structural formulae of azidophenylgalactosides.

giving information about the polarity of the environment in which it is situated.

The second class of compounds contains an arylazide moiety linked to the galactosyl portion of the molecule (Fig. 7.5). Irradiation of arylazides with visible light causes photolysis of the azido group to form molecular nitrogen and a highly reactive nitrene, which can then interact covalently with the macromolecule to which the nitrene-containing ligand is bound.

Dansylgalactosides Are Not Transported

Each of the dansylgalactosides shown in Figure 7.4 is a competitive inhibitor of lactose uptake in membrane vesicles prepared from strains of *E. coli* that contain a functional *lac* carrier protein (K_i values for Dns[0,2,3,4,5,] and [6]-Gal are approximately 550, 30, 12, 6, 3, and 5 μM, respectively). Strikingly, however, an exhaustive number of experiments show that the dansylgalactosides are not transported to any demonstrable extent (Reeves et al., 1973; Schuldiner et al., 1975a, b).

Binding Is Energy-Dependent

The fluorescent properties of the dansylgalactosides in aqueous solution resemble those of other dansyl derivatives, with emission maxima at approximately 540-550 nm and excitation maxima at approximately 340 nm. Importantly, these parameters do not change in the presence of membrane vesicles. When D-lactate is added to the cuvette, however, there is a marked increase in dansylgalactoside fluorescence, which is absolutely dependent upon the presence of functional *lac* carrier protein in the vesicles. Moreover, the fluorescence of 2'-(N-dansyl)aminoethyl 1-thio-β-D-*gluco*pyranoside (dansyl*gluco*side) is not altered under the same conditions, indicating that the effects are highly specific for the galactosyl configuration of the ligand. It is also very significant that the effect of D-lactate can be mimicked by the imposition of ionic diffusion potentials (interior negative) or by a lactose diffusion gradient (inside > outside).

Fluorescence enhancement observed with each dansylgalactoside under the appropriate conditions exhibits an emission maxima at 500 nm and excitation maxima at 345 and 292 nm. The blue shift in the emission maximum is equivalent to that observed when the dansylgalactosides are dissolved in 85 percent dioxane, indicating that the affected molecules find themselves in a hydrophobic environment. The appearance of the new peak in the excitation spectrum at 292 nm indicates, moreover, that the affected dansylgalactosides are excited by energy transfer from tryptophanyl residues in the membrane proteins.

The increase in dansylgalactoside fluorescence induced by D-lactate oxidation is blocked or rapidly reversed by addition of β-galactosides, sulfhydryl reagents, inhibitors of D-lactate oxidation, and proton conductors or other reagents that collapse the membrane potential. On the other hand, fluorescence enhancement induced by imposition of ionic diffusion potentials is blocked by β-galactosides, sulfhydryl reagents, and proton conductors, but not by respiratory poisons. Moreover, the degree to which dansylgalactoside fluorescence is increased under these conditions is dependent upon the magnitude of the applied ionic diffusion potential. Since D-lactate oxidation is intimately related to the generation of $\Delta\Psi$ (interior negative), it seems clear that the increase in dansylgalactoside fluorescence observed with D-lactate and artificially imposed ion gradients occurs by a similar mechanism. Finally, the increase in dansylgalactoside fluorescence observed in the presence of a lactose diffusion gradient is inhibited by β-galactosides and sulfhydryl reagents, both of which block the binding site of the carrier, but not by respiratory poisons or reagents that collapse $\Delta\Psi$. Thus, the fluorescence changes induced under these conditions occur by a mechanism which is independent of $\Delta\bar{\mu}_H+$.

Titration studies demonstrate that the vesicles bind 1-2 nm of each dansylgalactoside per mg of membrane protein, a value that is almost identical to that obtained from direct binding measurements with [³H] dansylgalactosides (Schuldiner and Kaback, 1975; Schuldiner et al., 1977). Assuming that one mole of dansylgalactoside is bound per mole of *lac* carrier protein and that the molecular weight of the *lac* carrier protein is approximately 30,000, 1-2 nm/mg membrane protein is equivalent to 3-6 percent of the membrane protein. This value is remarkably similar to that reported by Jones and Kennedy (1969), who used a completely independent method. In addition, these titration studies indicate that the affinity of the *lac* carrier for ligand is directly related to the length of the alkyl chain between the galactosyl and dansyl moieties of the probes. In addition, the affinity constants of the various dansylgalactosides, as determined by fluorimetric titration or flow dialysis (Reeves et al., 1973; Schuldiner et al., 1975a; Schuldiner, Weil, and Kaback, 1976; Schuldiner et al., 1977) and their apparent K_i values for lactose transport correlate very well.

There are at least three conceivable mechanisms by which energy might lead to dansylgalactoside binding in these experiments: (1) The carrier might be accessible to the external medium and binding could occur spontaneously. In this case, energy coupling would result in partial translocation of the bound ligand, resulting in its exposure to the hydrophobic interior of the membrane, and thus to the fluorescence changes observed. This possibility seems unlikely because no changes whatsoever in the emission or excitation spectra of the dansylgalactosides are observed unless the vesicles are "energized" and for reasons to be discussed below. (2) The carrier might be accessible to the external medium in the absence of energy coupling, and its affinity might be increased when energy is supplied. (3) The carrier might be inaccessible to the external medium, and energy coupling might cause a conformational change such that high-affinity binding sites appear on the external surface of the membrane. If it is postulated that the lac carrier protein (or part of it) is negatively charged, the appearance of binding sites on the exterior surface of the membrane can be easily conceptualized. Imposition of $\Delta\Psi$ (interior negative) would cause "movement" of the negatively charged carrier to the external surface of the membrane and binding of dansylgalactosides. It should be emphasized, however, that the last two possibilities are not mutually exclusive and that energy-coupling may well increase the accessibility and the affinity of the carrier simultaneously. In any case, the data suggest that energy is coupled to one of the intial steps in transport, and that facilitated diffusion therefore cannot represent the first step in the active transport of β-galactosides.

Although we have not been able to distinguish between the latter two alternatives, further evidence for the proposition that the fluorescence changes observed upon energization of the membrane are due to binding of the dansylgalactosides per se, rather than binding followed by translocation into the hydrophobic interior of the vesicle membrane, has been provided by the use of other techniques. Anisotropy (depolarization) of fluorescence can be used to assess binding specifically, since changes in this parameter reflect alterations in the rotation of molecules in solution. Studies with Dns^2-Gal and Dns^6-Gal (Schuldiner et al., 1975c) demonstrate that there is a marked increase in fluorescence anisotropy in vesicles containing the lac carrier protein during D-lactate oxidation. In the absence of electron donor, anisotropy values are minimal and identical in vesicles with or without the lac carrier protein when D-lactate is added. In addition to changes in anisotropy, fluorescence lifetime studies with Dns^2-Gal and Dns^6-Gal are also consistent with the proposition that changes in dansylgalactoside fluorescence observed on energization of the vesicles reflect binding of the fluorescent probes to the *lac* carrier protein (Schuldiner et al., 1975c). Finally, it can be calculated from the anisotropy and lifetime values that the rotational relaxation time of Dns^2-Gal increases dramatically when the probe is bound to the *lac* carrier protein.

High specific activity $[^3H]Dns^6$-Gal has been synthesized, and binding to membrane vesicles has been studied directly by flow dialysis (Schuldiner, Weil, and Kaback, 1976; Schuldiner et al., 1977). With vesicles containing the *lac* carrier protein, little, if any, binding is detected in the absence of D-lactate or reduced PMS. In the presence of these electron donors, binding is observed and the binding constant and number of binding sites are approximately 5 μM and 1.5 nm/mg membrane protein, respectively. Both values are in excellent agreement with those obtained by fluorescence titration. These results provide a strong indication that the enhancement of dansylgalactoside fluorescence observed on energization of membrane vesicles containing the β-galactoside transport system reflects binding of the probe to the *lac* carrier protein.

Finally, studies with photoreactive azidophenylgalactosides (Fig. 7.5) provide completely independent support for the conclusions derived from the dansylgalactoside experiments. 2-Nitro-4-azidophenyl-1-thio-β-D-galactopyranoside (APG_O) competitively inhibits lactose transport in ML 308-225 membrane vesicles with an apparent K_i of 75 μM (Rudnick, Weil, and Kaback, 1975a). The initial rate and steady-state level of $[^3H]APG_O$ accumulation are markedly stimulated by the addition of D-lactate to vesicles containing the β-galactoside transport system, and kinetic studies reveal an apparent K_m of 75 μM. Vesicles devoid of the β-galactoside transport system do not take up significant amounts of APG_O in the presence or absence of D-lactate. When exposed to visible light in the presence of D-lactate, APG_O irreversibly inactivates the β-galactoside transport system. Strikingly, APG_O-dependent photoinactivation is not observed in the absence of electron donors. Kinetic studies of the inactivation process yield a K_D of 77 μM, a value that is almost identical to the K_m and K_i values obtained with this compound. Moreover, lactose protects against APG_O photoinactivation, and significant inactivation of amino acid transport is not observed. It is clear, therefore, that this compound specifically inactivates the *lac* carrier protein.

Analogous studies with 2'-N(2-nitro-4-azidophenyl) aminoethyl-1-thio-β-D-galactopyranoside (Fig. 7.5) demonstrate that this compound behaves similarly with respect to photoinactivation of the β-galactoside transport system, with two important exceptions (Rudnick, Weil, and Kaback, 1975b). Like its analogue Dns^2-Gal, APG_2 is not accumulated by the vesicles in the presence of D-lactate or ascorbate-PMS and it exhibits a higher affinity for the *lac* carrier protein than APG_O (the K_i for competitive inhibition of lactose transport and the K_D for photoinactivation in the presence of D-lactate are 35 μM). In addition, it has been demonstrated that an artificially imposed $\Delta\Psi$ (interior negative) also leads to APG$_2$-dependent photoinactivation of the *lac* carrier protein.

Resolution of an Apparent Discrepancy

Although these studies strongly indicate that the accessibility, the affinity, or both, of the *lac* carrier protein are dependent upon energy-coupling, Kennedy et al. (1974) have reported that binding of β-D-galactopyrano-syl-1-thio-β-D-galactopyranoside (TDG) and *p*-nitrophenyl-*a*-D-galacto-pyranoside (NPG) to membrane particles prepared by ultrasonic disruption is not energy dependent. It should be emphasized, though, that the amount of binding observed by these workers is approximately 10-20 times less than the total amount of *lac* carrier protein (that is, M protein) present in the membrane, as determined by Jones and Kennedy (1969) and by titration studies with dansylgalactosides. In any case, a small amount of nonspecific binding of dansylgalactosides is detected by fluorescence anisotropy and lifetime studies, allowing the possibility that the techniques utilized thus far are not sufficiently sensitive to detect less than 10 percent of the binding observed under energized conditions. For this reason, high specific activity [^{3}H]NPG was synthesized and its binding to membrane vesicles investigated by means of flow dialysis (Rudnick, Schuldiner, and Kaback, 1976). These studies corroborate the findings of Kennedy et al. (1974), and they also confirm the observations discussed above. There is a small amount of NPG binding by vesicles containing the *lac* carrier protein (about 0.2 nmoles per mg of membrane protein at saturation), which is abolished by *p*-chloromercuribenzene-sulfonate (*p*-CMBS) but not by proton conductors, and this binding is not dependent upon the presence of D-lactate. During D-lactate oxidation, however, approximately 2.3 nmoles of NPG are bound per mg of membrane protein (an amount virtually identical to that observed with the dansylgalactosides), and all of the bound ligand is displaced by *p*-CMBS. In addition, the K_D for binding observed under both conditions is very similar (6-9 μM). Thus, there is a small amount of binding in the nonenergized state, but it is quite clear that the great bulk of the *lac* carrier protein is unable to bind ligand unless the membrane is energized.

Translocation and Accumulation of β-Galactosides

The observation that a small but significant amount of NPG binding to membrane vesicles containing the *lac* carrier protein occurs in the absence of an electrochemical gradient of protons or artificially applied ion diffusion potentials suggest that the *lac* carrier protein may exist in two forms that are in a state of dynamic equilibrium: (1) A high-affinity form that is accessible on the external surface of the membrane; and (2) a low-affinity, cryptic form. In the absence of D-lactate or reduced PMS, 90 percent or more of the carrier is in the low-affinity, cryptic form, and only 10 percent or less is in the high-affinity, accessible form. Upon generation of an electrochemical proton gradient across the vesicle membrane (interior negative and alkaline), one or more negatively charged groups in the low-affinity, cryptic form of the protein might be influenced, resulting in a conformational change and a shift in the equilibrium. According to such a model, active transport would occur by binding of ligand to the high-affinity form of the carrier on the external surface of the membrane, followed by conversion of the carrier to the low-affinity, cryptic form and release of ligand from the inner surface of the membrane. However, since the carrier may be negatively charged, translocation of ligand would require neutralization of the high-affinity form of the carrier on the external surface of the membrane, which might be accomplished by protonation of the negatively charged functional group or groups. The protein would then be uncharged and no longer under the influence of the electrochemical gradient. The protonated carrier-ligand complex would "relax" to the cryptic form, releasing protons and ligand on the inner surface of the membrane, which would regenerate the charged form of the carrier, and the cycle could be repeated. Clearly, this reaction sequence is consistent with many of the observations discussed above. It is also noteworthy that, according to this model, only the ternary complex would catalyze translocation, a stipulation that is necessary lest the carriers themselves dissipate the electrochemical gradient without performing work. It should also be emphasized that this formulation might also account for low rates of facilitated diffusion, without necessitating that the process represent the initial step in the active transport of β-galactosides.

Distance Measurements with the Lac Carrier Protein

As discussed above, the fluorescence changes observed with the dansylgalactosides are due specifically to binding of these ligands to the *lac* carrier protein. Since the spectral properties of the dansyl group are sensitive to solvent polarity, varying the distance between the galactosyl and dansyl moieties of the dansylgalactosides might be informative with respect to the environment in the immediate vicinity of the binding site.

Recently (Schuldiner et al., 1977), each dansylgalactoside homologue shown in Figure 7.1 was synthesized in radioactive form, and binding to membrane vesicles containing the *lac* carrier protein was measured directly by flow dialysis. The results were then compared with the D-lactate-induced fluorescence enhancement and fluorescence anisotropy observed with each dansylgalactoside and with the ability of N-methylpicolinium perchlorate (Shinitzky and Rivnay, 1977), a water-soluble quencher, to abolish the fluorescence of the bound homologues. The following observations have been clearly documented (Schuldiner et al., 1977): (1) The binding affinity of the *lac* carrier protein is directly related to the length of the alkyl chain linking the galactosyl and dansyl ends of the molecules. (2) The maximum number of binding sites observed for each homologue is essentially identical. (3) The increase in fluorescence observed when the probes are bound to the *lac* carrier protein changes markedly when the distance between the galactosyl and dansyl moieties is varied. As the linkage is lengthened from two to four carbons, fluorescence decreases by a factor of ten or more and then increases dramatically and progressively with Dns^5-Gal and Dns^6-Gal. These effects vary directly with Δp values of the bound homologues (polarization in the presence of D-lactate minus polarization in the absence of electron donor) and inversely with the ability of N-methylpicolinium perchlorate to quench the fluorescence of the bound probes.

In view of the specificity of dansylgalactoside binding and the fluorescence properties of the dansyl group, it seems clear that the galactosyl end of these molecules is anchored at the binding site of the *lac* carrier protein, while the dansyl end proceeds from a hydrophobic environment to an aqueous environment to a hydrophobic environment as the alkyl linkage is lengthened. Possible explanations of this behavior depend upon the flexibility and hydrophobicity of the alkyl linkage in the molecules. If it is assumed simplistically that the linkage maintains a linear configuration, the variations in fluorescence might reflect differences in the polarity of the microenvironment within the membrane or on the surface in the vicinity of the *lac* carrier protein. An alternative interpretation is that the alkyl linkage is both flexible and hydrophobic, in which case both parameters would vary directly with chain length. Under these circumstances, the dansyl moiety in Dns^2-Gal might reflect a hydrophobic environment in the binding site of the *lac* carrier protein, and as the dansyl end of the molecule is removed three and then four carbons from the binding site, it may become accessible to the aqueous solvent at the membrane interface. When the alkyl linkage is then elongated to five and subsequently six carbons, however, the molecules might become sufficiently flexible and hydrophobic that the dansyl moiety and part of the alkyl chain can adsorb to a hydrophobic site on the surface of the membrane, which may or may not comprise part of the *lac* carrier protein. Although it is impossible to distinguish absolutely between these two alternatives at the present time, it should be apparent that the latter interpretation is favored by the results of the anisotropy studies and the quenching studies with N-methylpicolinium perchlorate. Moreover, if this interpretation is correct, the binding site in the *lac* carrier protein is probably about 5-6 Å from the aqueous solvent at the surface of the membrane.

REFERENCES

Altendorf, K., H. Hirata, and F. M. Harold. 1975. Impairment and restoration of the energized state in membrane vesicles of a mutant of *Escherichia coli* lacking adenosine triphosphatase. *J. Biol. Chem.* 249:4587.

Altendorf, K., C. R. Müller, and H. Sandermann. 1977. β-Galactoside transport in *Escherichia coli:* reversible inhibition by aprotic solvents and its reconstitution in transport negative membrane vesicles. *Eur. J. Biochem.* 73:545.

Altendorf, K., and L. A. Staehelin. 1974. Orientation of membrane vesicles from *Escherichia coli* as detected by freeze-cleave electron microscopy. *J. Bacteriol.* 117:888.

Barnes, E. M., Jr., and H. R. Kaback. 1970. β-Galactoside transport in bacterial membrane preparations: energy coupling via membrane-bound D-lactic dehydrogenase and c-galactoside transport in *Escherichia coli* membrane

Barnes, E. M., Jr., and H. R. Kaback. 1971. Mechanisms of active transport in isolated membrane vesicles. I. The site of energy coupling between D-lactic dehydrogenase and β-galactoside transport in *Escherichia coli* membrane vesicles. *J. Biol. Chem.* 246:5518.

Boonstra, J., M. T. Huttunen, W. N. Konings, and H. R. Kaback. 1975. Anaerobic transport in *Escherichia coli* membrane vesicles. *J. Biol. Chem.* 250:6792.

Colowick, S. P., and F. C. Womack. 1969. Binding of diffusible molecules by macromolecules: rapid measurement by rate of dialysis. *J. Biol. Chem.* 244:774.

Cox, G. S., H. R. Kaback, and H. Weissbach. 1974. Defective transport in S-adenosyl methionine synthetase mutants of *Escherichia coli. Arch. Biochem. Biophys.* 161:610.

Cox, G. S., E. Thomas, H. R. Kaback, and H. Weissbach. 1973. Synthesis of cyclopropane fatty acids in isolated bacterial membranes. *Biophys.* 158:667.

Crane, R. K. 1977. In *Reviews on physiology, biochemistry, and pharmacology.* Springer-Verlag, Heidelberg, in press.

Eisenbach, M., S. Cooper, H. Garty, R. M. Johnstone, H. Rottenberg, and S. R. Caplan. 1977. Light driven sodium transport in sub-bacterial particles of *Halobacterium halobium. Biochim. Biophys. Acta,* in press.

Fogarty International Symposium. 1976. *Cellular regulation of transport and uptake of nutrients. J. Cell. Physiol.* 89:495.

Futai, M. 1973. Membrane D-lactate dehydrogenase from *Escherichia coli:* purification and properties. *Biochemistry* 12:2468.

Futai, M. 1974a. Stimulation of transport into *Escherichia coli:* membrane vesicles by internally generated reduced nicotinamide adenine dinucleotide. *J. Bacteriol.* 120:861.

Futai, M. 1974b. Orientation of membrane vesicles from *Escherichia coli* prepared by different procedures. *J. Membrane Biol.* 15:15.

Futai, M., and Y. Tanaka. 1975. Localization of D-lactate dehydrogenase in membrane vesicles prepared by using a French Press or ethylene diamine tetraacetate-lysozyme from *Escherichia coli. J. Bacteriol.* 124:470.

Greville, G. D. 1969. A scrutiny of Mitchell's chemiosmotic hypothesis of respiratory chain and photosynthetic phosphorylation. *Curr. Top. Bioenergetics* 3:1.

Grollman, E., G. Lee, F. S. Ambesi-Impiombato, M. F. Meldolesi, S. M. Aloj, H. G. Coon, H. R. Kaback, and L. D. Kohn. 1977. *Proc. Natl. Acad. Sci. USA,* in press.

Hare, J. B., K. Olden, and E. P. Kennedy. 1974. Heterogeneity of membrane vesicles from *Escherichia coli* and their subfractionation with antibody to ATPase. *Proc. Natl. Acad. Sci. USA* 71:4843.

Harold, F. M. 1972. Conservation and transformation of energy by bacterial membranes. *Bacteriol. Rev.* 36:172.

Hirata, H., K. Altendorf, and F. M. Harold. 1973. Role of an electrical potential in the coupling of metabolic energy to active transport by membrane vesicles of *Escherichia coli. Proc. Natl. Acad. Sci. USA* 70:1804.

Hirata, H., K. Altendorf, and F. M. Harold. 1974. Energy coupling in membrane vesicles of *Escherichia coli.* I. Accumulation of metabolites in response to an electrical potential. *J. Biol. Chem.* 249:2939.

Hirata, H., H. Sone, M. Yoshida, and Y. Kazawa. 1976. Solubilization and partial purification of alanine carrier from membranes of a thermophilic bacterium and its reconstitution into functional vesicles. *Biochem. Biophys. Res. Commun.* 69:665.

Hong, J.-s., and H. R. Kaback. 1972. Mutants of *Salmonella typhimurium* and *Escherichia coli* pleiotropically defective in active transport. *Proc. Natl. Acad. Sci. USA* 69:3336.

Jones, T. H. D., and E. P. Kennedy. 1969. Characterization of the membrane protein component of the lactose transport system of *Escherichia coli. J. Biol. Chem.* 244:5981.

Kaback, H. R. 1960. Uptake of amino acids by "ghosts" of mutant strains of *E. coli. Fed. Proc.* 19:130.

Kaback, H. R. 1970. Transport. *Annu. Rev. Biochem.* 39:561.

Kaback, H. R. 1971. Bacterial membranes. *Methods Enzymol.* 22:99.

Kaback, H. R. 1972. Transport across isolated bacterial cytoplasmic membranes. *Biochim. Biophys. Acta* 265:367.

Kaback, H. R. 1974a. Transport in isolated bacterial membrane vesicles. *Methods Enzymol.* 31:698.

Kaback, H. R. 1974b. Transport studies in bacterial membrane vesicles and cytoplasmic membrane vesicles devoid of soluble constituents catalyze the transport of many metabolites. *Science* 186:882.

Kaback, H. R. 1976. Molecular biology and energetics of membrane transport. *J. Cell. Physiol.* 89:575.

Kaback, H. R., and E. M. Barnes, Jr. 1971. Mechanisms of active transport in isolated membrane vesicles. *J. Biol. Chem.* 246:5523.

Kaback, H. R., and L. S. Milner. 1970. Relationship of a membrane-bound D-(-)-lactic dehydrogenase to amino acid transport in isolated bacterial membrane preparations. *Proc. Natl. Acad. Sci. USA* 66:1008.

Kennedy, E. P., M. K. Rumley, and J. S. Armstrong. 1974. Direct measurement of the binding of labeled sugars to the lactose permease M protein. *J. Biol. Chem.* 249:33.

Kohn, L. D., and H. R. Kaback. 1973. Mechanisms of active transport in isolated bacterial membrane vesicles. XV. Purification and properties of the membrane bound D-lactate dehydrogenase from *Escherichia coli. J. Biol. Chem.* 248:7012.

Konings, W. N. 1975. Localization of membrane proteins in membrane vesicles of *Bacillus subtilis. Arch. Biochem. Biophys.* 167:570.

Konings, W. N., E. M. Barnes, Jr., and H. R. Kaback. 1971. Mechanisms of active transport in isolated membrane vesicles. III. The coupling of reduced phenazine methosulfate to the concentrative uptake of β-galactosides and amino acids. *J. Biol. Chem.* 246:5857.

Konings, W. N., A. Bisschop, M. Voenhuis, and C. A. Vermeulen. 1973. New procedure for the isolation of membrane vesicles of *Bacillus subtilis* and an electron microscopy study of their ultra structure. *J. Bacteriol.* 116:1456.

Konings, W. N., and J. Boonstra. 1977a. Anaerobic electron transfer and active transport in bacteria. *Curr. Top. Membranes & Transport*, in press.

Konings, W. N., and J. Boonstra. 1977b. Active solute transport in bacterial membrane vesicles. *Adv. Microb. Physiol.*, in press.

Konings, W. N., and E. Freese. 1972. Amino acid transport in membrane vesicles of *Bacillus subtilis*. *J. Biol. Chem.* 247:2408.

Konings, W. N., and H. R. Kaback. 1973. Anaerobic transport in *Escherichia coli* membrane vesicles. *Proc. Natl. Acad. Sci. USA* 70:3376.

Lanyi, Y., R. Renthal, and R. I. MacDonald. 1976. Light-induced glutamate transport in halo bacterium halobium envelope vesicles. II. Evidence that the driving force is a light-dependent sodium gradient. *Biochemistry* 15:1603.

Lombardi, F. J., and H. R. Kaback. 1972. Mechanisms of active transport in isolated bacterial membrane vesicles. *J. Biol. Chem.* 247:7844.

Mitchell, P. 1966. Chemiosmotic coupling in oxidative and photosynthetic phosphorylation. *Biol. Rev.* 41:445.

Mitchell, P. 1967. Translocations through natural membranes. *Adv. Enzymol.* 29:33.

Mitchell, P. 1968. *Chemiosmotic coupling in oxidative phosphorylation and photosynthetic phosphorylation.* Glynn Res. Ltd., Bodmin, England.

Mitchell, P. 1970. In E. E. Bittar, ed., *Membranes and ion transport.* Vol. I, p. 192. Wiley (Interscience), New York.

Mitchell, P. 1970. In H. P. Charles and B. C. J. G. Knight, eds., *Organization and control in procaryotic and eukaryotic cells.* Symposium of the Society of General Microbiology. Vol. 20, p. 121. Cambridge University Press, London and New York.

Mitchell, P. 1973. Performance and conservation of osmotic wash by proton-coupled solute porter systems. *J. Bioenerg.* 4:63.

Padan, E., D. Zilberstein, and H. Rottenberg. 1976. The proton electrochemical gradient in *Escherichia coli* cells. *Eur. J. Biochem.* 63:533.

Ramos, S., and H. R. Kaback. 1977a. The electrochemical proton gradient in *Escherichia coli* membrane vesicles. *Biochemistry* 16:848.

Ramos, S., and H. R. Kaback. 1977b. The relationship between the electrochemical proton gradient and active transport in *Escherichia coli* membrane vesicles. *Biochemistry* 16:854.

Ramos, S., S. Schuldiner, and H. R. Kaback. 1976. The electrochemical gradient or protons and its relationship to active transport in *Escherichia coli* membrane vesicles. *Proc. Natl. Acad. Sci. USA* 73:1892.

Ramos, S., S. Schuldiner, and H. R. Kaback. 1977. Use of flow dialysis for determinations of delta pH and active transport. *Methods in Enzymology*, in press.

Reeves, J. P., J.-s. Hong, and H. R. Kaback. 1973. Reconstitution of D-lactate-dependent transport in membrane vesicles from a D-lactate dehydrogenase mutant of *Escherichia coli*. *Proc. Natl. Acad. Sci. USA* 70:1917.

Reeves, J. P., F. J. Lombardi, and H. R. Kaback. 1972. Mechanisms of active transport in isolated bacterial membrane vesicles. *J. Biol. Chem.* 247:6204.

Reeves, J. P., E. Shecter, R. Weil, and H. R. Kaback. 1973. Dansyl-galactoside, a fluorescent probe of active transport in bacterial membrane vesicles. *Proc. Natl. Acad. Sci. USA* 70:2722.

Rosen, B. P., and J. S. McClees. 1974. Active transport of calcium in inverted membrane vesicles of *Escherichia coli. Proc. Natl. Acad. Sci. USA* 71:5042.

Rudnick, G., S. Schuldiner, and H. R. Kaback. 1976. Equilibrium between two forms of the lac carrier protein in energized and non-energized membrane vesicles from *Escherichia coli. Biochemistry* 15:5126.

Rudnick, G., R. Weil, and H. R. Kaback. 1975a. Photo inactivation of the β-galactoside transport system in *Escherichia coli* membrane vesicles with 2-nitro-4-azidophenyl-1-thio-β-D-galactopyranoside. *J. Biol. Chem.* 250:1371.

Rudnick, G., R. Weil, and H. R. Kaback. 1975b. Photo inactivation of the β-galactoside transport system in *Escherichia coli:* membrane vesicles with an impermeant azidophenylgalactoside. *J. Biol. Chem.* 250:6847.

Schuldiner, S., and H. R. Kaback. 1975. Membrane potential and active transport in membrane vesicles from *Escherichia coli. Biochemistry* 14:5451.

Schuldiner, S., and H. R. Kaback. 1977. Modified transport substrates as membrane probes. *Biochim. Biophys. Acta*, in press.

Schuldiner, S., G. Kerwar, R. Weil, and H. R. Kaback. 1975a. Energy-dependent binding of dansyl galactosides to the β-galactoside carrier protein. *J. Biol. Chem.* 250:1361.

Schuldiner, S., H. Kung, R. Weil, and H. R. Kaback. 1975b. Differentiation between binding and transport of dansyl galactosides in *Escherichia coli. J. Biol. Chem.* 250:3679.

Schuldiner, S., G. Rudnick, R. Weil, and H. R. Kaback. 1976. Mechanism of β-galactoside transport in *Escherichia coli* membrane. *Trends in Biochem. Sci.* 1:41.

Schuldiner, S., R. D. Spencer, G. Weber, R. Weil, and H. R. Kaback. 1975c. Lifetime and rotational relaxation time of dansyl galactoside bound to the lac carrier protein. *J. Biol. Chem.* 250:8893.

Schuldiner, S., R. Weil, and H. R. Kaback. 1976. Energy-dependent binding of dansyl galactoside to the lac carrier protein: direct binding measurements. *Proc. Natl. Acad. Sci. USA* 73:109.

Schuldiner, S., R. Weil, D. Robertson, and H. R. Kaback. 1977. Microenvironment of the binding site in the lac carrier protein. *Proc. Natl. Acad. Sci. USA*, in press.

Shaw, L., F. Grau, H. R. Kaback, J.-s. Hong, and C. T. Walsh. 1975. Vinylglycolate resistance in *Escherichia coli. J. Bacteriol.* 121:1047.

Shinitzky, M., and B. Rivnay. 1977. Degree of exposure of membrane proteins determined by fluorescence quenching. *Biochemistry* 16:982.

Short, S., D. White, and H. R. Kaback. 1972a. Mechanisms of active transport in isolated bacterial membrane vesicles. *J. Biol. Chem.* 247:7452.

Short, S., D. White, and H. R. Kaback. 1972b. Active transport in isolated bacterial membrane vesicles. *J. Biol. Chem.* 247:298.

Short, S. A., H. R. Kaback, T. Hawkins, and L. D. Kohn. 1975. Immunochemical properties of the membrane-bound D-lactate dehydrogenase from

Escherichia coli. J. Biol. Chem. 250:4285.

Short, S. A., H. R. Kaback, and L. D. Kohn. 1974. D-Lactate dehydrogenase binding in *Escherichia coli dld*-membrane vesicles reconstituted for active transport. *Proc. Natl. Acad. Sci. USA* 71:1461.

Short, S. A., H. R. Kaback, and L. D. Kohn. 1975. Localization of D-lactate dehydrogenase in native and reconstituted *Escherichia coli* membrane vesicles. *J. Biol. Chem.* 250:4291.

Stock, J., and S. Roseman. 1971. A sodium-dependent sugar co-transport system in bacteria. *Biochem. Biophys. Res. Commun.* 44:132.

Stroobant, P., and H. R. Kaback. 1975. Ubequinone-mediated coupling of NADH dehydrogenase to active transport in membrane vesicles from *Escherichia coli. Proc. Natl. Acad. Sci. USA* 72:3970.

Thomas, E., H. Weissbach, and H. R. Kaback. 1972. Further studies on metabolism of phosphatidic acid of isolated *E coli* membrane vesicles. *Arch. Biochem. Biophys.* 150:797.

Thomas, E., H. Weissbach, and H. R. Kaback. 1973. Studies on the metabolism of ATP on isolated bacterial membranes: solubilization and phosphorylation of a protein component of the diglyceride kinase system. *Arch. Biochem. Biophys.* 157:327.

Thompson, J., and R. A. MacLoed. 1974. Specific electron donor-energized transport of a-amino isobutyric acid and K^+ into intact cells of a marine pseudomonad. *J. Bacteriol.* 117:1055.

Tokuda, H., and H. R. Kaback. 1977. Sodium-dependent methyl 1-thio-β-D-galactopyranoside transport in membrane vesicles isolated from Salmonella typhimurium. *Biochemistry* 16:2130.

Walsh, C. T., and H. R. Kaback. 1973. Vinylglycolic acid. *J. Biol. Chem.* 248:5456.

Walsh, C. T., and H. R. Kaback. 1974. Membrane transport as a potential target for antibiotic action. *Ann. N.Y. Acad. Sci.* 235:519.

Weiner, J. H. 1974. The localization of glycerol-3-phosphate dehydrogenase in *Escherichia coli. J. Membrane Biol.* 15:1.

Weissbach, H., E. Thomas, and H. R. Kaback. 1971. Membrane-bound phosphatidic acid. *Arch. Biochem. Biophys.* 147:249.

Wolfson, E. B., and T. A. Krulwich. 1974. Requirement of a functional respiration-coupled D-fructose transport system for induction of phosphoenol pyruvate: D-fructose phosphotransferase activity. *Proc. Natl. Acad. Sci. USA* 71:1739.

PART TWO

Asymmetry in Transport

8 Asymmetry and the Mechanism of the Red Cell Na-K Pump, Determined by Ouabain Binding

JOSEPH F. HOFFMAN

This chapter concerns the active transport of Na and K in red blood cells, with particular emphasis on the sidedness of action of various ligands as they affect the operation of the pump. The energy-requiring process of active transport is oriented, in the sense that Na and K are each moved uphill against their respective electrochemical potential gradients. The notion that the underlying molecular device must possess structural as well as functional asymmetry emerges from thermodynamic analysis of the nature of the coupling between uphill transport and metabolism (Katchalsky, 1970). Therefore, the properties defining the sidedness of the pump should provide useful insight into its molecular mechanism.

With human red cells taken as the prototypical case, some of the general side-dependent features of the Na-K pump are summarized in Figure 8.1. Cells normally contain high $[K_i]$ and low $[Na_i]$ inside, while outside $[K_o]$ is low and $[Na_o]$ is high. The pump operates, of course, to move Na_i out and K_o in (Harris, 1941; Danowski, 1941), and to the extent that the pump can hold the sum $[K_i + Na_i]$ constant, in compensation for changes in Na_i and K_i that result from diffusion leaks, the cell is able to control its volume (Tosteson and Hoffman, 1960). While Na-K exchange by the pump can occur in the absence of K_i, of Na_o, or of both, there is an obligatory dependence of K influx on Na_i and Na efflux on K_o (Harris and Maizels, 1951; Glynn, 1956, 1962; Whittam, 1962). On the other hand, there are competitive effects between Na_o and K_o (Post et al., 1960), as well as between K_i and Na_i (Hoffman, 1962a; Knight and Welt, 1974). The Na-K pump is also sensitive to the tonicity of the suspending medium and reaches maximum activation when the medium is isotonic (Hoffman, 1962a). This contrasts with Na,K-ATPase, where the enzyme

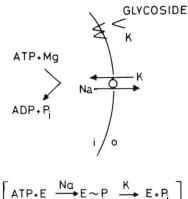

$$\left[\text{ATP} \cdot \text{E} \xrightarrow{\text{Na}} \text{E} \sim \text{P} \xrightarrow{\text{K}} \text{E} \cdot \text{P}_i \right]$$

FIGURE 8.1. Model showing the side-assignments of different ligands that interact with the Na-K pump. The membrane separates inside (i) from outside (o). The pump moves Na_i out and K_o in, at the expense of internal Mg-ATP. Cardiac glycosides, such as ouabain, inhibit the pump from the outside, and the binding of the glycoside to the membrane is antagonized by K_o. The chemical reaction indicates that the pump (E) is phosphorylated (E $\sim$ P) in the presence of Na and is dephosphorylated by K. (From Hoffman, 1972.)

is as active at low as it is at high tonicity. This indicates that the tonicity affects the pump complex asymmetrically to alter its molecular volume or its surround.

The energy source for the pump is ATP_i, and the pump requires Mg_i (Gardos, 1954; Hoffman, 1962b); the products of the reaction, ADP and orthophosphate (P_i), remain inside the cell throughout the transport cycle (Whittam and Ager, 1964). The stoichiometry of the pump appears to be such that for each molecule of ATP_i split, three Na_i are exchanged for two K_o (Post and Jolly, 1957; Sen and Post, 1964; Whittam and Ager, 1965; Garrahan and Glynn, 1967d). The events of phosphorylation of the pump (E), which have been proposed to underlie the translocation of the ions (Albers et al., 1968; Post et al., 1969), are indicated in the equation in Figure 8.1, where phosphorylation of the pump (E $\sim$ P) is catalyzed by Na and where dephosphorylation is catalyzed by K.

The molecular weight of the pump in red cells has been estimated to be between 250,000 and 400,000, the former estimate obtained by irradiation inactivation analysis (Kepner and Macey, 1968) and the latter by centrifugation after solubilization (Dunham and Hoffman, 1970), in accord with values estimated in various ways on other types of source material (see Jørgensen, 1975). The fact that antibodies prepared against a purified preparation of pig kidney microsomal Na,K-ATPase inhibits the Na-K pump in human red cells (Jørgensen, et al., 1973) indicates that the physiological role, as well as the antigenic determinants, of the pump complex are the same regardless of source. This conclusion is supported by evidence that purified Na,K-ATPase preparations incorporated into liposomes have the capacity to actively transport Na and K (Hilden et al., 1974). In highly purified preparations it appears that each unit molecular weight of pump complex can bind one molecule of ATP or one molecule of a cardiac glycoside and can contain one or two Na-dependent phosphoprotein (E $\sim$ P) sites (see Jørgensen, 1975). Each pump complex also contains two major protein components: one a polypeptide with a molecular weight of about 100,000 daltons and the other a glyco-protein of about 40,000 daltons. It is evident that only the large polypeptide can be phosphorylated or can bind ouabain (see Jørgensen, 1975). Although recent evidence indicates that two of the 100,000-dalton copies and at least one (but perhaps two) of the glycoproteins are involved (Jørgensen, 1975; Kyte, 1975; Giotta, 1976), the substructure of the pump is not yet known in sufficient detail to account in molecular terms for the mechanism of action of various ligands that determine or affect the pump's activity.

Cardiotonic steroids, such as ouabain, are known to inhibit the pump (Schatzmann, 1953) only from the outside of the membrane (Caldwell and Keynes, 1959; Hoffman, 1966; Perrone and Blostein, 1973), with the form of the inhibition being noncompetitive or allosteric (Hoffman, 1966). As indicated in Figure 8.1, the binding of glycoside to the membrane is antagonized by K_O (Solomon, et al., 1956; Glynn, 1957; Hoffman, 1966). There are approximately 275 glycoside (ouabain) binding sites per human red cell (Ingram, 1970), and when these are fully occupied all pump activity ceases. If each ouabain binding site is taken to represent a surrogate pump, then it can be estimated that each pump transports about 6,000 ions per minute; and since the range of variation in this turnover number is rather narrow it would appear that differences in pump rates shown by different types of cells reflect, at least to a first approximation, differences in the surface density of pumps rather than differences in their primary properties (Dunham and Hoffman, 1970).

It is of interest now to consider the sidedness of action of certain determinants of glycoside binding as a way of analyzing different conformational states of the pump. In order to study sidedness it is necessary to use a membrane preparation, such as resealed human red-cell ghosts, where the internal and external compartments are not only separated from each other but their composition can be varied independently. While many of the determinants of glycoside binding have been worked out in microsomal or porous ghost preparations (see Albers et al., 1968; Schwartz et al., 1975; Skou, 1975; Hoffman, 1969) it is not possible to know the side-dependencies of the various ligands without determining their effects under conditions where the sidedness relationship has been established. Perhaps the most compelling evidence that conformational changes of the pump can occur comes from studies concerned with the stimulating effects that nucleotide triphosphates (NTP) and inorganic phosphate (P_i) have on the rate at which glycosides bind and inhibit. While the overt effects of NTP and P_i were first worked out on microsomal preparations (Schwartz et al., 1968; Lindenmayer, et al., 1968) it was with resealed ghosts that it was established that *inside* NTP (Bodemann and Hoffman, 1976a) and *inside* P_i (Lisko, et al., 1972; Bodemann and Hoffman, 1976b) together with inside magnesium were responsible for the promotion of glycoside binding to the *outside*.

The model shown in Figure 8.2 summarizes these effects where the reasonable assumption is made that NTP or P_i act by binding to the inward facing aspect of the pump complex (Hoffman, 1969). The glycoside binding site can exist in two forms, A and B. It is occluded in the A form and open to the outside in the B form. Since glycosides, G, can only bind to the B form to give BG, the availability of the B form defines the rate of glycoside binding and thereby the rate of pump inactivation. This is because the equilibrium constant, at least in human red cells, for the reaction $B + G \rightleftharpoons BG$, where G is ouabain, is greater than 10^{10}, indicating the essential irreversibility of the BG form; therefore it is only possible to study differences in binding rates and not in equilibrium. Changes in the set of the $A \rightleftharpoons B$ equilibrium are seen as changes in the conformation of the pump complex. Thus, ATP binding on the inside favors the B form by shifting the equilibrium to the right; whereas K_o in antagonizing glycoside binding (see below) shifts the equilibrium to the left. Just how the conformational changes associated with the glycoside binding site are related to the translocation of Na and K is not specified in the model. Nevertheless, it appears that all of the determinants of glycoside binding can be rationalized with this model in terms of their influence on the $A \rightleftharpoons B$ equilibrium.

Mention should be made of the fact that at least two types of thiol groups appear to be involved in the pump complex and that these are located on opposite sides of the membrane (Hoffman, 1972). This conclusion is based on studies using different sulfhydryl reagents and comparing their effects when they could interact only from the outside of the membrane (intact cells or resealed ghosts) with their effects when they were accessible to both sides of the membrane (porous ghosts). These agents act from either side to inhibit the pump or Na,K-ATPase activity. However, ouabain binding is only inhibited when the thiol groups on the inside are attacked. Thus, inside thiol groups either control the interaction of inside ligands, such as ATP, with the pump or are involved directly with changes in pump conformation. In either instance, it is evident that the inside thiol groups can influence the $A \rightleftharpoons B$ equilibrium.

We come now to survey in some detail the side-dependent actions of Na and K on the rate of ouabain binding to human red cells. These actions are of special interest since the *sided* effects of Na and K

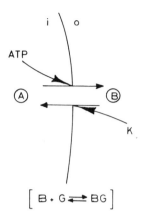

$$\left[\text{B} \cdot \text{G} \rightleftarrows \text{BG} \right]$$

FIGURE 8.2. Model of the glycoside (ouabain) binding site of the membrane. The binding site is either open to the outside (B form) or occluded (A form), but glycosides (G) combine only with the B form. Different conformational states of the pump are presumed to define the set of the A$\rightleftarrows$B equilibrium. K_o shifts the equilibrium to the left, and nucleotide triphosphates such as ATP_i shift the equilibrium to the right. (From Hoffman, 1972.)

have been found to be different from the effects Na and K have on un-sided preparations, such as microsomes or porous ghosts. In the present work reconstituted ghosts were used in order to separately set and control the concentrations of Na and K on the two sides of the membrane (Bode-mann and Hoffman, 1976a). We will consider the effects of Na and K, first on the outside and then on the inside, on the rate of ATP-promoted ouabain binding. Thus, when $[Na_o]$ is varied at constant $[K_o]$, $[K_i]$, and $[Na_i]$, the rate at which ouabain binds, as shown in Figure 8.3, is increased as $[Na_o]$ is increased. This effect of Na_o has also been observed with in-tact cells (Schatzmann, 1965; Beauge and Adragna, 1971; Sachs, 1974) and presumably represents a competitive action of Na_o with K_o. Since, as already mentioned in connection with Figure 8.1, K_o antagonizes the binding of ouabain to the membrane (Glynn, 1957; Hoffman, 1966), Na_o in effect reduces the membrane's affinity for K_o and thereby stimulates the rate of ouabain binding. This type of interaction implies that the effect of Na_o should be decreased as $[K_o]$ is increased and that its effect should disappear when the pump becomes saturated with K_o. That this is so is shown in Table 8.1. Therefore, the results indicate that, since Na_o is known to competitively inhibit the activation of the pump by K_o (Post et al., 1960), the effects that Na_o and K_o have on the ouabain binding rate are opposite to their effects on the Na-K pump (Bodemann and Hoffman, 1976a). This is to say that the ouabain binding rate is inversely related to the pumping rate, a conclusion supported by results summarized below, where $[Na_i]$ is varied at constant $[K_o]$. In terms of the model shown in

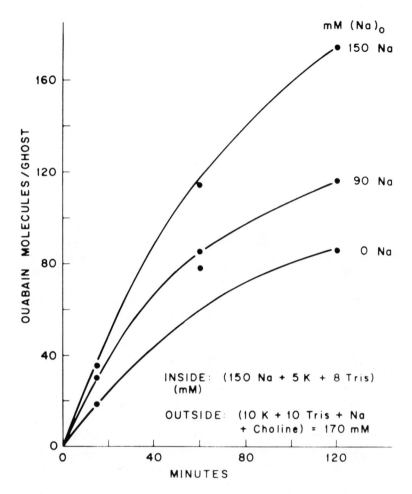

FIGURE 8.3. The effect of varying $[Na_O]$ on the rate of ATP_i-promoted ouabain binding to reconstituted human red-cell ghosts. The uptake of ouabain was estimated by the use of $[^3H]$ ouabain. (From Bodemann and Hoffman, 1976a, which should be consulted for methods used and experimental details.)

Figure 8.2, K_O favors the A form and Na_O, in antagonizing the effect of K_O, favors a shift of the A $\rightleftharpoons$ B equilibrium to the right.

The effect of altering $[Na_i]$ when the concentrations of K_i, Na_O, and K_O are all held constant is shown in Figure 8.4, where it is apparent that the rate of binding of ouabain is *decreased* as $[Na_i]$ is increased. This action of Na_i is opposite to that of Na_O and corresponds to the effect that Na_i has in activating the Na-K pump, reaching saturation at about 40 mM Na_i (Post and Jolly, 1957). That this action of Na_i is dependent upon the activity of the Na-K pump is shown by the results presented in Figure

TABLE 8.1. Effect of Na_O and K_O on ouabain binding. (Data from Bodemann and Hoffman, 1976a.)

K_O (mM)	Na_O (mM)	Molecules ouabain bound per ghost in 60 min
1	3	194
1	40	243
1	150	321
15	3	133
15	40	136
15	150	140

Note: The competitive effect of Na_O on K_O alters ouabain binding to reconstituted human red-cell ghosts. ATP-promoted ouabain binding was measured with [^{3}H]-ouabain. The ghosts were suspended in media containing the indicated concentrations of Na and K such that the sum, Na + K + choline, was 165 mM. The measure of the rate of ouabain binding was taken as the number of ouabain molecules bound per ghost after exposure to [^{3}H]-ouabain for one hour. Bodemann and Hoffman (1976a) should be consulted for experimental details.

8.5, where [Na_i] and [K_O] have been varied such that the pump is only operative under the condition where there is 40 mM Na_i and 6 mM K_O. Thus, there is an inverse relationship between the rate of ouabain binding and the rate of active transport. And not only is the ouabain binding rate dependent on the coupled action of Na_i and K_O, but for K_O to antagonize the binding of ouabain, Na_i (or K_i, see below) must also be present (Bodemann and Hoffman, 1976a).

Since K_i is known (Hoffman, 1962a; Knight and Welt, 1974; Simons, 1974) to alter (lower) the pump's affinity for Na_i, it might be expected that increased [K_i] at constant [Na_i], [K_O] and [Na_O] would result in increased rates of binding of ouabain. But, contrary to expectation, increasing [K_i] decreases the rate of ouabain binding in a manner similar to the effect of Na_i, as shown in Figure 8.4, under circumstances where the activity of the Na-K pump was found to be unaffected (Bodemann and Hoffman, 1976a). This action of K_i correlates with its effect on activating K-K exchange (Simons, 1974) as demonstrated by measuring the ouabain-sensitive K efflux as a function of [K_i]. Since K-K exchange requires K_O as well as K_i, in parallel to the pump's requirement for Na_i and K_O, it was possible to establish the relationship between K-K exchange and the rate of ouabain binding. This is shown in Figure 8.6, where it is apparent that, in analogy to the situation with Na-K exchange (Fig. 8.5), the ouabain binding rate is slower the faster the rate of K-K exchange. It can also

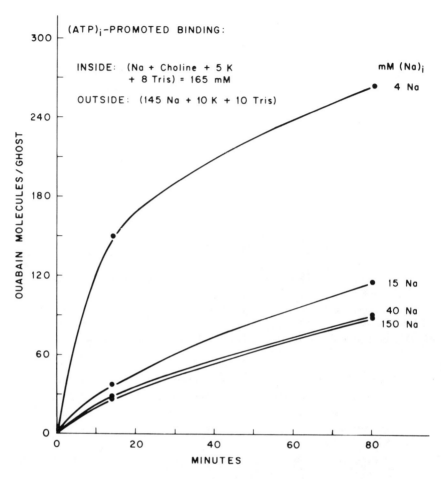

FIGURE 8.4. The effect of varying $[Na_i]$ on the rate of ATP_i-promoted ouabain binding to reconstituted red-cell ghosts. (From Bodemann and Hoffman, 1976a.)

be stated again that in order for K_o to antagonize the binding of ouabain either K_i or Na_i must be present.

The conclusion that the rate of ouabain binding is affected (albeit inversely) by the transport rate of the pump, whether it is operating to carry out Na-K or K-K exchange, does not appear to apply to Na-Na exchange. It will be recalled that Na-Na exchange requires Na on both sides of the membrane; it is inhibited by the presence of K_o; and, since it is ouabain-sensitive, it is considered, like K-K exchange, to represent a partial reaction of the pump, at least in human red cells (Garrahan and Glynn, 1967a, b, c). It is also known that the Na-Na exchange occurs at a rate proportional to the intracellular concentration of ADP (Glynn and Hoffman, 1971). An indication that ouabain binding is essentially inde-

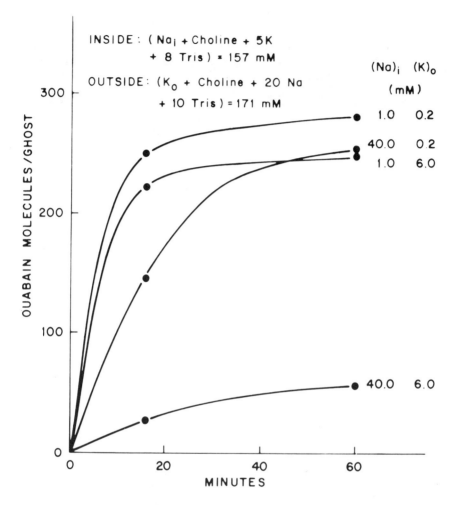

FIGURE 8.5. Na-K exchange and the rate of ATP$_i$-promoted ouabain binding to reconstituted red-cell ghosts. Ghosts prepared to contain either high or low [Na$_i$] were suspended in media containing either high or low [K$_o$]. The Na-K pump is activated when [Na$_i$] and [K$_o$] are both high. (From Bodemann and Hoffman, 1976a.)

pendent of Na-Na exchange can be seen in Figure 8.5, where the rate of ouabain binding had little effect when [Na$_i$] was changed from 1 to 40 mM in the presence of 20 mM Na$_o$ and 0.2 mM K$_o$. When this relationship was studied in more detail and the conditions for Na-Na exchange were varied systematically, for instance, by varying [Na$_i$] at constant [Na$_o$], or by varying [Na$_o$] at constant [Na$_i$], or by altering the availability of ADP$_i$, the ouabain binding rate was always found to be maximum and unaffected by the rate of Na-Na exchange (Bodemann and

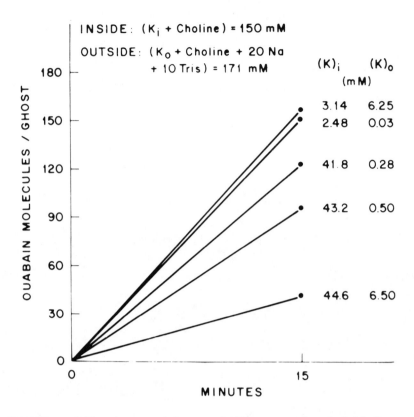

FIGURE 8.6. K-K exchange and the rate of ATP_i-promoted ouabain binding to reconstituted red-cell ghosts. Ghosts prepared to contain varying $[K_i]$ were suspended in media containing varying $[K_o]$. The rate of K-K exchange increases as $[K_i]$ together with $[K_o]$ is increased. (From Bodemann and Hoffman, 1976a.)

Hoffman, 1976a). In addition, Sachs (1974) found that the rate of ouabain binding to intact human red cells in the absence of both K_i and K_o was the same whether or not Na_o was present. These findings are also consistent with the fact that Na, in the absence of K, has little effect on the rate at which ouabain binds to porous ghosts (Hoffman, 1969), but are in marked contrast to the effect of Na in microsomal preparations. The binding of ouabain to microsomal preparations of Na,K-ATPase are almost completely dependent upon the addition of Na (Schwartz, Matsui, and Laughter, 1968; Lindenmayer et al., 1968). If this is an authentic effect of Na alone, acting independent of the possible presence of K, then this difference between ghosts and microsomal preparations must be accounted for before generalizations can be made about the properties of different transport systems. And to the extent that sidedness remains unspecified there is also an uncertainty with regard to the mechanism of action of any added ligand.

One way of explaining the fact that the ouabain binding rate can be altered by K_O, but not by Na_O in the absence of K_O, is to suppose that the pump can assume at least two different conformations, either of which can face the outside but only one of which can influence the affinity of the B site for ouabain. It is of interest to relate these different conformations to those that have been associated with the different phosphorylation states of the pump (Fahn et al., 1966a, b; Albers et al., 1968) since something is known about the sensitivity of phosphorylation to different ligands. These different forms, E_1 and E_2, can be represented by rewriting the equation in Figure 8.1 such that the pump cycles in the sequence, $E \rightleftharpoons E_1P \rightleftharpoons E_2P \rightleftharpoons E$. E_2P would then be the conformation that would be associated with the B form, and since E_2P is known to be K-sensitive (dephosphorylated by K) and ADP-insensitive (Post et al., 1969) a role for K_O is provided while the pump is carrying out either Na-K or K-K exchange. The formation of E_1P, or some equivalent form such as ADP·E_1P (see Bodemann and Hoffman, 1976a), would be the form of the pump that would be associated with Na-Na exchange (Glynn and Hoffman, 1971), because it is insensitive to K and because its formation is reversed by ADP (Post, et al., 1969). While this formulation is reasonable on the basis of the evidence available, it still has to be shown that it is K_O, that is, outside K, that affects E_2P, just as it is still necessary to establish experimentally the specific relationship between the different phosphorylation states and the transport of ions in the different modes.

Since the original suggestion (Fahn et al., 1966a, b) was that the formation of E_1P and its conversion to E_2P was sensitive to and dependent upon $[Mg_i]$ it was of interest to see whether the sided effects of Na on ouabain binding were changed by varying $[Mg_i]$, as well as whether or not there was an effect on Na-K compared to Na-Na exchange when $[Mg_i]$ was altered (Bodemann and Hoffman, 1976c). Obviously, such studies provide a way of evaluating changes in the pump's transport capacity as they might relate to the different postulated conformational states of the pump.

The results presented in Figure 8.7 show that the rate of ouabain binding depends upon $[Mg_i]$ and increases as $[Mg_i]$ is increased. Since K is present in the external medium the ouabain binding rate at high values of $[Mg_i]$ is decreased by increasing $[Na_i]$, the same as that already described (Fig. 8.4). However, this effect of $[Na_i]$ is lost when the $[Mg_i]$ is low (0.8 mM) and, in addition, increasing $[Na_O]$ is now seen to increase the rate of ouabain binding, in contrast to the situation when $[Mg_i]$ is high (Table 8.2). These effects of Na_i and Na_O at low $[Mg_i]$ are examined in more detail in Figure 8.8, where it is apparent that even small changes in $[Na_O]$ alter the binding rate of ouabain and that, as before, changes in $[Na_i]$ are

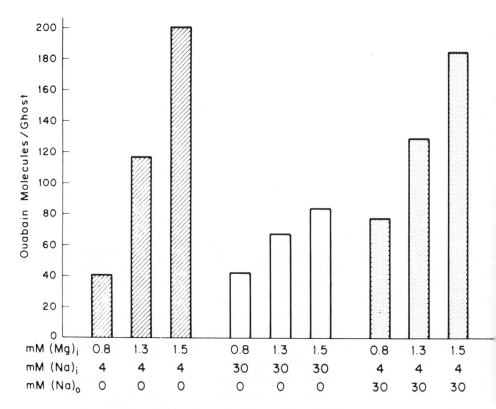

mM $(Mg)_i$	0.8	1.3	1.5	0.8	1.3	1.5	0.8	1.3	1.5
mM $(Na)_i$	4	4	4	30	30	30	4	4	4
mM $(Na)_o$	0	0	0	0	0	0	30	30	30

FIGURE 8.7. The effect of altering $[Mg_i]$ on the rate of ATP_i-promoted binding to reconstituted red-cell ghosts. The effects of changing $[Na_i]$ and $[Na_o]$ are also shown. In all circumstances the medium contained 10 mM K. The measure of the ouabain binding rate is taken as the number of molecules of ouabain bound per ghost during a 30-minute exposure to [³H] ouabain. (From Bodemann and Hoffman, 1976c.)

without effect. The results are consistent with the idea that the effects of Na and K are limited to the outside at low $[Mg_i]$, whereas cross-membrane effects operate at high $[Mg_i]$. Thus, when the $[Mg_i]$ is high, the effect of K_o in altering the ouabain binding rate is found to be dependent upon the presence of either Na_i (Fig. 8.5) or K_i (Fig. 8.6), but this coupled action is completely lost at low $[Mg_i]$, since K_o can now act independent of Na_i or K_i to directly antagonize the binding of ouabain to the membrane (Bodemann and Hoffman, 1976c). The change in the effectiveness of Na_o in antagonizing the action of K_o at low compared to high $[Mg_i]$ indicates that the affinity of the pump for Na_o or K_o (or both) can be influenced directly by Mg_i.

This switching of affinities from the inside to the outside surface of the membrane by alterations in [Mg] can also be demonstrated in porous

TABLE 8.2. Effect of [Mg_i] on Na-K and Na-Na exchange. (Modified from Bodemann and Hoffman, 1976c.)

Experiment	Mg_i (mM)	$^ok_{Na}$ (hr^{-1})	
		Na-K	Na-Na
1	0.8	0.35	0.35
	2.5	0.77	0.52
2	0.8	0.38	0.13
	2.5	0.88	0.24
3	0.8	0.21	0.11
	1.5	0.48	0.12

Note: The outward rate constant, $^ok_{Na}$, in units of reciprocal hours, was measured in ghosts containing a Mg + EDTA (1.5 mM) buffer system together with ATP. The ouabain-sensitive efflux of Na that occurred in the presence or absence of K_O was taken as a measure of the extent of Na-K and Na-Na exchange, respectively. For reasons given in Bodemann and Hoffman (1976c) the concentration of free Mg_i is not precisely known, but is estimated, using the procedure outlined by Wolf (1973), to be about 2×10^{-6} M when Mg_i is 0.8 mM, about 5×10^{-6} M when Mg is 1.5 mM, and about 10^{-3} M when Mg is 2.5 mM. Bodemann and Hoffman (1976c) should be consulted for complete experimental details.

ghost systems. This is shown in Figure 8.9, where it is apparent that when the concentration of Na, in the presence of K, is increased, the rate of ouabain binding is affected in opposite ways, dependent upon the concentration of Mg. From the foregoing discussion it is clear that when the addition of Na results in a decrease in the ouabain binding rate it is due to *inside* Na and is a consequence of the coupled action of Na_i and K_O. Since lowering the concentration of Mg eliminates the linkage of Na_i to K_O, the increase in the ouabain binding rate that occurs upon the addition of Na is due to Na acting on the *outside* of the membrane, presumably by decreasing the affinity of the pump for external K. But the reason these results are included in this discussion is to emphasize the importance of the previously described studies concerned with specifying sidedness of action. Without this type of information, considerable uncertainty would obtain in interpreting the various effects of the different ligands, as illustrated in Figure 8.9.

We turn now to consider the effects of altering [Mg_i] on Na-Na exchange compared to Na-K exchange. These studies were carried out using

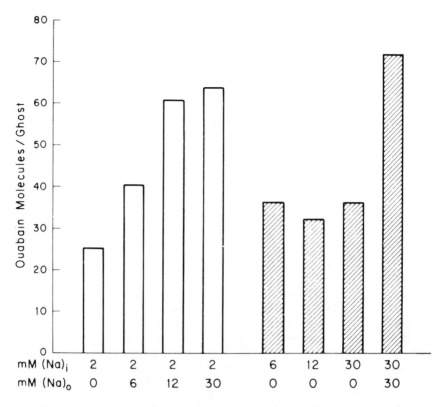

FIGURE 8.8. The effects of changing $[Na_i]$ and $[Na_o]$ at a low $[Mg_i]$ on the rate of ATP_i-promoted ouabain binding to reconstituted red-cell ghosts. These studies are similar in kind to those presented in Figure 8.7, but examine in more detail the ouabain binding properties when $[Mg_i]$ is 0.8 mM. K_o in all cases was 6 mM. The measure of the ouabain binding rate is taken as the number of ouabain molecules bound per ghost during a 30-minute exposure to $[^3H]$ ouabain. (From Bodemann and Hoffman, 1976c.)

reconstituted ghosts that had been prepared to contain—in addition to different concentrations of Mg_i buffer (EDTA)—ATP_i and Na_i as well. The ouabain-sensitive efflux of Na measured in the presence and absence of K_o was taken to indicate the extent of Na-K and Na-Na exchange, respectively. It is apparent from the results presented in Table 8.2 that Na-K exchange is more sensitive to changes in $[Mg_i]$ than Na-Na exchange, since increased $[Mg_i]$ was found to stimulate Na-K exchange to a much greater extent than Na-Na exchange (an average increase of 126 percent compared to 72 percent, respectively, in experiments 1 and 2). In fact, in experiment 3 it is evident that Na-K exchange was increased under circumstances where Na-Na exchange remained unaffected. Since Fahn et al. (1966a, b) suggested that *low* Mg in the presence of Na is

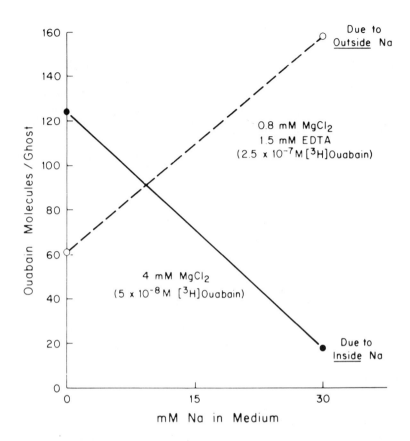

FIGURE 8.9. The effect of varying [Na] at two different values of [Mg] on the rate of ATP-promoted ouabain binding to porous ghosts. The measure of the ouabain binding rate is taken as the number of ouabain molecules bound per ghost after a 30-minute exposure to [³H] ouabain. See text for discussion. (From Bodemann and Hoffman, 1976c.)

associated with the reversible formation of E_1P (or ATP·E going to E_1P or to ADP·E_1P) and *high* Mg is associated with the conversion of E_1P to E_2P, the results provide evidence that the conformations of the pump depicted by the E_1P and E_2P forms underlie the Na-Na and Na-K transport modes, as discussed before, with regard to the ouabain binding rate and the role of K_O (Bodemann and Hoffman, 1976c). Thus it would appear that Mg_i is directly involved not only with determining the sidedness of action of different ligands but also with defining different configurational states of the pump complex concerned with ion translocation.

Brief mention should be made of Ca and the way it interacts with the Na-K pump. The effects of Ca are markedly sided, since the pump is inhibited by Ca_i but not by Ca_O (Hoffman, 1962b). And although Ca_i

appears to act by preventing dephosphorylation of the pump by K (Knauf et al., 1974) it is not clear how this relates to the effect that Ca has on ATP-promoted ouabain binding. Because ATP_i can be used to pump Ca_i out of resealed ghosts (Schatzmann and Vincenzi, 1969), the effect of Ca on ATP-promoted ouabain binding was studied in porous ghosts; while it is reasonable to assume that the observed effects of Ca are due to Ca_i, this needs to be established in a sided preparation. In any event, Ca is most effective in inhibiting the ouabain binding rate when Na and K are present together, even though the binding rate is depressed by Ca in the presence of either ion alone or in their absence (Bodemann and Hoffman, 1976c). Presumably the A $\rightleftharpoons$ B equilibrium, in being shifted to the left in a situation where K is prevented from exerting its normal action, implies that the conformational change associated with the B form is not locked to the phosphorylated form of the pump complex. An alternative explanation might be that Ca in this situation changes the dissociation constant of the B site for ouabain (see Tobin, et al., 1974; Hansen, 1974).

While the main theme of this review has been to look at various asymmetries of the pump in an attempt to gain some insight as to how it might work, limitations of space necessarily preclude a comprehensive survey. It would appear that before the pump mechanism can be specified in any detail it is necessary to correlate the events of transphosphorylation with the translocation of Na and K. One way of approaching this issue is to study in a sided system the side-dependent effects of various ligands, such as Na and K, on the phosphorylation of the pump, in an analogous fashion to that described above for ouabain binding, but carried out in concert with an assessment of the unidirectional fluxes of Na and K. Given the relation between ion flow and energy coupling, it would still be necessary to know how the pump transferred ions across the membrane, and this would appear to require information about the tertiary structure of the complex. And whether the transfer occurs, for instance, by a flip-flop (Repke et al., 1975) or half-site mechanism (Lazdunski, 1972), or by an internal transfer mechanism (Stein, et al., 1973), it would still require information on whether or not the Na sites and the K sites were always present (a sequential mechanism) or interconvertible (a ping-pong mechanism). Recent evidence, based on kinetic analysis of cross-membrane effects of changes in affinities of the membrane for Na and K, favors (but see Albers, 1976) a sequential mechanism (Hoffman and Tosteson, 1971; Garrahan and Garay, 1976; Whittam and Chipperfield, 1975; Robinson, 1974). But what is really missing is direct information that defines the detailed interaction of Na and K with the pump throughout its cycling; without this it would be difficult to specify the mechanism of active transport at the molecular level.

By way of summary, it is important to again emphasize the utility of a knowledge of sidedness in interpreting the mode or locus of action of any added ligand in the pump reaction. It would seem that there is an indeterminancy that attends the analysis of results derived from studies on systems that lack sidedness, especially when the ligands involved, such as Na and K, are known to interact with each other on each side of the membrane as well as to affect one another across it. Therefore the usefulness of reaction schemes (say of transphosphorylation as affected by Na and K) would obviously be strengthened to the extent that the supposed interaction had been independently established under circumstances where the sidedness was defined.

ACKNOWLEDGMENTS

The work reported in this paper was supported in part by U.S. Public Health Service grants HL 09906 and AM 17433.

REFERENCES

Albers, R. W. 1976. The (sodium plus potassium)-transport ATPase. In A. Martonosi, ed., *The enzymes of biological membranes.* Vol. 3, pp. 283-301. Plenum Press, New York.

Albers, R. W., G. J. Koval, and G. J. Siegel. 1968. Studies on the interaction of ouabain and other cardioactive steroids with sodium-potassium-activated adenosine triphosphatase. *Mol. Pharmacol.* 4:325-336.

Baker, P. F., and A. J. Stone. 1966. A kinetic method for investigating hypothetical models of the sodium pump. *Biochim. Biophys. Acta.* 126:321-324.

Beauge, L. A., and N. Adragna. 1971. The kinetics of ouabain inhibition and the partition of rubidium influx in human red blood cells. *J. Gen. Physiol.* 57:576-592.

Bodemann, H. H., and J. F. Hoffman. 1976a. Side-dependent effects of internal versus external Na and K on ouabain binding to reconstituted human red blood cell ghosts. *J. Gen. Physiol.* 67:497-525.

Bodemann, H. H., and J. F. Hoffman. 1976b. Comparison of the side-dependent effects of Na and K on orthophosphate, UTP, and ATP-promoted ouabain binding to reconstituted human red blood cell ghosts. *J. Gen. Physiol.* 67:527-545.

Bodemann, H. H., and J. F. Hoffman. 1976c. Effects of Mg and Ca on the side dependencies of Na and K on ouabain binding to red cell ghosts and the control of Na transport by internal Mg. *J. Gen. Physiol.* 67:547-561.

Caldwell, P. C., and R. D. Keynes. 1959. The effect of ouabain on the efflux of sodium from a squid giant axon. *J. Physiol.* (Lond.) 148:8P.

Danowski, T. S. 1941. The transfer of potassium across the human blood cell membrane. *J. Biol. Chem.* 139:693-705.

Dunham, P. B., and J. F. Hoffman. 1970. Partial purification of the ouabain-binding component and of Na,K-ATPase from human red cell membranes. *Proc. Natl. Acad. Sci. USA* 66:936-943.

Fahn, S., G.J. Koval, and R. W. Albers. 1966a. Sodium-potassium-activated adenosine triphosphatase of *Electrophorus* electric organ. I. An associated sodium-activated transphosphorylation. *J. Biol. Chem.* 241:1882-1889.

Fahn, S., M. R. Hurley, G. J. Koval, and R. W. Albers. 1966b. Sodium-potassium-activated adenosine triphosphatase of *Electrophorus* electric organ. II. Effects of N-ethylmaleimide and other sulfhydryl reagents. *J. Biol. Chem.* 241:1890-1895.

Gardos, G. 1954. Akkumulation der Kaliumionen durch menschliche Blutkörperchen. *Acta physiol. hung.* 6:191-199.

Garrahan, P. J., and R. P. Garay. 1976. The distinction between sequential and simultaneous models for sodium and potassium transport. *Curr. Top. Membranes & Transport* 8:29-97.

Garrahan, P. J., and I. M. Glynn. 1967a. The behaviour of the sodium pump in red cells in the absence of external potassium. *J. Physiol.* (Lond.) 192: 159-174.

Garrahan, P. J., and I. M. Glynn. 1967b. The sensitivity of the sodium pump to external sodium. *J. Physiol.* (Lond.) 192:175-188.

Garrahan, P. J., and I. M. Glynn. 1967c. Factors affecting the relative magnitudes of the Na:K and Na:Na exchange catalysed by the sodium pump. *J. Physiol.* (Lond.) 192:189-216.

Garrahan, P. J., and I. M. Glynn. 1967d. The stoichiometry of the sodium pump. *J. Physiol.* (Lond.) 192:217-235.

Giotta, G. J. 1976. Quarternary structure of $(Na^+ + K^+)$-dependent adenosine triphosphatase. *J. Biol. Chem.* 251:1247-1252.

Glynn, I. M. 1956. Sodium and potassium movements in human red cells. *J. Physiol.* (Lond.) 134:278-310.

Glynn, I. M. 1957. The action of cardiac glycosides on sodium and potassium movements in human red cell. *J. Physiol.* (Lond.) 136:148-173.

Glynn, I. M. 1962. Activation of adenosine triphosphate activity in a cell membrane by external potassium and internal sodium. *J. Physiol.* (Lond.) 160:18P.

Glynn, I. M., and J. F. Hoffman. 1971. Nucleotide requirements for sodium-sodium exchange catalysed by the sodium pump in human red cells. *J. Physiol.* (Lond.) 218:229-256.

Hansen, O. 1974. The influence of monovalent cation and Ca on g-strophanthin binding to (Na + K)-activated ATPase. *Ann. N.Y. Acad. Sci.* 242:635-645.

Harris, E. J., and M. Maizels. 1951. The permeability of human erythrocytes to Na. *J. Physiol.* 113:506-524.

Harris, J.E. 1941. The influence of the metabolism of human erythrocytes on their potassium content. *J. Biol. Chem.* 141:579-595.

Hegyvary, C., and R. L. Post. 1971. Binding of adenosine triphosphate to sodium and potassium ion-stimulated adenosine triphosphatase. *J. Biol. Chem.* 246:5235-5240.

Hilden, S., H. M. Rhee, and L. E. Hokin. 1974. Sodium transport by phospholipid vesicles containing purified sodium and potassium ion-activated adenosine triphosphatase. *J. Biol. Chem.* 249:7432-7440.

Hoffman, J. F. 1962a. The active transport of sodium by ghosts of human red blood cells. *J. Gen. Physiol.* 45:837-859.

Hoffman, J. F. 1962b. Cation transport and structure of the red cell plasma membrane. *Circulation* 26:1201-1213.

Hoffman, J. F. 1966. The red cell membrane and the transport of sodium and potassium. *Am. J. Med.* 41:666-668.

Hoffman, J. F. 1969. The interaction between tritiated ouabain and the Na-K pump in red blood cells. *J. Gen. Physiol.* 54:343s-350s.

Hoffman, J. F. 1972. Sidedness of the red cell Na:K pump. In L. Bolis, R. D. Keynes, and W. Wilbrandt, eds., *Role of membranes in secretory processes,* pp. 203-214. North Holland Publishing Co., Amsterdam.

Hoffman, P. G., and D. C. Tosteson. 1971. Active sodium and potassium transport in high potassium and low potassium sheep red cells. *J. Gen. Physiol.* 58: 438-466.

Ingram, C. J. 1970. Ouabain binding to human red blood cells. Ph.D. Dissertation. Yale University.

Jorgensen, P. L. 1975. Purification and characterization of (Na$^+$ + K$^+$)-ATPase. V. Conformational changes in the enzyme transitions between the Na-form and the K-form studied with tryptic digestion as a tool. *Biochim. Biophys. Acta.* 401:399-415.

Jorgensen, P. L., D. Hansen, I. M. Glynn, and J. D. Cavieres. 1973. Antibodies to pig kidney (Na$^+$ + K$^+$)-ATPase inhibit the Na$^+$ pump in human red cells provided they have access to the inner surface of the cell membranes. *Biochim. Biophys. Acta.* 291:795-800.

Karlish, S. J. D., D. W. Yates, and I. M. Glynn. 1976. Transient kinetics of (Na + K)-ATPase studied with a fluorescent substrate. *Nature* 263:251-253.

Katchalsky, A. 1970. A thermodynamic consideration of active transport. In L. Bolis, A. Katchalsky, R. D. Keynes, W. R. Loewenstein, and B. A. Pethica, eds., *Permeability and function of biological membranes,* pp. 20-35. North Holland Publishing Co., Amsterdam.

Kepner, G. R., and R. I. Macey. 1968. Membrane enzyme systems: molecular size determinations by radiation inactivation. *Biochim. Biophys. Acta.* 163:188-203.

Knauf, P. A., F. Proverbio, and J. F. Hoffman. 1974. Electrophoretic separation of different phosphoproteins associated with Ca-ATPase and Na,K-ATPase in human red cell ghosts. *J. Gen. Physiol.* 63:324-326.

Knight, A. B., and L. G. Welt. 1974. Intracellular potassium: a determinant of the sodium-potassium pump rate. *J. Gen. Physiol.* 63:351-373.

Kyte, J. 1975. Structural studies of sodium and potassium ion-activated adenosine triphosphatase. *J. Biol. Chem.* 250:7443-7449.

Lazdunski, M. 1972. Flip-flop mechanisms and half-site enzymes. In B. Horecker and E. Stadtman, eds., *Current topics in cellular regulation.* Vol. 6, pp. 267-310. Academic Press, New York.

Lindenmayer, G. E., A. H. Laughter, and A. Schwartz. 1968. Incorporation of inorganic phosphate-32 into a Na$^+$,K$^+$-ATPase preparation: stimulation by ouabain. *Arch. Biochem. Biophys.* 127:187-192.

Lisko, V. K., M. K. Malysheva, and T. I. Grevizirskaya. 1972. The interaction of the (Na,K)-ATPase of erythrocyte ghosts with ouabain. *Biochim. Biophys. Acta.* 288:103-106.

Perrone, J. R., and R. Blostein. 1973. Asymmetric interaction of inside-out and right-side out erythrocyte membrane vesicles with ouabain. *Biochim. Biophys. Acta.* 291:680-689.

Post, R. L., and Ph. C. Jolly. 1957. The linkage of sodium, potassium and ammonium active transport across the human erythrocyte membrane. *Biochim. Biophys. Acta.* 25:118-128.

Post, R. L., S. Kume, T. Tobin, B. Orcutt, and A. K. Sen. 1969. Flexibility of an active center in sodium-plus-potassium adenosine triphosphatase. *J. Gen. Physiol.* 54:306s-326s.

Post, R. L., C. R. Merritt, C. R. Kinsolving, and C. D. Albright. 1960. Membrane adenosine triphosphatase as a participant in the active transport of sodium and potassium in the human erythrocyte. *J. Biol. Chem.* 235:1796-1802.

Repke, K. R. H., R. Schön, and F. Dittrich. 1975. Energy relationship in flip-flop model of (Na,K)-ATPase transport and ATP synthesis function. *Proc. Fed. Europ. Biochem. Soc.* 35:241-253.

Robinson, J. D. 1974. Cation interactions with different functional states of the Na,K-ATPase. *Ann. N.Y. Acad. Sci.* 242:185-202.

Sachs, J. R. 1974. Interaction of external K, Na and cardiactive steroids with the Na-K pump of the human red blood cell. *J. Gen. Physiol.* 63:123-143.

Schatzmann, H. J. 1953. Herzglykoside als Hemmstoffe für den aktiven Kalium und Natrium transport durch die erythrocytenmenbran. *Helv. Physiol. Pharmacol. Acta.* 11:346-354.

Schatzmann, H. J. 1965. The role of Na and K in the ouabain-inhibition of the Na + K-activated membrane adenosine triphosphatase. *Biochim. Biophys. Acta.* 94:89-96.

Schatzmann, H. J., and F. F. Vincenzi. 1969. Calcium movements across the membrane of human red cells. *J. Physiol.* (Lond.) 201:369-395.

Schwartz, A., G. E. Lindenmayer, and J. C. Allen. 1975. The sodium-potassium adenosine triphosphatase: pharmacological, physiological and biochemical aspects. *Pharmacol. Rev.* 27:3-134.

Schwartz, A., H. Matsui, and A. H. Laughter. 1968. Tritiated digoxin binding to $(Na^+ + K^+)$-activated adenosine triphosphatase: possible allosteric site. *Science* 160:323-325.

Sen, A. K., and R. L. Post. 1964. Stoichiometry and localization of ATP dependent Na and K transport in the erythrocyte. *J. Biol. Chem.* 239:345-352.

Simons, T. J. B. 1974. Potassium:potassium exchange catalysed by the sodium pump in human red cells. *J. Physiol.* (Lond.) 237:123-155.

Skou, J. C. 1975. The (Na + K) activated enzyme and its relationship to transport of sodium and potassium. *Q. Rev. Biophys.* 7:401-434.

Solomon, A. K. 1952. The permeability of the human erythrocyte to sodium and potassium. *J. Gen. Physiol.* 36:57-110.

Solomon, A. K., T. J. Gill, and G. L. Gold. 1956. The kinetics of cardiac glycoside inhibition of potassium transport in human erythrocytes. *J. Gen. Physiol.* 40:327-350.

Stein, W. D., W. R. Lieb, S. J. D. Karlish, and Y. Eilam. 1973. A model for the active transport of sodium and potassium ions as mediated by a tetrameric enzyme. *Proc. Natl. Acad. Sci. USA* 70:275-278.

Tobin, T., T. Akera, and T. M. Brody. 1974. Studies on the two phosphoenzyme conformations of Na + K-ATPase. *Ann. N.Y. Acad. Sci.* 242:120-132.

Tosteson, D. C., and J. F. Hoffman. 1960. Regulation of cell volume by active cation transport in high and low potassium sheep red cells. *J. Gen. Physiol.* 44:169-194.

Whittam, R. 1962. The asymmetrical stimulation of a membrane adenosine triphosphate in relation to active cation transport. *Biochem. J.* 84:110.

Whittam, R., and M. E. Ager. 1964. Vectional aspects of adenosine-triphosphatase activity in erythrocyte membranes. *Biochem. J.* 93:337-348.

Whittam, R., and M. E. Ager. 1965. The connexion between active cation transport and metabolism in erythrocytes. *Biochem. J.* 97:214-227.

Whittam, R., and A. R. Chipperfield. 1975. The reaction mechanism of the sodium pump. *Biochim. Biophys. Acta.* 415:149-171.

Wolf, H. U. 1973. Divalent metal ion buffers with low pH-sensitivity. *Experientia* 29:241-249.

9 Simultaneous or Consecutive Occupancy by Sodium and Potassium Ions of Their Membrane Pump

ROBERT L. POST

Sodium and potassium consecutively occupy sodium-, potassium-dependent adenosine triphosphatase, Na,K-ATPase (EC 3.6.1.3), during the reaction sequence of an experiment that will be described in this chapter. Sodium and potassium ions were present simultaneously throughout the experiment at constant concentrations and the sequence of changes was initiated by a transient pulse of ATP. The experiment was designed in response to comments by Skou (1975, p. 422) and more specifically by Garrahan and Garay (1976, p. 69), who write: "Consecutive schemes have been used not only to interpret results of biochemical experiments but also as working hypotheses to design experiments. This way of thinking obviously leads to sequential [consecutive] experiments, that is, to experiments in which activating cations are added one after the other and which, therefore, unavoidably give sequential [consecutive] results."[1] The new simultaneous experiment showed consecutive results.

In a larger context there is a controversy between a school of simultanists and a school of consecutivists. The simultanists maintain that simultaneous occupancy of Na,K-ATPase by sodium and potassium is important for rate-determining steps; the consecutivists emphasize alternating occupancy by sodium and potassium. In general the simultanists rely on the kinetics of the whole system and the consecutivists on partial or transient reactions. In principle, there is no obligatory incompatibility be-

[1] In this quotation I have replaced "sequential" in the original by "consecutive," in order to avoid confusion with the usage of enzyme kineticists, who say "A *sequential* mechanism will be one in which all the substrates must be present on the enzyme before any products can leave" (Plowman, 1972).

tween the two approaches, as illustrated by the following primitive reaction sequences:

$$E \longrightarrow E \cdot K \longrightarrow Na \cdot E \cdot K \longrightarrow Na \cdot E \longrightarrow E \qquad (1)$$
$$E \longrightarrow Na \cdot E \longrightarrow Na \cdot E \cdot K \longrightarrow E \cdot K \longrightarrow E \qquad (2)$$

where E is the free transport enzyme and Na and K are bound sodium and potassium, respectively. The field has been favored by eight review articles in the past two years, and almost all have commented on the simultanist-consecutivist controversy (Albers, 1976; De Weer, 1975; Garrahan and Garay, 1976; Glynn and Karlish, 1975; Jørgensen, 1975; Schwartz, Lindenmeyer, and Allen, 1975; Skou, 1975; Whittam and Chipperfield, 1975).

The Na-K pump in the plasma membranes of most animal cells actively moves sodium ions outward and potassium ions inward, using energy from the hydrolysis of ATP. In preparations of broken plasma membranes, the activity of the pump persists, appearing as an ATPase dependent upon the combined presence of sodium, potassium, and magnesium ions, but transport is not accomplished. Reconstitution of purified Na,K-ATPase into tight phospholipid vesicles restores the transport function. Restoration shows that the enzyme in broken membranes no doubt still translocates ions from one face of the membrane to the other, but in vain, since the incipient concentration gradients produced in this way immediately dissipate through the breaks. Both the Na-K pump and Na,K-ATPase are inhibited specifically by cardioactive steroids such as ouabain, strophanthin, or the digitalis glycosides.

In each cycle, three intracellular Na ions go outward, two extracellular K ions go inward, and the terminal phosphate bond of one molecule of intracellular ATP is hydrolyzed to yield intracellular ADP and inorganic phosphate with catalysis by intracellular Mg (Fig. 9.1). The inequality of charge transfer is real; the pump is electrogenic. However, the stoichiometry is not rigid, since Na is pumped out at about 10 percent of capacity even in the absence of extracellular K. Net outward transport is specific for Na but net inward transport accepts congeners of K, namely Li, NH_4, Rb, Cs, or Tl.

Na,K-ATPase is an intrinsic enzyme of plasma membranes. In its most purified state it contains at least two peptides and phospholipid. One peptide of about 100,000 daltons carries the active site for phosphorylation and probably an active site for ouabain binding. The other peptide, of about 50,000 daltons, is a glycopeptide. The enzyme accepts a phosphate group on the beta-carboxyl group of a specific aspartate residue either from ATP or from inorganic phosphate. Mg is required in both cases. In the presence of Na, uniquely among monovalent inorganic cations, the enzyme is phosphorylated from ATP. In the absence of Na, uniquely among monovalent inorganic cations, the enzyme is phosphorylated reversibly from inorganic phosphate. The phosphoenzyme from inorganic phosphate shows four reactive states, of which two

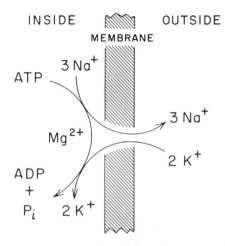

FIGURE 9.1. Stoichiometry and sidedness of the sodium and potassium ion pump of plasma membranes of animal cells. (From Post, Toda, Kume, and Taniguchi, 1975.)

appear to participate in the normal reaction sequence (Fig. 9.2). Potassium-sensitive phosphoenzyme (E_2P) is the prominent (about 90 percent) product of phosphorylation from ATP at concentrations of Na below 0.1 M. Its rate of dephosphorylation at 0° is moderate ($k=0.1\ s^{-1}$). Addition of a low concentration of K increases the rate of dephosphorylation beyond the range of estimation ($k > 1\ s^{-1}$) (see also Mårdh and Zetterqvist, 1974). Although potassium-sensitive phosphoenzyme is a product of phosphorylation from ATP, it equilibrates with inorganic phosphate and is insensitive to addition of ADP. Potassium-complexed phosphoenzyme ($K \cdot E_2P$) appears when the enzyme is incubated in the presence of inorganic phosphate, Mg, and K. It can be distinguished from other reactive states of the phosphoenzyme by its extremely rapid rate of equilibration with inorganic phosphate ($k > 1\ s^{-1}$). The maximal level of this phosphoenzyme is ordinarily about 20 percent of phosphorylation capacity, probably because of an unfavorable equilibrium with a complex of the enzyme with non-covalently bound inorganic phosphate.

Experimental Design

The experiment was designed to estimate relative occupancy of the enzyme by sodium or potassium by taking the level of phosphorylation from inorganic phosphate as an index. Occupancy by Na prevents phosphorylation from inorganic phosphate and occupancy by K permits this phosphorylation (Post, Toda, and Rogers, 1975). The experiment started with the enzyme exposed to both Na and K, but with the concentration of Na in sufficient excess to occupy the enzyme completely. Occupancy

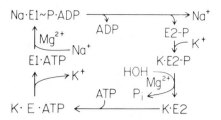

FIGURE 9.2. A reaction sequence for phosphorylation and dephosphorylation of Na, K-ATPase. E represents the enzyme. Ligands inside the loop are intracellular and those outside are extracellular. Enzyme forms on the left side show one pattern of tryptic digestion and those on the right show another (Jørgensen, 1975b). $E_1 \sim P$ is a form that transfers its phosphate group reversibly to ADP and E_2-P is a form that transfers its phosphate group reversibly to water. The stoichiometry of Na and K is not shown. (Modified slightly from Post, 1977.)

by Na or K was estimated by incubating the enzyme with radioactive inorganic phosphate and Mg for 20 minutes or longer at 0° (compare Post, 1977). At concentrations of 80 mM for Na and 0.2 mM for K the steady-state level of phosphoenzyme was 0.3 percent of capacity. Replacement of Na by K gave a level of 26 percent. Higher concentrations of P_i, Mg, or K do not significantly increase this level (Post, Toda, and Rogers, 1975). Another precaution was the use of acetate salts in place of the more usual chloride salts since chloride is an inhibitor of phosphorylation from inorganic phosphate (Post, Toda, and Rogers, 1975).

The reaction was initiated by addition of the smallest effective amount of unlabeled ATP. Preliminary experiments showed that an amount of ATP 1.8-fold greater than the amount of enzyme was the minimum amount that would produce maximal phosphorylation from ATP in the absence of K and in the presence of 1 mM ouabain and 0.2 mg/ml of oligomycin, which were added to partially inhibit spontaneous dephosphorylation.

Addition of ATP induced phosphorylation from P_i. In order to characterize the phosphoenzyme as specifically potassium-complexed phosphoenzyme, $K \cdot E_2P$, an excess of a chelator of magnesium ions was added to it in order to prevent further phosphorylation and expose the apparent monomolecular rate of release of inorganic phosphate from the enzyme. Potassium-complexed phosphoenzyme uniquely loses its phosphate group in less than one second in response to such a test (Post, Toda, and Rogers, 1975). A second means of characterization was the rate of attack of excess Na on $K \cdot E_2P$. Preliminary experiments showed that such an attack required tens of seconds or even a few minutes at 0° and 32 mM Na (Post, 1977), and that the rate was still conveniently slow even at 0.2 M Na. Accordingly, the potassium-complexed phosphoenzyme was tested for stability at two concentrations of Na.

In order to compare the potassium-complexed phosphoenzyme obtained indirectly from ATP with that formed directly from inorganic phosphate in the absence of ATP, a control experiment was done in which the enzyme was incubated first only with K and then the reaction was started by addition of inorganic phosphate, Mg, and the excess Na. The same tests of addition of a chelator or a higher concentration of Na were also applied to this phosphoenzyme.

Methods

The methods are given in detail by Post, Toda, and Rogers (1975). Membranes were isolated from a homogenate of guinea pig kidney by repeated centrifugation and resuspension. Na,K-ATPase was partially purified by soaking the membranes overnight at about 6° in 1.3 M urea. The specific activity was about 3 μmol (mg protein)$^{-1}$ (min)$^{-1}$ at 37° and the activity independent of Na and K was about 10 percent of the dependent activity. The phosphorylation reaction was quenched with acid and the denatured membranes were collected by centrifugation. They were digested with pepsin to release phosphopeptides overlapping the active site of phosphorylation. These were isolated by high voltage paper electrophoresis and counted. [^{32}P]ATP in the supernatant after acidification was estimated by adsorption on charcoal.

Results

When the enzyme was incubated with excess Na, with K, Mg, and radioactive inorganic phosphate, no phosphoenzyme appeared (Fig. 9.3). This result showed that at this concentration Na completely occupied the enzyme, preventing phosphorylation from inorganic phosphate. When 2 nmole of unlabeled ATP was added to 1 nmole of this enzyme, 34 percent of the enzyme (estimated by extrapolation) became labeled from inorganic phosphate in less than 10 seconds. This phosphoenzyme was unstable and disappeared with a rate constant of 0.011 s^{-1} (Fig. 9.3). (In another similar experiment with 100 mM Na and 0.3 mM K the level of phosphoenzyme was 1 percent of capacity after 15 minutes.) Characterization of the phosphoenzyme by addition of excess chelator showed almost complete dephosphorylation in less than 4 seconds. An increase in the concentration of Na from 80 mM to 215 mM increased the rate constant for disappearance to 0.040 s^{-1}.

In order to compare these results of indirect phosphorylation via ATP with those of direct phosphorylation from inorganic phosphate in the absence of ATP under the same conditions and with the same enzyme preparation, the enzyme was incubated with K first and then phosphorylation was initiated by addition of the radioactive inorganic phosphate, Mg, and Na, all in the same solution. The results were indistinguishable (Fig. 9.3).

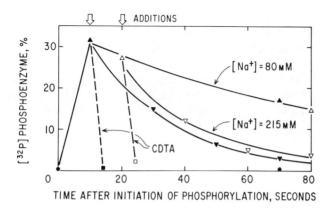

FIGURE 9.3. Comparison of potassium-complexed phosphoenzyme formed directly from labeled inorganic phosphate, P_i, with that formed indirectly by antecedent phosphorylation from unlabeled ATP and subsequent exchange. In 0.8 ml the direct reaction mixture (open symbols) contained 2.9 mg of membrane protein from guinea pig kidney, 15 μmol of imidazole with 20 μmol of 2-(N-morpholino) propane sulfonic acid at pH 7.0, and 1.6 μmol of KCl. The membrane protein contained about 1.0 μmol of enzyme. At zero time, phosphorylation was initiated by addition of 0.2 ml of 0.2 mM KCl containing 1 μmol of $^{32}P_i$, 1 μmol of $MgCl_2$ and 80 μmol of sodium acetate (Δ). At 20 seconds the reactivity of the phosphoenzyme was tested by addition of 0.1 ml containing 10 μmol of (Tris)$_3$ (1, 2-cyclohexylenedinitrilo)tetraacetate (CDTA [□]), or 1.6 nmol of sodium acetate (∇). The indirect reaction mixture (closed symbols) was the same, except that: (1) The initial 0.8 ml contained, additionally, 64 μmol of sodium acetate. (2) The 0.2 ml contained 16 μmol of sodium acetate and was added 10 seconds before zero time. The other reagents were the same. No phosphoenzyme appeared (•). (3) At zero time, 2 nmol of unlabeled (Tris)$_4$ ATP was added in 20 μl (▲). The reactivity was tested by additions at 10 seconds instead of 20 seconds (■, ▼). At the times indicated on the horizontal axis, the reaction was stopped with acid and radioactive phosphopeptides overlapping the active site were isolated. For the scale of the vertical axis, 100 percent phosphoenzyme was estimated after an interval of 15 minutes with 1 mM $_{32}P_i$, 1 mM $MgCl_2$, and 0.25 mM ouabain. When ouabain was replaced with 40 mM potassium acetate, the level of phosphoenzyme was 23 percent.

In order to find out the rate of disappearance of the ATP the experiment was repeated with the ^{32}P label in the ATP and not in the P_i. A different enzyme preparation was used. Addition of 2.25 nmole of ATP to 0.90 nmole of enzyme under conditions otherwise identical to those in Fig. 9.3 produced a diphasic disappearance of the ATP. Within 5 seconds, 57 percent of the ATP disappeared abruptly and the rest disappeared with a rate constant of 0.14 s^{-1}. At the same time, the level of phosphoenzyme from [^{32}P]ATP was estimated; about 25 percent of the initial amount was found at zero time by extrapolation. This amount disappeared with the same rate constant as ATP during the slow phase. The molar ratio of ATP to [^{32}P]phosphoenzyme during this phase was about 4.0. This phosphoenzyme probably represents a form in equilibrium with ATP such as E_1P in Figure 9.2. About 2 percent of the enzyme was still phosphorylated from ATP 20 seconds after the addition of the ATP, and about the same fraction of the initial amount of ATP remained. In the absence of K, 66 percent of the enzyme was phosphorylated from ATP at 5 seconds and 37 percent of the ATP remained.

Discussion

It is well known that Na,K-ATPase has many properties in the presence of sodium ions different from those it has in the presence of potassium ions (Glynn and Karlish, 1975). In particular, partial tryptic digestion shows a difference in conformation between a specific sodium-form and a specific potassium-form. These forms are interconvertible upon addition of nucleotides with or without Mg (Jørgensen, 1975b). The experiment in this paper has employed one of these properties, namely, phosphorylation from inorganic phosphate, as an index of the occupancy of the enzyme by sodium or potassium. In the simultaneous presence of both sodium and potassium at constant concentrations, the addition of an almost stoichiometric amount of ATP immediately converted the enzyme from a sodium-form to a form that was as fully occupied by potassium as when the enzyme was incubated with potassium alone and the rest of the reaction mixture, less ATP, was added later. Following hydrolysis of the ATP, the enzyme slowly reverted to the sodium-form. Thus, the experiment showed an alternating occupancy by Na or K in the simultaneous presence of constant concentrations of both. Karlish, Yates, and Glynn (1976) have recently observed a corresponding sequence of states by means of transient kinetics of nucleotide binding, but in their experiments sodium and potassium ions were not equilibrated with the enzyme and then held at constant concentrations thereafter.

The reaction sequence in our experiment may seem too slow to participate in the normal function of the enzyme. However, release of K from the enzyme is probably greatly accelerated by the binding of ATP to a low-affinity site at which it is not a phosphate donor (Glynn and Karlish, 1975). The level of phosphoenzyme when sodium was added to the potassium-enzyme, 34 percent, was higher than the level when potassium simply replaced sodium in a steady state, 26 percent. This result appeared in each of four similar experiments. It may reflect an allosteric action of sodium, as proposed by Cavieres and Ellory (1975). It may indicate a state in which sodium and potassium are bound simultaneously.

Garrahan and Garay (1976) have argued at length not only for simultaneous occupancy by sodium and potassium but also for simultaneous accessibility to inside and outside binding sites. They favor a bi-substrate, random, quasi-equilibrium mechanism without binding interaction between substrates. They have considered intracellular sodium and extracellular potassium as substrates. As products they have considered extracellular sodium and intracellular potassium. Their argument is based on their observations of the kinetics of transport in intact cells. In their experiments, changes in the ion concentrations on one side of the membrane affected the kinetics with respect to the ions on the other side of the membrane in only a very simple way. The effect was simply to change the maximum rate without changing the half-maximal concentration. Thus, the apparent ion affinity on one side was not affected by the composition of the solution on the other side. This invariance between the affinity on one side and the composition on the other side led Garrahan and Garay to reject a ping-pong mechanism, in which an enzyme is occupied by the first substrate, undergoes a conformational change, and releases the first product; the changed enzyme then accepts the second substrate, undergoes a reversal of the conformational change, and releases the second product. Application of this model to Na,K-ATPase would give an alternation of an inward conformation, in which an ion translocation site communicates with an intracellular solution, with an outward conformation, in which an ion translocation site communicates with the extracellular solution (Post, 1977).

There is a limitation to Garrahan and Garay's experiments. In all of their experiments the enzyme was saturated with sodium or potassium ions with respect to the intracellular solution. Unfortunately, definitive experiments require that the enzyme be unsaturated with respect to both extracellular and intracellular solutions. With the enzyme saturated with respect to the intracellular solution, even the rejected ping-pong mechanism can closely approximate the observed kinetics for net transport, provided that suitable kinetic constants are chosen. All that is required is that the apparent monomolecular rate-constant for release of sodium into either solution should be the same as the corresponding constant for potassium.

Recently Chipperfield and Whittam (1976) have reported experiments in which product concentrations of intracellular potassium and extracellular sodium were kept as low as possible. In these experiments, changes in the level of intracellular sodium modified the apparent affinity for extracellular potassium. The half-maximal concentration for extracellular potassium increased about as much as the maximum rate with respect to extracellular potassium when the intracellular sodium concentration was increased. These experiments showed that under stringent conditions the fundamental relationship upon which Garrahan and Garay founded their hypothesis is not preserved.

What future discovery might harmonize presently discordant interpretations? Perhaps it would be enough to show that addition of extracellular K to the enzyme increases the unidirectional rate constant for dissociation of translocated Na into the extracellular solution or vice versa for intracellular Na with respect to translocated K, as suggested by equation (1) or (2).

ACKNOWLEDGMENTS

I am grateful to Dr. John R. Sachs for a talk, "On the reaction mechanism of the Na-K pump," given to the Red Cell Club on November 26, 1976, in New Haven. This work was supported by a grant, 5RO1 HL-01974, from the National Heart, Lung, and Blood Institute and by a grant, 5PO1 AM-07462, from the National Institute of Arthritis and Metabolic Diseases.

REFERENCES

Albers, R. W. 1976. The (sodium plus potassium)-transport ATPase. In A. N. Martonosi, ed., *The enzymes of biological membranes.* Vol. 3, pp. 283-301. Plenum, New York.

Cavieres, J. D., and J. C. Ellory. 1975. Allosteric inhibition of the sodium pump by external sodium. *Nature* 255: 338-340.

Chipperfield, A. R., and R. Whittam. 1976. The connexion between the ion-binding sites of the sodium pump. *J. Physiol.* 260: 371-385.

De Weer, P. 1975. Aspects of the recovery processes in nerve. In C. C. Hunt, ed., *MTP international reviews of science, physiology.* Series 1, vol. 3, pp. 231-278. University Park Press, Baltimore.

Garrahan, P. J., and R. P. Garay. 1976. The distinction between sequential and simultaneous models for sodium and potassium transport. *Curr. Top. Membranes & Transport* 8: 29-97.

Glynn, I. M., and S. J. D. Karlish. 1975. The sodium pump. *Annu. Rev. Physiol.* 37: 13-55.

Jørgensen, P. L. 1975a. Isolation and characterization of the components of the sodium pump. *Q. Rev. Biophys.* 7: 239-274.

Jørgensen, P. L. 1975b. Purification and characterization of (Na^+, K^+)-ATPase. V. Conformational changes in the enzyme: transitions between the Na-form and the K-form studied with tryptic digestion as a tool. *Biochim. Biophys. Acta* 401: 399-415.

Karlish, S. J. D., D. W. Yates, and I. M. Glynn. 1976. Transient kinetics of $(Na^+ + K^+)$-ATPase studied with a fluorescent substrate. *Nature* 263: 251-253.

Mårdh, S., and Ö. Zetterqvist. 1974. Phosphorylation and dephosphorylation reactions of bovine brain $(Na^+ - K^+)$-stimulated ATP phosphohydrolase studied by a rapid-mixing technique. *Biochim. Biophys. Acta* 350: 473-483.

Plowman, K. M. 1972. *Enzyme kinetics.* McGraw-Hill, New York.

Post, R. L. 1977. Titration of sodium against potassium for occupancy of the sodium, potassium ion transport adenosine triphosphatase. In G. Semenza and E. Carafoli, eds., *Biochemistry of membrane transport.* FEBS Symposium no. 42, pp. 352-362. Springer Verlag, New York.

Post, R. L., G. Toda, S. Kume, and K. Taniguchi. 1975. Synthesis of adenosine triphosphate by way of potassium-sensitive phosphoenzyme of sodium, potassium adenosine triphosphatase. *J. Supramol. Struct.* 3: 479-497.

Post, R.L., G. Toda, and F.N. Rogers. 1975. Phosphorylation by inorganic phosphate of sodium plus potassium ion transport adenosine triphosphatase. *J. Biol. Chem.* 250: 691-701.

Schwartz, A., G. E. Lindenmayer, and J. C. Allen. 1975. The sodium-potassium adenosine triphosphatase: pharmacological, physiological, and biochemical aspects. *Pharmacol. Rev.* 27: 3-134.

Skou, J. C. 1975. The (Na$^+$ + K$^+$) activated enzyme system and its relationship to transport of sodium and potassium. *Q. Rev. Biophys.* 7: 401-434.

Whittam, R., and A. R. Chipperfield. 1975. The reaction mechanism of the sodium pump. *Biochim. Biophys. Acta* 415: 149-171.

10 Proton Translocation in Submitochondrial Particles and Reconstituted Segments of the Respiratory Chain

PETER C. HINKLE

Peter Mitchell's chemiosmotic theory (1961, 1966) is now used by a majority of researchers in the fields of oxidative phosphorylation, photophosphorylation, and bacterial transport. The chemiosmotic view of the mitochondrial inner membrane is illustrated in Figure 10.1. The components of the respiratory chain are arranged across the mitochondrial inner membrane so that electron transfer down the chain brings about proton translocation across the membrane, generating a membrane potential ($\Delta\Psi$) and proton concentration gradient (ΔpH), which constitute the electrochemical proton gradient. This gradient in turn is proposed to drive ATP synthesis by the F_1-F_0 ATPase complex. Other transport reactions are also driven by the electrochemical proton gradient; for example, calcium ion uptake is driven by the membrane potential and sodium ion efflux and phosphate uptake are driven by the pH gradient. Many of the details of this scheme remain to be worked out but there is considerable evidence for the central role of proton translocation in the coupling of oxidation reactions to phosphorylation of ADP.

I have studied the properties of proton translocation in inverted submitochondrial vesicles and in isolated complexes of the respiratory chain reconstituted in phospholipid vesicles. These systems are significantly simpler than whole mitochondria and do not require substrate transport. In addition, they are not as susceptible to interference from the many transport systems present in the mitochondrial membrane or the endogenous substrates and ions of mitochondria.

Proton Translocation in Submitochondrial Vesicles

Submitochondrial vesicles formed from beef heart mitochondria by sonication have an inverted orientation but are well coupled, showing more

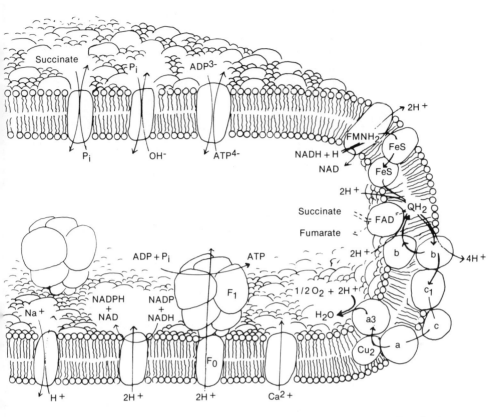

FIGURE 10.1. Diagram of Mitchell's chemiosmotic hypothesis for mitochondrial oxidative phosphorylation. A simplified version of the respiratory chain from NADH to oxygen is shown on the right, driving protons outward. The F_1-F_0 ATPase complex and the NAD-NADP transhydrogenase use the electrochemical proton gradient to drive a chemical reaction. Examples of the substrate and ion transport systems are also shown.

than five-fold stimulation of respiration by uncouplers. In our studies of the stoichiometry of proton translocation driven by the respiratory chain, two protons were taken up by the vesicles for each electron pair traversing a coupling region of the chain (Hinkle and Horstman, 1971). When NADH regenerated with ethanol and alcohol dehydrogenase was the substrate, six protons were transported per oxygen atom reduced. When succinate or reduced naphthoquinone sulfonate (plus N-methyl-phenazine as mediator to internal cytochrome c) was used, four and two protons were transported, respectively, confirming earlier work by Mitchell and Moyle using rat liver mitochondria (1967). We also studied the stoichiometry of proton translocation driven by ATP hydrolysis and found two protons were taken up by the submitochondrial vesicles per ATP hydrolyzed (Thayer and Hinkle, 1973).

Recently Brand, Reynafarje, and Lehninger (1976) have found higher coupling ratios using rat liver mitochondria treated with N-ethylmaleimide to inhibit phosphate transport. Phosphate was not present in our studies, however, and the reason for the discrepancy between our work and that from Lehninger's laboratory is not clear.

ATP Synthesis Driven by an Artificial Electrochemical Proton Gradient

Jagendorf and Uribe (1966) reported that a pH gradient in chloroplasts, formed by incubation at pH 4 and rapidly adding base to bring the external pH to 8, caused the synthesis of ATP from ADP and P_i. ATP synthesis has also been demonstrated in mitochondria driven by a base-to-acid pH jump (Reid, Moyle, and Mitchell, 1966) and by a potassium ion gradient on addition of valinomycin (Cockrell, Harris, and Pressman, 1967). William Thayer and I decided to investigate ATP synthesis driven by artificial ion gradients in inverted submitochondrial vesicles in more detail than had been done in previous studies in mitochondria and chloroplasts (1975a, b). We found the optimum conditions to demonstrate ATP synthesis were quite different from those for chloroplasts. A pH gradient, acid inside, was created by incubation in acid followed by base. The pH of the acid stage could not be as low as in chloroplasts, however, because the submitochondrial vesicles were damaged more easily. To supplement the pH gradient, we used a potassium ion gradient with high concentration outside and we added valinomycin to create potassium ion permeability, forming a membrane potential that was positive inside. The combined pH gradient and membrane potential created a burst of ATP synthesis that lasted about six seconds. The important measurement to make, however, in order to characterize the relationship between the electrochemical proton gradient and ATP synthesis, is not the yield of ATP in such a burst, but the initial rate of phosphorylation. To measure the initial rate of ATP synthesis we used a continuous flow rapid-mixing device to mix the acid and base stages and then stop the reaction. The initial rate driven by the artificial electrochemical proton gradient was found to be faster than phosphorylation driven by NADH respiration (see Fig. 10.2). This observation established that the connection between the electrochemical proton gradient and ATP synthesis was not a minor side pathway but was as rapid as ATP synthesis during oxidative phosphorylation.

In the course of these studies we also observed an interesting aspect of the initial kinetics of oxidative phosphorylation, one which, strangely, no one had bothered to measure before. This was that when oxidative phosphorylation was initiated by adding NADH or another substrate for the respiratory chain, or by adding oxygen, there was a lag in the rate of phosphorylation of about one second. When ADP and P_i were added to submitochondrial vesicles that were already oxidizing NADH, however,

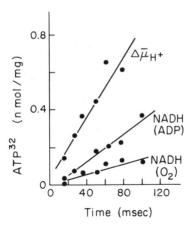

FIGURE 10.2. Initial rates of ATP synthesis by submitochondrial particles driven by an artificial electrochemical proton gradient $(\Delta \bar{\mu}_{H}+)$ or by NADH oxidation initiated by addition of ADP + P_i or O_2, as shown. The measurements were made with a double-mixer continuous flow apparatus. (For details, see Thayer and Hinkle, 1975b.)

no lag was observed. It seems likely that we had observed the time required for the respiratory chain to generate the high level of the proton gradient necessary for rapid ATP synthesis.

Another property of proton-driven ATP synthesis that we studied was the relationship between the magnitude of the electrochemical proton gradient and the rate of ATP synthesis. When there was a decrease in the size of the pH change and potassium ion concentration change that were used to create the electrochemical proton gradient, the rate of ATP synthesis decreased. The membrane potential and pH gradient components appear to have very similar effects, as shown in Figure 10.3.

Reconstitution of Proton Translocation by Segments of the Respiratory Chain

The development by Kagawa and Racker (1971) of a method to reconstitute enzyme complexes into phospholipid membranes has allowed studies of the mechanism of oxidative phosphorylation with isolated enzymes. The reconstitution method consists of mixing the enzyme with liposomes and cholate and removing the cholate by dialysis. In some cases (Racker, 1973), simply sonicating a mixture of liposomes and an enzyme also produces vesicles in which transport is mediated by the enzyme. A new method using a freeze-thaw step followed by brief sonication has been reported recently (Kasahara and Hinkle, 1977).

We have now succeeded in applying the cholate dialysis technique to

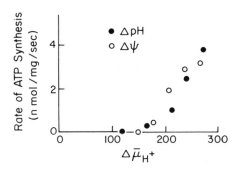

FIGURE 10.3. Initial rate of ATP synthesis by submitochondrial particles as a function of the magnitude of the electrochemical proton gradient in millivolts. The gradient was varied by changing the initial pH (ΔpH) or potassium ion concentration ($\Delta\Psi$) used for creation of the artificial electrochemical proton gradient. (See Thayer and Hinkle, 1975b.)

all three coupling regions of the respiratory chain: NADH-CoQ reductase (Ragan and Hinkle, 1975), reduced CoQ-cytochrome c reductase (Hinkle and Leung, 1974; Leung and Hinkle, 1975), and cytochrome c oxidase (Hinkle, Kim, and Racker, 1972; Hinkle, 1973), in addition to the original reconstitution of the isolated mitochondrial ATPase complex, F_1-F_0, by Kagawa and Racker (1971). Each of the complexes of the respiratory chain showed electrogenic proton translocation in the reconstituted system when assayed with artificial electron acceptors and donors, as illustrated in Figure 10.4. Possible mechanisms of electron transfer and coupled proton translocation are shown, although the details are controversial. Until more is known about the structure of the respiratory chain in the region from NADH to CoQ and from CoQ to cytochrome c, the exact mechanism of proton transport in these systems must be speculative. However, the fact that the isolated complexes, which are the minimum units that will catalyze electron transfer, bring about proton translocation when they are incorporated into liposomes shows that proton movements are directly associated with the respiratory chain. No special proton pumps, coupling factors, or subunits of the ATPase complex are needed for proton translocation by the electron transfer components.

Mitchell's chemiosmotic hypothesis has withstood the tests that we have attempted in these studies of proton translocation and I feel that the central role of the electrochemical proton gradient in coupling oxidation and phosphorylation reactions is well established. A more detailed picture of the mechanisms of proton translocation by the respiratory chain and the ATPase complex will probably come largely from studies of the structure of the components in the membrane.

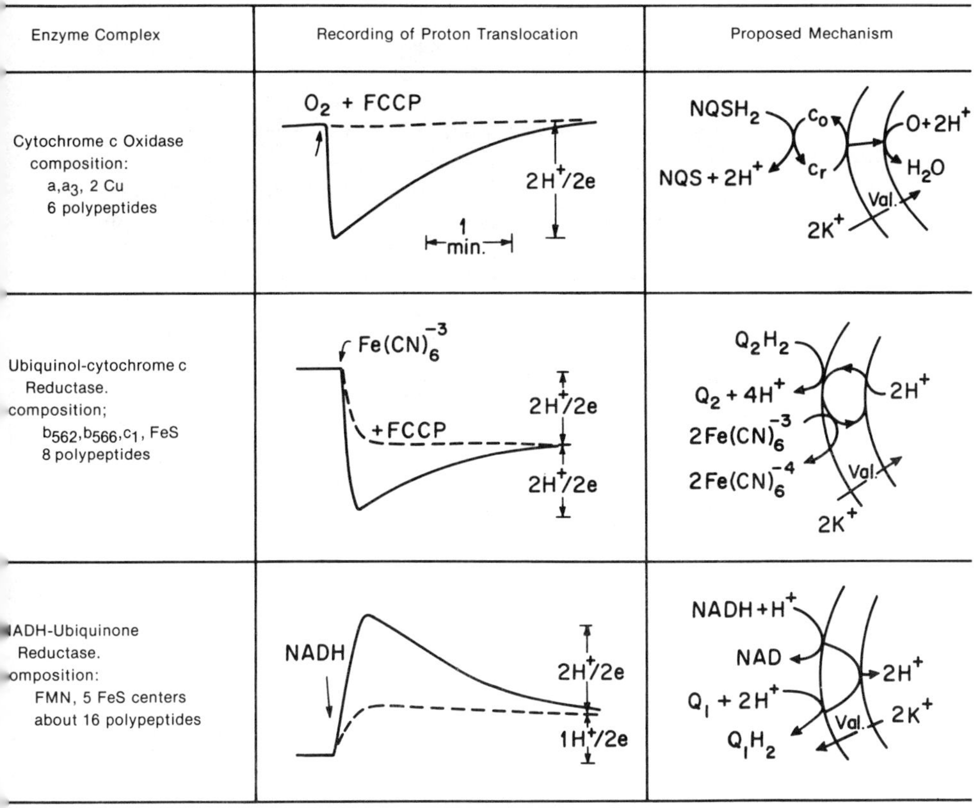

Enzyme Complex	Recording of Proton Translocation	Proposed Mechanism
Cytochrome c Oxidase composition: a, a_3, 2 Cu 6 polypeptides	O_2 + FCCP $2H^+/2e$ $\vdash$1 min.$\dashv$	$NQSH_2$, $NQS + 2H^+$, C_o, C_r, $O+2H^+$, H_2O, Val., $2K^+$
Ubiquinol-cytochrome c Reductase. composition; b_{562}, b_{566}, c_1, FeS 8 polypeptides	$Fe(CN)_6^{-3}$ $+FCCP$ $2H^+/2e$ $2H^+/2e$	Q_2H_2 $Q_2 + 4H^+$ $2Fe(CN)_6^{-3}$ $2Fe(CN)_6^{-4}$ $2H^+$ Val. $2K^+$
NADH-Ubiquinone Reductase. composition: FMN, 5 FeS centers about 16 polypeptides	NADH $2H^+/2e$ $1H^+/2e$	$NADH+H^+$ NAD $Q_I + 2H^+$ Q_IH_2 $2H^+$ Val. $2K^+$

FIGURE 10.4. Proton translocation by reconstituted segments of the respiratory chain. Traces from the recording pH meter are shown. The oxidation reactions were initiated by addition of a small pulse of the appropriate substrate. After a few seconds the substrate is exhausted and electron transfer ceases, allowing the protons to diffuse back across the membrane. The dashed lines show parallel experiments in the presence of the uncoupler FCCP. The mechanisms shown at the right are not certain, but are consistent with the results. Abbreviations: NQS, naphthoquinone Sulfonate; Q_2, analogue of Ubiquinone with two isoprenoid units; C_o and C_r oxidized and reduced cytochrome c; Val., valinomycin; FCCP, carbonylcyanide-p-trifluoromethoxyphenylhydrazone.

REFERENCES

Brand, M. D., B. Reynafarje, and A. L. Lehninger. 1976. Stoichiometric relationship between energy-dependent proton ejection and electron transport in mitochondria. Proc. Natl. Acad. Sci. USA 73: 437.

Cockrell, R. S., E. J. Harris, and B. C. Pressman. 1967. Synthesis of ATP driven by a potassium gradient in mitochondria. Nature 215: 1487.

Hinkle, P. C. 1973. Electron transfer across membranes and energy coupling.

Fed. Proc. 32: 1988.

Hinkle, P. C., and L. L. Horstman. 1971. Respiration-driven proton transport in submitochondrial particles. *J. Biol. Chem.* 246: 6024.

Hinkle, P. C., J. J. Kim, and E. Racker. 1972. Ion transport and respiratory control in vesicles formed from cytochrome oxidase and phospholipids. *J. Biol. Chem.* 247: 1338.

Hinkle, P. C., and K. H. Leung. 1974. Ion transport and respiratory control in vesicles formed from co-enzyme-Q cytochrome C reductase and phospholipids. In G. F. Azzone, M. E. Klingenberg, E. Quagliariello, and N. Siliprandi, eds., *Membrane proteins in transport and phosphorylation*, p. 73. Elsevier, New York.

Jagendorf, A. T., and E. Uribe. 1966. ATP formation caused by acid-base transition of spinach chloroplasts. *Proc. Natl. Acad. Sci. USA* 55: 170.

Kagawa, Y., and E. Racker. 1971. Partial resolution of the enzymes catalyzing oxidative phosphorylation. *J. Biol. Chem.* 246: 5477.

Kasahara, M., and P. C. Hinkle. 1977. Reconstitution and purification of the d-glucose transport protein from human erythrocytes. In G. Semenza and E. Carafoli, eds., *Biochemistry of membrane transport.* FEBS Symposium no. 42, pp. 346-350. Springer Verlag, New York.

Leung, K. H., and P. C. Hinkle. 1975. Reconstitution of ion transport and respiratory control in vesicles formed from reduced coenzyme Q-cytochrome *c* reductase and phospholipids. *J. Biol. Chem.* 250: 8467.

Mitchell, P. 1961. Coupling of phosphorylation to electron and hydrogen transfer by a chemi-osmotic type of mechanism. *Nature* 191: 144.

Mitchell, P. 1966. Chemiosmotic coupling in oxidative and photosynthetic phosphorylation. *Biol. Rev.* 41: 445.

Mitchell, P., and J. Moyle. 1967. Respiration-driven proton translocation in rat liver mitochondria. *Biochem. J.* 105: 1147.

Racker, E. 1973. A new procedure for the reconstitution of biologically active phospholipid vesicles. *Biophys. Res. Commun.* 55: 224.

Ragan, C. I., and P. C. Hinkle. 1975. Ion transport and respiratory control in vesicles formed from reduced nicotinamide adenine dinucleotide coenzyme Q reductase and phospholipids. *J. Biol. Chem.* 250: 8472.

Reid, R. A., J. Moyle, and P. Mitchell. 1966. Synthesis of adenosine triphosphate by a proton motive force in rat liver mitochondria. *Nature* 212: 257.

Thayer, W. S., and P. C. Hinkle. 1973. Stoichiometry of adenosine triphosphate-driven proton translocation in bovine heart submitochondrial particles. *J. Biol. Chem.* 248: 5395.

Thayer, W. S., and P. C. Hinkle. 1975a. Synthesis of adenosine triphosphate by an artificially imposed electrochemical proton gradient in bovine heart submitochondrial particles. *J. Biol. Chem.* 250: 5330.

Thayer, W. S., and P. C. Hinkle. 1975b. Kinetics of adenosine triphosphate synthesis in bovine heart submitochondrial particles. *J. Biol. Chem.* 250: 5336.

11   Studies on the Molecular Mechanism of Anion Transport across the Red Blood Cell Membrane

HERMANN PASSOW AND LAILA ZAKI

The elucidation of the molecular events that take place when an anion crosses the red blood cell membrane requires:

1. the demonstration of the existence of a specific anion transport system and the delineation of that system
2. the biochemical identification of the membrane constituents involved in the transport process
3. the assignment of specific functions to the identified constituents
4. the reconstitution of the transport system from the isolated constituents in artificial membrane systems

This chapter is concerned with points 2 and 3. Some comments on point 1 will serve as an introduction.

Existence and Delineation

The existence of an anion transport system in the red cell is demonstrated fairly easily by comparing anion transport in liposomes made from red-cell lipids with transport in the intact cells. Although such liposomes are permeable to anions, their permeability is many orders of magnitude smaller than the permeability of the membrane of the intact red cell. This indicates the presence in the intact cell membrane of one or several constituents that facilitate anion penetration.

It is difficult to delineate a specific anion transport system. There are two problems:

(1) The anion flux as measured with ^{36}Cl or $^{35}SO_4$ is composed of two components, one that contributes to the conductance of the membrane and another that does not (Harris and Pressman, 1967; Scarpa, Cecchetto, and Azzone, 1970). In Cl transport, the component that con-

tributes conductance is about $1/10^4$ of the total transport (Hunter, 1971; Wieth et al., 1973); in SO_4 transport it is about 1/5 of the total flux at pH 7.8 and negligibly small at pH 6.5 (Passow et al., 1975). It is not yet clear whether the transport that contributes to the conductance and the transport that does not contribute to it are on parallel and independent pathways or whether they represent two different aspects of the same mechanism. Whatever the explanation, it is possible to adjust the experimental conditions such that one measures predominantly that flux component that does not contribute to the conductance, and this chapter is concerned only with that component.

(2) Another problem that cannot be resolved so casually is the question of whether all anion species penetrate by the same mechanism or there are different mechanisms for ions like sulfate or the halides. The penetration rates for these anion species are different by a factor of 10^4 to 10^5. The pH (Gunn et al., 1973; Passow and Wood, 1973) and temperature (Brahm and Wieth, 1976) dependences are also different, and halide transport shows the peculiar feature of "acceleration" (Passow and Wood, 1973), which has not been observed with divalent anions (unpublished results). The work of Deuticke (1976) and Rice and Steck (1976) also suggests that certain organic anions penetrate by a transport mechanism that, in certain respects, is distinct from that of either SO_4 or halides. It is necessary, therefore, to keep in mind that this chapter is confined to sulfate transport and that the results do not necessarily apply to anion transport in general.

Involvement of a Specific Membrane Protein in Anion Transport

The identification of the membrane proteins that are involved in the active transport of K, Na, and Ca was greatly facilitated by the fact that these transport processes are linked to the hydrolysis of ATP. Thus, even in preparations of broken membranes, the isolation and purification of the protein can easily be followed by measuring the activation of ATP hydrolysis by the cations to be transported. But anions are passively transported across the red blood cell membrane; hence, one cannot rely on the convenient technique used to identify the cation transport protein. The obvious alternative would be to label with a tightly binding specific inhibitor. Unfortunately, no inhibitor is available that inhibits anion transport with a degree of specificity similar to that with which ouabain inhibits Na-K transport.

Several years ago we found that amino-reactive reagents are covalently binding inhibitors of anion transport (Passow, 1969; Passow and Schnell, 1969). These reagents are at best group-specific, and they combine more or less readily with any accessible amino group. Our work with pronase (Passow, 1971), and that of Obaid, Rega, and Garrahan (1972) with maleic anhydride have suggested that lipid-amino groups are not involved in anion transport. However, even though this limits the search to protein-amino groups, it is still a formidable task to identify those groups that could possibly be involved in transport.

In 1971 Knauf and Rothstein reported that the nonpenetrating isothiocyanate derivative of a stilbene disulfonic acid (SITS[1]) was a powerful inhibitor of anion transport. This discovery proved to be important, since it limited the range of candidates for the anion transport protein to those few membrane proteins that are exposed to the outer cell surface. SITS was used in two parallel approaches. Cabantchik and Rothstein (1972, 1974) tried to modify the molecule in order to obtain an inhibitor that was specific enough to label one membrane constituent only. We tried to improve the selectivity by combining the application of SITS with that of another inhibitor of anion transport, DNFB, and with an enzyme, trypsin (Passow et al., 1975; Lepke and Passow, 1976).

The work done in our laboratory was based on the assumption that common binding sites for two different inhibitors of anion transport are likely to be involved in that transport process. The two inhibitors chosen were DNFB and SITS, or DNFB and some other anion transport inhibiting disulfonic acid. For the identification of common binding sites, we first exposed the cells to one of the disulfonic acids. After removal of the unreacted material by washing, we dinitrophenylated the cells with [14C] DNFB, isolated the labeled membranes, and separated the membrane proteins on SDS-gel electropherograms. We compared the labeling pattern with that obtained in cells that had been dinitrophenylated in the absence of a disulfonic acid, and we looked for a band at which dinitrophenylation was prevented by the preexposure to the nonradioactive acid.

Figure 11.1 shows an experiment where SITS was used as the disulfonic acid. The easily penetrating DNFB labeled virtually all major membrane proteins. However, in the SITS-treated membranes, the protein in band 3 remained largely unlabeled. This suggested immediately that band 3 was a candidate for the transport protein. There was, however, the problem that small numbers of other binding sites might have been overlooked. To reduce this source of uncertainty we tried to remove from the membrane as many proteins as possible and to look again for common binding sites for SITS and DNFB. Most membrane proteins are associated with the inner membrane surface. Therefore, we incorporated some trypsin into red-cell ghosts and resealed them. Under these conditions the ghosts became fragmented. Gel electrophoresis of the isolated membranes showed that virtually all of the membrane proteins were affected by the enzyme and that most of the resulting small peptides

[1]Abbreviations: SITS, 4-isothiocyanato-4'-acetamido stilbene-2,2'-disulfonic acid; DAS, 4,4'-diacetamido stilbene-2,2'-disulfonic acid; DIDS, 4,4'-diisothiocyanato stilbene-2,2'-disulfonic acid; H$_2$DIDS, dihydro-DIDS; APMB, 2-(4'-aminophenyl)-6-methyl benzenethiazol-3',7-disulfonic acid; DNFB, 1-fluoro-2,4-dinitrobenzene; SDS, sodium dodecyl sulfate.

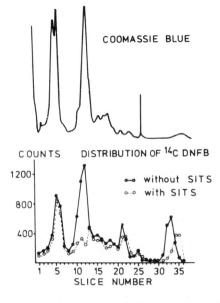

FIGURE 11.1. Common binding sites for DNFB and SITS on the protein in band 3. The peak of band 3 is located at slice no. 12. The lower tracings ("Distribution of ^{14}C DNFB") represent the distribution of ^{14}C DNP residues on polyacrylamide-gel electropherograms of red-cell membranes isolated from cells that had been dinitrophenylated (30 minutes at pH 7.6, 37° C, 0.14 mM DNFB) with or without prior exposure to SITS (30 minutes at pH 7.6, 25° C, 0.5 mM SITS; excess SITS was removed prior to dinitrophenylation). The distribution pattern of the Coomassie blue stain (upper tracing, "Coomassie blue") is independent of pretreatment with SITS. (From Passow et al., 1974.)

became detached from the membrane. There remained three, sometimes four, bands. The molecular weights of the two most prominent bands are close to 58,000 daltons and 48,000 daltons.[2] Only these two fragments carried the common binding sites for SITS and DNFB and hence were derived from the protein in band 3. Anion transport could still be measured in the enzymatically modified ghosts and was still inhibitable by a variety of substances that produced different types of inhibition. Since the binding sites on the peptides derived from band 3 were the only ones on the enzymatically cleaned membrane that bound two different

[2] The determinations of the molecular weight of the protein in band 3 differ considerably from laboratory to laboratory (range 88,000-106,000 daltons; for references, see Passow et al., 1977). Our recent estimates were obtained by adding standard proteins labeled with [^{14}C] DNFB to the unknown samples. With this technique we obtained values of 106,000 daltons, which are at the upper limit of the indicated range. In this context it should be borne in mind that numerical values as derived from SDS-polyacrylamide gel electropherograms are useful estimates rather than true absolute values.

inhibitors of anion transport, we suggested that the protein in band 3 was involved in that transport (Passow et al., 1975; Zaki et al., 1975; Lepke and Passow, 1976). (We should like to add, parenthetically, that the experiments with internal trypsin and external SITS confirmed previous findings indicating that the protein in band 3 spans the membrane [Bretscher, 1971; Jenkins and Tanner, 1975; for reviews see Steck, 1974; Juliano, 1973].)

Although we strongly suggested the involvement of the protein in band 3 in anion transport, we also expressed certain reservations:

(1) Our analytical techniques only permit the study of the labeling of the so-called "major" membrane proteins, that is, the more abundant protein species. The less abundant "minor" membrane proteins, which include the transport ATPases for Ca and alkali ions, acetylcholinesterase, and many other proteins, are totally disregarded. Since isothiocyanates are not at all specific for anion transport sites, it is not impossible that one of these minor proteins is the transport protein. The magnitude of the problem can be easily visualized by labeling the membrane proteins with diazo sulfanilic acid, a reagent that is less selective than the isothiocyanates and therefore more powerful in demonstrating the presence of minor membrane proteins (Bender, Garan, and Berg, 1971). In untreated and trypsin-treated ghosts, we find a continuous spectrum of labeled proteins with molecular weights ranging from more than 100,000 to less than 10,000 (Lepke and Passow, 1976).

(2) The gel electropherograms that showed the common binding sites on the peptides derived by tryptic digestion from the protein in band 3 indicated that such sites exist on both fragments. Thus, if we assume that the protein in band 3 does actually participate in transport, it remains to be seen whether or not all of the sites are involved and exert the same function.

The question of an involvement of small numbers of common binding sites within or outside band 3 leads immediately to the question whether or not we can obtain an estimate of the minimal number of sites required for anion transport at the observed rate. We would like to compare this number with the number of common sites for SITS and DNFB that can be correlated experimentally with anion transport. We found that about 1×10^6 of the sites in band 3 could possibly be correlated with transport, while several hundred thousand additional sites in that band could not (Zaki et al., 1975). The question of the minimal numbers of sites required to account for the observed transport rates will be postponed briefly while we turn to comment on the work that went on in Toronto while we did the experiments we just described in Frankfurt.

After publishing some inconclusive results (1972) with a diiodo derivative of a stilbene disulfonic acid, Cabantchik and Rothstein (1974) reported that the diisothiocyanate derivative of SITS, called DIDS, labeled one membrane constituent only, band 3. They described a linear relationship between inhibition and binding and found that inhibition was complete when about 300,000 sites on band 3 were occupied. In their work, they measured binding by means of $^3H_2DIDS$, and inhibition by means of DIDS, and they plotted the two sets of data, obtained in different experiments, against each other. Although they had some evidence that H_2DIDS and DIDS behaved alike, later work in our laboratory showed that the reactivity of H_2DIDS is lower than that of DIDS and that the observed straight-line relationship was due to several compensating errors. A reinvestigation (Lepke, Fasold, Pring, and Passow, 1976) revealed that for each, DIDS and H_2DIDS, there still exists a linear relationship between inhibition and binding. The total number of sites occupied in band 3 at complete inhibition is, however, about 1.1×10^6 and 1.2×10^6 molecules/cell for DIDS and H_2DIDS, respectively. On the basis of these data it is possible to calculate turnover numbers. For chloride transport one obtains about $2\text{-}3 \times 10^4$ per second per H_2DIDS binding site and for sulfate transport 0.07 per second per site (both at $25°C$).

If one could demonstrate that these numbers represent an upper limit for any realistic model of anion transport across the red-cell membrane, it would be justifiable to assume that all of the $^3H_2DIDS$ binding sites on the protein in band 3 must participate in the transport process. Unfortunately, it is not even clear whether or not the higher of the two numbers represents such a limiting value. It has been shown that the turnover number for valinomycin-mediated K transport across lipid bilayers is 1×10^4 per second (Läuger, 1972) and hence close to the turnover number calculated for Cl transport in the red-cell membrane. K transport across the bilayer is governed both by a relatively slow rate of reaction between the carrier molecule and the K ion to be transported and by the diffusion through the whole thickness of the lipid bilayer. It is unlikely that anion transport across the red blood cell membrane is accomplished by rotation in toto of a molecule of molecular weight 100,000 (see Cherry, 1976). If such a molecule does in fact play a role, one must assume that oscillations of a side chain accomplish the transition across the permeability barrier. The amplitude of such oscillations would be much smaller than the thickness of the lipid bilayer, and the rate of a single oscillatory movement could exceed that for diffusion of a lipid-soluble carrier across the whole thickness of the rather viscous bilayer. In addition, the combination between the side chain and the anion to be transported would not necessarily involve major conformational changes but could be due to purely electrostatic binding (this is suggested by the unpublished observation that the enthalpy changes associated with sulfate binding to the transport system are very low, in the range of 8-12 KJ/mole) and hence be a very fast process. Thus, it is not at all inconceivable that anion transport may be carried out with turnover numbers that exceed those for valinomycin-mediated K transport by one or two orders of magnitude. If so, 1/10 or 1/100 of the H_2DIDS binding sites on the red-cell membrane could account for the observed transport rate. Such small numbers could easily be overlooked on the gels.

In this context it should be pointed out that the H_2DIDS binding sites on band 3 do not seem to form a homogeneous population. The proteolytic dissection of the protein in band 3 by chymotrypsin, pronase, and papain suggests a nonuniform resistance of the H_2DIDS carrying protein to enzymatic cleavage (Passow et al., 1977). For the present purposes, experiments with chymotrypsin are pertinent. Exposure of intact red cells to increasing concentrations of chymotrypsin shows that part of the protein in band 3 is split into one peptide of 70,000 daltons[3] and another of 36,000 daltons. The rest remains undigested or, at least, does not undergo major changes in its molecular weight even at very high concentrations of the enzyme and after long times of exposure. Although a continuation of enzymatic degradation for very extended periods of time might also lead to a further digestion of some or all of the more resistant fraction of protein in band 3, it is quite obvious that the response of that protein is not uniform. This indicates differences in accessibility or susceptibility to the enzyme, which might be due to differences in primary structure, conformation, or arrangement within the membrane structure (Fig. 11.2a, b).

The effect of chymotrypsin on the protein in band 3 is not associated with an inhibition of anion transport. It is possible, therefore, to study the relationship between inhibition and H_2DIDS binding to the enzymatically modified membrane. Inhibition of sulfate equilibrium exchange is linearly related to binding both to the surviving 106,000-dalton protein and to the 70,000-dalton fragment (Fig. 11.3). Thus, we are unable to decide whether the surviving 106,000-dalton peptide or the 70,000-dalton peptide is involved in anion transport, or whether both are. In any event, further work will be required to decide unequivocally whether all, some, or none of the H_2DIDS binding sites in band 3 are involved in anion transport.

Similarly inconclusive results were obtained from work with pronase-treated red cells. Low concentrations of pronase split part of the protein in band 3 into fragments of 70,000 and 36,000 daltons. The rest of the protein remains unaltered, at least so far as SDS-gel electrophoresis can tell. There is little inhibition of anion transport. At higher concentrations of pronase, the enzyme digests both the 70,000-dalton peptide and the

---

[3] On our gels, the location of this band is indistinguishable from that of plasma albumin, the molecular weight of which is 67,000 daltons. Thus, this band is possibly, but not certainly, somewhat lower than 70,000 daltons, as determined for the major split products produced by pronase or papain.

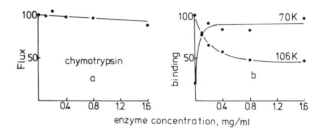

FIGURE 11.2. *Left (a):* Effect of increasing concentrations of chymotrypsin on sulfate equilibrium exchange in human red cells. $^{35}SO_4$ efflux was measured at 37° C in cells equilibrated in a medium containing 5 mM Na_2SO_4, 20 mM Na phosphate, and 122.5 mM NaCl, pH 7.4. Hematocrit 10 percent. Prior to the flux measurements the cells were exposed to chymotrypsin for one hour at 37° C and at the concentrations indicated on the abscissa. The enzymatic reaction was interrupted by repeated washes in albumin-containing media of the composition indicated above. Efflux was measured in the absence of the enzyme. Ordinate: $^{35}SO_4$ efflux as a percent of control. Abscissa: enzyme concentration, mg/ml.

Right (b): $^3H_2$ DIDS binding as measured after termination of the enzymatic reaction. *106K* and *70K* denote binding to the surviving 106,000-dalton peptides and the newly formed 70,000-dalton peptides, respectively. To measure binding, an aliquot of the cells that had been equilibrated in the medium described above for flux measurements was removed and exposed to a large excess of $^3H_2$ DIDS (25 μM) for 90 minutes at 37° C. Subsequently, the unreacted $^3H_2$ DIDS was removed by washing, membranes were prepared, dissolved in SDS, subjected to gel electrophoresis, sliced, and counted in a liquid scintillation counter. Ordinate: H_2DIDS binding as a percent of binding to control. Abscissa: enzyme concentration, mg/ml.

surviving 106,000-dalton peptide. This is accompanied by a considerable inhibition of anion transport. If one titrates the number of H_2DIDS binding sites on the 106,000- and the 70,000-dalton peptides one observes that the number of available sites decreases linearly with increasing inhibition. Thus, it is not possible to decide whether the surviving 106,000-dalton protein or the 70,000-dalton fragment is involved in anion transport (Figs. 11.4, 11.5).

In spite of this uncertainty, it seems worthwhile to pursue the idea that at least some of the binding sites in band 3 play some role. However, before discussing further data, it may be worthwhile to illustrate the situation by a story that originated at the Weizmann Institute and which is told by Eigen and Winkler in their recent book (1975). A graduate student had trained an army of cockroaches to follow his orders. When he gave the order "turn left," all turned left. When he ordered "turn right," all turned to the right. After he had demonstrated this achievement to his professor he took one of the poor creatures and tore off all its legs. Then he put the legless roach back into the line. Again he com-

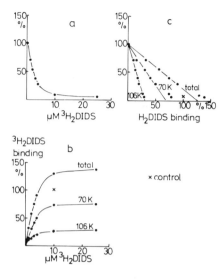

FIGURE 11.3. (a) Effect of H_2DIDS on $^{35}SO_4$ efflux from chymotrypsin-treated red cells. Prior to the exposure to H_2DIDS at the concentrations indicated on the abscissa (90 minutes at pH 7.4), the cells had been treated with 0.3 mg/ml chymotrypsin for one hour. The flux measurements were performed after removal of reversibly bound H_2DIDS by washing in albumin-containing media. All reactions and the flux measurements were carried out at 37° C in a medium containing 5 mM Na_2SO_4, 122.5 mM NaCl, and 20 mM Tris-Cl pH 7.4. Ordinate: sulfate efflux as percent of efflux from chymotrypsin-treated control. This control value was indistinguishable from efflux from intact red cells that had not been exposed to the enzyme.

(b) Irreversible binding of H_2DIDS to chymotrypsin-treated red cells as a function of H_2DIDS concentration in the medium. Same experiment as in a. Control: H_2DIDS binding to red cells that had not been exposed to chymotrypsin (1.23 x 10^6 molecules/cell). All other binding data (represented on the ordinate) are expressed in percent of this value. Total, 106 K, and 70 K refer, respectively, to binding to the membrane as a whole, to the chymotrypsin-resistant 106,000-dalton peptide, and to the 70,000-dalton fragment of chymotryptic digestion. Abscissa: as in a.

(c) Relationship between sulfate efflux and irreversible H_2DIDS binding to chymotrypsin-treated red cells. Data from a and b plotted against each other. Ordinate: SO_4 efflux. Abscissa: H_2DIDS binding. Both axes are expressed in percent of control value.

manded: "turn right," "turn left." All cockroaches did correctly what they were supposed to do, except, of course, the cockroach without legs. "What do you conclude from this experiment?" asked his professor. "I think," the student said, "I have demonstrated that cockroaches hear with their legs." The point of the story is that cockroaches do, in fact, seem to hear with their legs. Thus, in spite of the low level of sophistica-

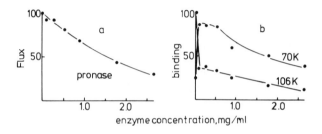

FIGURE 11.4. (a) Effect of increasing concentrations of pronase on sulfate equilibrium exchange in human red cells. $^{35}SO_4$ efflux was measured under the same conditions as described for chymotrypsin in the legend to Figure 11.2a. Ordinate: $^{35}SO_4$ efflux as a percent of control. Abscissa: enzyme concentration, mg/ml.

(b) $^3H_2$DIDS binding as measured after termination of the enzymatic reaction. Same experimental conditions as in the experiment with chymotrypsin described in the legend to Figure 11.2b. Ordinate: H_2DIDS binding as a percent of binding to control. Abscissa: enzyme concentration, mg/ml.

tion of our current attempts to assign a role in anion transport to the protein in band 3, it may eventually turn out that band 3 is in fact involved. The experiments described below on the sidedness of action of APMB on anion transport and dinitrophenylation of H_2DIDS binding sites on the protein in band 3 do indeed strongly suggest such involvement.

Function of Membrane Constituents in Anion Transport

What is the function in anion transport of the common binding sites for DIDS and DNFB on the protein in band 3 and transport anions? Before answering this question, two preliminary problems have to be resolved: (1) Are these sites mobile or fixed? (2) Are they transfer or modifier sites? That is, are they sites that bind anions and then translocate them across the membrane, or are they sites that do not combine with the anions to be transported but participate, nevertheless, in the control of binding or translocation?

The investigation of these two problems will also shed some light onto the additional question whether or not the common binding sites for DIDS and DNFB are the only sites involved in anion transport.

The first of the two problems was studied by investigating the sidedness of action of nonpenetrating chemical modifiers of sulfate transport in resealed red-cell ghosts. If anion transport were controlled by fixed sites located only at the outer membrane surface, a nonpenetrating inhibitor could exert its action only at that surface but not at the inner membrane surface. Hence, the observation of an asymmetric inhibition would suggest that the receptor sites for that inhibitor are fixed and asymmetrically distributed. Asymmetric inhibition is not easily compatible with a reaction of the inhibitor with a mobile carrier molecule. If

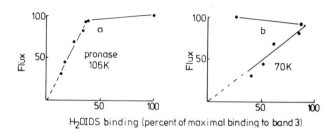

H₂DIDS binding (percent of maximal binding to band 3)

FIGURE 11.5. Effect of pronase treatment on sulfate equilibrium exchange and the number or titratable $^3H_2$DIDS binding sites on the 106,000-dalton protein (*106K*) and on the 70,000-dalton fragment (*70K*). Intact red cells were exposed to eight different concentrations of pronase, with a range of 0-2 mg/ml, and incubated for 60 min at 37° C. After removal of pronase, each of the eight batches of cells was subdivided into two batches. One of these batches was used for flux measurements, the other for titration of $^3H_2$DIDS binding sites, as described in the legend to Figure 11.2 *b*. The rate constant for sulfate equilibrium exchange (as a percent of control) is plotted against the number of titrated $^3H_2$DIDS binding sites in molecules per cell.

there were valinomycin-like mobile carriers, one would expect them to move to that surface which is in contact with the nonpenetrating inhibitor until all carrier molecules are inactivated or until all inhibitor molecules are consumed by the reaction with the carrier. This situation is completely symmetrical. The affinity of a carrier molecule to an inhibitor should depend on the environment in which the reaction takes place. For this reason, it cannot be ruled out that a carrier molecule changes its affinity when it crosses the membrane, and thus gives rise to an asymmetry of inhibition. Nevertheless, an inhibition confined to only one surface would represent a fairly strong argument against a mode of action that involves the inactivation of a mobile carrier.

Our first experiments (Lepke and Passow, 1973; Schnell et al., 1973) were done with phlorizin. We prepared red-cell ghosts and observed the effects of external or internal phlorizin. However, before applying this agent to the study of anion transport we checked whether or not we could observe the expected symmetrical inhibition in a transport process that clearly shows the kinetics of a carrier system. For this purpose we studied the sidedness of inhibition of sugar transport. Figure 11.6 shows that, in accord with expectation, the permeability to D-xylose is inhibited by phlorizin at either surface. In contrast, the inhibition of anion transport is not symmetrical (Fig. 11.7). Phlorizin inhibits if applied to the outer cell surface but is ineffective at the inner surface. This finding suggests a reaction with fixed sites at the outer membrane surface.

To study the sidedness of a compound that is structurally similar to SITS or DIDS, we chose DAS, the diacetamido derivative of a stilbene disulfonic acid. DAS does not form a covalent bond with the membrane.

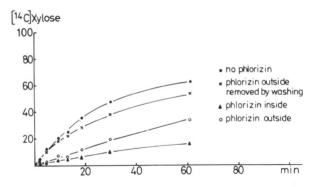

FIGURE 11.6. Sidedness of action of phlorizin on D-xylose efflux from human red-cell ghosts. Ghosts and external media contained 10 mM D-xylose dissolved in a solution consisting of 10 mM K_2SO_4, 100 mM KCl, and 20 mM Tris-Cl pH 7.6. The D-xylose in the ghosts was labeled with ^{14}C. At zero time, ghosts and medium were mixed and the time course of appearance of ^{14}C in the supernatant was followed. Temperature: 0° C. Ordinate: ^{14}C in the supernatant, as a percent of the equilibrium value. Abscissa: time in minutes. "No phlorizin": control, neither internal nor external phlorizin present. "Phlorizin inside" and "phlorizin outside" refer to internal or external phlorizin at a concentration of 2.0 mM, respectively. "Phlorizin outside removed by washing": to demonstrate that the binding of the agent is completely reversible, one batch of resealed ghosts was incubated for 5 minutes in the presence of external phlorizin, subsequently washed in phlorizin-free media, and then subjected to flux measurements as the other batches of cells.

Like phlorizin, it inhibits only at the outer membrane surface (Fig. 11.8). Like SITS, and unlike phlorizin, it has common binding sites with DNFB on the protein in band 3. If we assume that the common binding sites on band 3 are in fact involved in anion transport, we have to postulate that phlorizin and DAS inhibit at different sites.

A third inhibitor, another disulfonic acid, called 2-(4'-aminophenyl)-6-methylbenzenethiazol-3',7-disulfonic acid (APMB) inhibits reversibly at either surface with nearly equal strength. This suggests the involvement of a third set of sites that are not accessible to DAS or phlorizin (Fig. 11.9).

The binding sites for APMB could be either mobile or fixed. They could be mobile for the reasons explained above. They coud be fixed if we assume that the same sites are present at either surface. Could one discriminate between the two possibilities? The following considerations would suggest that this can be done. If we dinitrophenylated in the presence of external APMB, we would protect all binding sites on a mobile carrier against inactivation. Also, if we dinitrophenylated in the presence of internal APMB, we would obtain the same protection. If the sites were fixed, however, external APMB could not prevent the dinitrophenylation

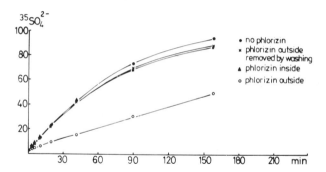

FIGURE 11.7. Sidedness of action of phlorizin on sulfate efflux from red cell ghosts. Same experimental conditions as in Figure 11.6, except that the sulfate inside the ghosts was labeled with $^{35}SO_4$ and the flux was measured at 37° C. Ordinate: $^{35}SO_4$ in the supernatant as a percent of the equilibrium value. Abscissa: time in minutes.

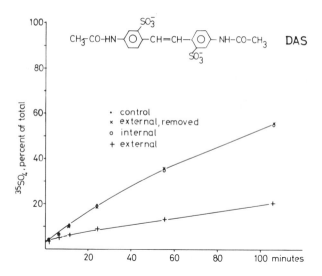

FIGURE 11.8. Sidedness of action of 4,4'-diacetamido-stilben-2,2'-disulfonic acid (DAS, 2 mM) on sulfate efflux from human red-cell ghosts. Ghosts and media contained 5.0 mM Na_2SO_4, 131.0 mM NaCl, 20 mM Na-phosphate, pH 7.3, otherwise same experimental conditions as in Figure 11.7. Temperature: 30° C. Ordinate: $^{35}SO_4$ in the supernatant. Abscissa: time in minutes.

of the internal sites and internal APMB could not prevent the dinitro-phenylation of external binding sites. Hence, exposure of the cells to DNFB in the presence of either internal or external APMB should lead to a protection of transport if we deal with a carrier system. It should show inhibition if we have fixed sites on both surfaces.

·

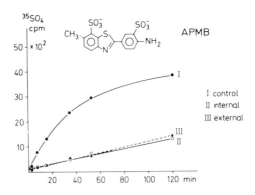

FIGURE 11.9. Sidedness of action of 2-(4'-aminophenyl)-6-methyl-benzenethi-azol-3,7'-disulfonic acid (APMB, 7.5 mM) on sulfate efflux from human red-cell ghosts. Composition of ghosts and media as described in the legend to Figure 11.8. Temperature: 30° C. Ordinate: $^{35}SO_4$ in the supernatant as a percent of the equilibrium value. Abscissa: time in minutes.

The actual performance of such experiments requires a removal of APMB and unreacted DNFB prior to the flux measurements with $^{35}SO_4$. For this purpose it is necessary to use ghosts that have been hemolyzed and sealed twice. During the first hemolysis step, APMB is incorporated. After the first sealing period, the ghosts are exposed to DNFB. Subsequently, the ghosts are washed to remove unreacted DNFB (which easily penetrates) and hemolyzed again. This leads to a release of the entrapped APMB. After resealing of these ghosts, SO_4 flux is measured to assess the effect of internal APMB on the action of DNFB. Control experiments have shown that the sulfate permeability of twice-sealed ghosts is virtually identical to that of once-sealed ghosts. Moreover, to make all experiments strictly comparable, the parallel runs with external APMB in the presence or absence of DNFB and the control runs without added inhibitors are also performed with twice-sealed ghosts. Figure 11.10 shows that both internal and external APMB protect the sulfate transport system against dinitrophenylation. The protection is not complete but still considerable in view of the fact that the reversibly binding APMB cannot indefinitely block the irreversible reaction of DNFB. These findings are obviously compatible with the assumption that the common binding sites for APMB and DNFB are mobile and able to cross the permeability barrier for APMB.

To test such an assumption further we applied a procedure that was similar to, but not identical to, a procedure used by Rothstein, Cabantchik, and Knauf (1976) in a study of the effects of pyridoxal phosphate on sulfate transport. We titrated the common binding sites for APMB and DNFB with external H_2DIDS and demonstrated that internal and ex-

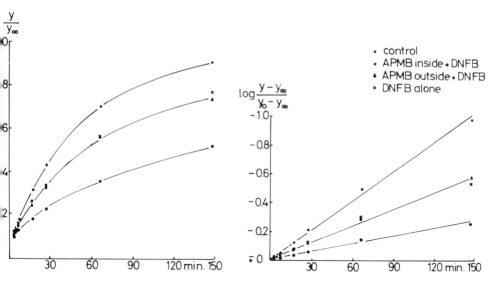

FIGURE 11.10 Effects of internal and external APMB on irreversible modification of the sulfate transport system by DNFB. Resealed ghosts were dinitrophenylated for 30 minutes at 37° C, pH 7.4, in the absence of APMB ("DNFB alone") or in the presence of 10 mM internal APMB ("APMB inside plus DNFB") or 10 mM external APMB ("APMB outside plus DNFB"). The control was incubated in the absence of both DNFB and APMB. The medium inside and outside the ghosts contained 5 mM Na_2SO_4, 122.5 mM NaCl, 20 mM Tris-Cl pH 7.4. After dinitrophenylation, internal and external APMB as well as unreacted DNFB were removed as described in the text and sulfate efflux was measured. y, y_o, $y\infty$ = radioactivity in the supernatant at times t, 0, and infinity, respectively. Abscissa: time in minutes.

ternal APMB protect about equal numbers of H_2DIDS binding sites against dinitrophenylation.

In these experiments the ghosts were again dinitrophenylated in the presence or absence of external or internal APMB. After removal of DNFB and APMB by washing, rehemolysis, and subsequent resealing in the absence of added inhibitors, each batch of ghosts was subdivided into two portions. One portion was used for measuring sulfate transport, the other for H_2DIDS binding to the protein in band 3. H_2DIDS binding was estimated after exposure to a large excess of the agent (25 μM) for a long time (90 minutes) at an elevated temperature (37°C) in order to ensure a saturation of all H_2DIDS binding sites on the protein in band 3. Table 11.1 shows the results of such an experiment: in the control, a high rate of sulfate transport is associated with a high capacity of the ghosts to bind externally applied H_2DIDS. In ghosts that are dinitrophenylated in the absence of either external or internal APMB we find a low rate of SO_4

TABLE 11.1. Protection against dinitrophenylation of sulfate transport system and H$_2$DIDS binding sites on the protein in band 3 by internal and external APMB.

Condition	Sulfate transport, ^ok_S $(\times 10^2 \text{min}^{-1})$	H$_2$DIDS binding to band 3 $(\times 10^{-6} \text{molec/cell})$
Control	1.78	1.90
DNFB + external APMB	1.13	0.94
DNFB + internal APMB	1.05	0.93
DNFB, no APMB	0.48	0.38

Note: Ghosts were dinitrophenylated in the presence or absence of APMB (internal or external), washed, rehemolyzed, sealed, divided into two batches, one for flux measurements, the other for measurements of H$_2$DIDS binding capacity.

transport, accompanied by a small number of surviving H$_2$DIDS binding sites. If APMB is present during dinitrophenylation, there is considerable protection against dinitrophenylation of the transport system as well as of the H$_2$DIDS binding sites that can be titrated by external H$_2$DIDS. Experiments of this type would agree with the view that anion transport and H$_2$DIDS binding are associated with mobile sites. At the same time they provide the strongest evidence yet available for the participation of the protein in band 3 in anion transport.

Although the conclusions concerning the motility of the H$_2$DIDS binding sites may seem reasonable, it is necessary to keep in mind that alternative explanations cannot be ruled out. The fact that internal APMB inhibits anion transport indicates that the compound does not penetrate readily through the membrane. Direct determinations of APMB retention by red-cell ghosts have confirmed this result. However, due to some binding of APMB to hemoglobin and, possibly, to certain membrane constituents, these direct determinations were not very accurate. APMB penetration has subsequently been studied with a more sensitive technique and it has been discovered that some leakage does in fact occur, amounting to about 5-10 percent per hour at 37°C. One could argue that the escaping APMB is immediately diluted by the large volume of the external medium surrounding the ghosts and hence could neither produce inhibition nor afford protection against dinitrophenylation by combination with a fixed binding site at the outer membrane surface. A more detailed consideration of the situation (Fig. 11.11) indicates, however, that such an inference may not necessarily be correct. In Figure 11.11 a fixed binding site is assumed to be located at the outer surface of the barrier that the anions cross by means of a carrier system; this barrier is

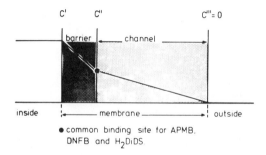

C'   C"   C'''= 0

barrier   channel

inside   membrane   outside

● common binding site for APMB,
  DNFB and H₂DIDS

FIGURE 11.11. Effect of an unstirred layer ("channel") on the diffusion of APMB across a barrier. $\Delta C = c'' - c'''$. For explanation see text.

responsible for anion exchange without contribution to the conductance of the membrane. This carrier system could consist of a side chain of some protein in band 3 and it could oscillate with an amplitude of a few angstroms. It could be located at the bottom of an invagination in the protein (*channel* in Fig. 11.11) that communicates at the other end with the external medium. This could give rise to an unstirred layer of a thickness Δx with a maximum thickness of about 40 Å. However, a semiquantitative assessment of the situation (see appendix) shows that unrestricted diffusion of APMB through a water-filled, convection-free channel of length Δx can certainly not explain the findings. The situation would be different, however, if one stipulated that the diffusion of APMB in the invagination occurred about 10^8 times more slowly than in a large volume of aqueous solution or, in other words, that the common binding sites for APMB, DNFB, and H₂DIDS were located inside the membrane. If these buried sites were accessible to APMB and DNFB from either surface but to H₂DIDS (and the related compound DAS) from the outer surface only, local accumulation of APMB within the membrane could explain the inhibition of anion transport and the protection against dinitrophenylation of common binding sites for APMB and H₂DIDS by internal (and, of course, external) APMB.

We shall now turn to a discussion of the possible function of the H₂DIDS binding sites. Figure 11.12 represents the relationship between inhibition of sulfate transport and APMB concentration in the external medium as measured at two different sulfate concentrations. The concentrations were chosen such that at a K_I for APMB of about 1 mM and a K_M for sulfate of about 30 mM one could expect to discover competition between sulfate and APMB for common sites on the transport system. However, no competition could be detected. In experiments with intact cells, using the technique of Cass and Dalmark (1973), the concentration range was extended to much higher sulfate concentrations. Under these conditions we found some 15-20 percent reduction of the inhibition by

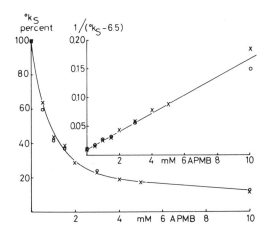

FIGURE 11.12. Effect of varying concentrations of external APMB on rate constant for $^{35}SO_4$ efflux from red-cell ghosts as measured at 15 mM (crosses) and 115 mM (circles). Na_2SO_4 in ghosts and medium. In addition to the sodium sulfate, 20 mM Na-phosphate, pH 7.4, was present in both experiments. If the sulfate concentration range is increased, a slight reduction of the inhibition by APMB can be demonstrated.

APMB.[4] The smallness of the effect argues against a participation of binding sites for APMB, which are common for this agent and H_2DIDS and DNFB, in SO_4 binding and translocation. Perhaps these sites act as modifier sites. The existence of such modifier sites has been inferred from work on concentration and pH dependence of anion transport (Schnell, 1975; Dalmark, 1975); and recent theories of Gunn (1972) and Dalmark (1975) even postulate that these sites are mobile like the transfer sites. The relationship between these modifier sites and those demonstrated to be involved in reaction with inhibitors is still unclear, but it is encouraging that similar thoughts are evolving from studies with and without inhibitors. Nevertheless the experiments on the competition between APMB and SO_4 are not conclusive enough to rule out reaction with a site that, after combination with APMB, blocks the access to the anion transport system without being a constituent of that system itself. Since APMB has common binding sites with H_2DIDS, it remains equally undecided whether or not the inhibition by H_2DIDS is due to reaction with

[4] The data and a more detailed discussion of this subject may be found in Passow, Pring, Legrum-Schuhmann, and Zaki (1977). In this paper it is shown that under certain specific assumptions about the mode of self-inhibition by sulfate, some competition between APMB and SO_4 for a transfer site cannot be ruled out.

those sites on the protein in band 3 that participate in the actual transfer of the anions across the red-cell membrane.

Conclusions

This chapter has assessed current attempts to elucidate the molecular mechanism of anion transport across the red-cell membrane. This assessment is of necessity confined to a few selected topics and illustrated primarily by experimental results obtained in our laboratory. With these limitations in mind, the following conclusions can be derived:

(1) Although the existence of facilitated diffusion can be easily demonstrated, it remains to be determined whether the two forms of anion transport—with and without contribution to the conductance of the membrane—proceed via two parallel and independent pathways or whether the two forms represent different aspects of the same process. It is also an open question whether the halides and sulfate penetrate via different pathways or share parts or all of a common pathway. (The following remarks deal with sulfate transport as measured under conditions where the contribution to the conductance is small.)

(2) Attempts to identify membrane constituents involved in sulfate transport by means of labeling with inhibitors like DNFB or isothiocyanate derivatives suggest that among the "major" membrane proteins the protein in band 3 is a likely candidate. The turnover numbers per binding site show, however, that not all of the labeled sites on this band are required to account for the observed transport rate. Thus, the observed effects on transport are not necessarily related to the observed binding but may be due to interaction with small numbers of binding sites within band 3 or on one of the so-called "minor" membrane proteins outside band 3. Enzymatic treatment of the red cells suggests that the protein in band 3 is not uniformly resistant to proteolysis and that at least two different peptides with H_2DIDS binding sites can be obtained. Attempts to correlate subfractions of H_2DIDS binding sites within band 3 with anion transport has not, as yet, been successful. Serious attempts to look for small numbers of binding sites outside band 3 have not yet been made.

(3) Studies of the possible function of the common binding sites for inhibitors like H_2DIDS, APMB, and DNFB show that they are unlikely to serve as binding sites for the anions to be transported. The agents could exert their effects by combination with membrane constituents that do or do not participate in anion translocation across the membrane. In either case, the binding sites are likely to be modifier sites rather than transfer sites. The experiments on the sidedness of action of APMB are compatible with the assumption that the common binding sites for this agent, DNFB, and H_2DIDS are mobile and possibly capable of crossing the membrane. An alternative explanation involving a fixed site that is buried

inside the membrane can be excluded. Regardless of the interpretation, the experiments on this subject provide the strongest evidence yet available for the presence of these binding sites on or near the anion transport system.

These conclusions show that anion transport is complex and far from being understood. Nevertheless, the kinetic and biochemical studies described in this chapter and in publications from other laboratories show that we are now able to deal with questions that could not be approached a few years ago. The rapid rate at which the subject has developed permits the hope that many of the points that are currently still in doubt will be clarified.

Appendix

Neglecting diffusion potentials and assuming a diffusion coefficient, D, for APMB of about 1 cm^2/day (for unrestricted diffusion in water), it is possible to estimate by means of Fick's law the concentration difference, Δc, that is required between the barrier containing the carrier system and the mouth of the invagination to obtain the observed efflux of APMB, j:

$$\Delta c = \frac{j \Delta x}{FD}$$

At 10 mM APMB inside, $j = 24$ mM/l ghosts/day; $F = nzA = 1.5 \times 10^5$ cm^2/l ghosts (n = number of H$_2$DIDS binding sites/cell [1.23 x 10^6); z = number of ghosts/l packed ghosts [10^{13}]; A = surface area of one channel [10 Å x 10 Å]; F = total diffusional area in 1 l of ghosts). Inserting the indicated numerical values yields 7.7 x 10^8 mM/l instead of 1-10 mM/l as needed to account for the observed inhibition or protection by internal APMB.

ACKNOWLEDGMENTS

We thank Dr. Martin Pring and Dr. Philip Wood for reading and commenting on the manuscript. The collaboration of Mss. Barbara Legrum-Schuhmann and Sigrid Lepke is gratefully acknowledged. We are indebted to Prof. H. Fasold for the supply of ^{3}H$_2$ DIDS and for much valuable advice.

REFERENCES

Bender, W. W., H. Garan, and H. C. Berg. 1971. Proteins of the human erythrocyte membrane as modified by pronase. *J. Mol. Biol.* 58: 783-797.
Brahm, J. 1977. Temperature-dependent changes of chloride transport kinetics in human red cells. *J. Gen. Physiol.* 70: 283-306.
Bretscher, M. S. 1971. Human erythrocyte membranes: specific labeling of surface proteins. *J. Mol. Biol.* 58: 775-781.

Cabantchik, Z. I., and A. Rothstein. 1972. The nature of the membrane sites controlling anion permeability of human red blood cells as determined by studies with disulfonic stilbene derivatives. *J. Membrane Biol.* 10: 311-330.

Cabantchik, Z. I., and A. Rothstein. 1974. Membrane proteins related to anion permeability of human red blood cells. I. Localization of disulfonic stilbene binding sites in proteins involved in permeation. *J. Membrane Biol.* 15: 207-226.

Cass, A., and M. Dalmark. 1973. Equilibrium dialysis of ions in nystatin-treated cells. *Nature New Biol.* 244: 47-49.

Cherry, R. J., A. Burkli, M. Busslinger, and G. Schneider. 1977. Rotational diffusion of proteins in membranes. FEBS Symposium on the Biochemistry of Membrane Transport, Zürich, July 18-23, 1976, pp. 86-95. Springer-Verlag, Berlin.

Dalmark, M. 1975. Chloride transport in human red cells. *J. Physiol.* (Lond.) 250: 39-64.

Deuticke, B. 1976. Stereospecific, SH-dependent lactate transfer in mammalian erythrocytes. FEBS Symposium on the Biochemistry of Membrane Transport, Zürich, July 18-23. Abstract No. P340.

Eigen, M., and R. Winkler. 1975. *Das ewige Spiel. Naturgesetze steuern den Zufall.* Piper, Munich.

Gunn, R. B. 1972. A titrable carrier model for both mono- and di-valent anion transport in human red blood cells. In M. Rørth and P. Astrup, eds., *Oxygen affinity of hemoglobin and red cell acid-base status*, pp. 823-827. Munksgaard, Copenhagen.

Gunn, R. B., M. Dalmark, D. C. Tosteson, and J. O. Wieth. 1973. Characteristics of chloride transport in human red blood cells. *J. Gen. Physiol.* 61: 185-206.

Harris, E. J., and B. C. Pressman. 1967. Obligate cation exchanges in red cells. *Nature* 216: 918-920.

Hunter, M. J. 1971. A quantitative estimate of the non-exchange-restricted chloride permeability of the human red cell. *J. Physiol.* 218: 49P-50P.

Jenkins, R. E., and M. J. A. Tanner. 1975. The major human erythrocyte membrane protein: evidence for an S-shaped structure which traverses the membrane twice and contains a duplicated set of sites. *Biochem. J.* 147: 393-399.

Juliano, R. L. 1973. The proteins of the erythrocyte membrane. *Biochim. Biophys. Acta* 300: 341-378.

Knauf, P. A., and A. Rothstein. 1971. Chemical modification of membranes. I. Effects of sulfhydryl and amino reactive reagents on anion and cation permeability of the human red blood cell. *J. Gen. Physiol.* 58: 190-210.

Läuger, P. 1972. Carrier-mediated ion transport. *Science* 178: 24-30.

Lepke, S., H. Fasold, M. Pring, and H. Passow. 1976. A study of the relationship between inhibition of anion exchange and binding to the red blood cell membrane of 4,4'-diisothiocyano stilbene-2,2'-disulfonic acid (DIDS) and its dihydroderivative (H_2DIDS). *J. Membrane Biol.* 29: 147-177.

Lepke, S., and H. Passow. 1973. Asymmetric inhibition by phlorizin of sulfate movements across the red blood cell membrane. *Biochim. Biophys. Acta* 298: 529-533.

Lepke, S., and H. Passow. 1976. The effect of incorporated trypsin on membrane proteins and anion transport in human red blood cell ghosts. *Biochim. Bio-*

phys. *Acta* 455: 353-370.

Obaid, A. L., A. F. Rega, and P. J. Garrahan. 1972. The effects of maleic anhydride on the ionic permeability of red cells. *J. Membrane Biol.* 9: 385-401.

Passow, H. 1969. Passive ion permeability of the erythrocyte membrane. *Progr. Biophys. Mol. Biol.* 19: 424-467. Pergamon, Oxford.

Passow, H. 1971. Effects of pronase on passive ion permeability of the human red blood cell. *J. Membrane Biol.* 6: 233-258.

Passow, H., H. Fasold, S. Lepke, M. Pring, and B. Schuhmann, 1977. Chemical and enzymatic modification of membrane proteins and anion transport in human red blood cells. In M. W. Miller and A. E. Shamoo, eds., *Membrane toxicity.* Proceedings of the Ninth Rochester International Conference on Environmental Toxicity, pp. 353-379. Plenum Press, New York.

Passow, H., H. Fasold, L. Zaki, B. Schuhmann, and S. Lepke. 1975. Membrane proteins and anion exchange in human erythrocytes. In G. Gardós and I. Szász, eds., *Biomembranes: structure and function.* Proceedings of the Ninth Meeting of the Federation of European Biochemical Societies, Budapest 1974, pp. 197-214.

Passow, H., M. Pring, B. Legrum-Schuhmann, and L. Zaki. 1977. FEBS Symposium on the Biochemistry of Membrane Transport. Zurich, July 18-23, 1976, pp. 306-315. Springer-Verlag, Berlin.

Passow, H., and K. F. Schnell. 1969. Chemical modifiers of passive ion permeability of the erythrocyte membrane. *Experientia* 25: 460-468.

Passow, H., and P. G. Wood. 1973. Current concepts of the mechanism of anion permeability. In B.A. Callingham, ed., *Drugs and transport processes,* pp. 149-171. Macmillan, London.

Rice, W. R., and T. L. Steck. 1976. Pyruvate flux into resealed ghosts from human erythrocytes. *Biochim. Biophys. Acta* 433: 39-53.

Rothstein, A., Z. I. Cabantchik, and P. Knauf. 1976. Mechanism of anion transport in red blood cells: role of membrane proteins. *Fed. Proc.* 35: 3-10.

Scarpa, A., A. Cecchetto, and G. F. Azzone. 1970. The mechanism of anion translocation and pH equilibration in erythrocytes. *Biochim. Biophys. Acta* 219: 179-188.

Schnell, K. F. 1975. *Untersuchungen zum Mechanismus des Sulfat-transportes durch die Erythrozytenmembran.* Habilitationsschrift, Regensburg.

Schnell, K. F., S. Gerhard, S. Lepke, and H. Passow. 1973. Asymmetric inhibition by phlorizin of halide movements across the red blood cell membrane. *Biochim. Biophys. Acta* 318: 474-477.

Steck, T. L. 1974. The organization of the proteins in the human red blood cell membrane. *J. Cell Biol.* 62: 1-19.

Wieth, J. O., M. Dalmark, R. B. Gunn, and D. C. Tosteson. 1973. The transfer of monovalent inorganic anions through the red cell membrane. In E. Gerlach, K. Moser, E. Deutsch, and W. Wilmanns, eds., *Erythrocytes, thrombocytes, leucocytes,* pp. 71-76. Georg Thieme Publishers, Stuttgart.

Zaki, L., H. Fasold, B. Schuhmann, and H. Passow. 1975. Chemical modification of membrane proteins in relation to inhibition of anion exchange in human red blood cells. *J. Cell. Physiol.* 86: 471-494.

PART THREE

Polar Faces in Epithelia

12 The Double-Membrane Model for Transepithelial Ion Transport: Are Homocellular and Transcellular Ion Transport Related?

STANLEY G. SCHULTZ

The membranes of epithelial cells are confronted with at least two distinct responsibilities. On the one hand they must maintain an intracellular ionic composition compatible with normal metabolic processes and one that is generally similar to that found in nonepithelial cells. And, at the same time, they must be capable of mediating specific and often quite varied absorptive and secretory functions. A question of central importance to our understanding of epithelial physiology is: Are these two membrane functions related? In other words, are the same mechanisms responsible for *homocellular* and *transcellular* ion transport?

In 1958, Koefoed-Johnsen and Ussing proposed a double-membrane (KJU) model for Na transport by isolated frog skin; the model provided a major conceptual framework for the design and interpretation of studies on a wide variety of epithelia (Ussing, 1960; Keynes, 1969). According to this now-classic model, which is illustrated in Figure 12.1, Na enters the epithelial cell across a barrier (*O.c.m.*) that is impermeable to K, and is subsequently extruded from the cell across a second barrer (*I.c.m.*) by means of an active Na-K exchange "pump" similar to that found in virtually all cells from higher animals. This second barrier was presumed to be impermeable to Na but permeable to K. Thus, in the steady state, K pumped into the cell in exchange for Na simply recycles across the second (inner or serosal) barrier. A central feature of this model is that the same mechanism is responsible for maintaining the high intracellular K and low intracellular Na characteristic of most animal cells, *as well as* for maintaining active transepithelial Na transport. By localizing the Na-K exchange pump to *one of only two* membranes, homocellular and transcellular Na transport were inextricably interwoven.

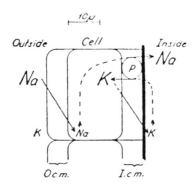

FIGURE 12.1. The double-membrane model for Na transport by isolated frog skin, proposed by Koefoed-Johnsen and Ussing (1958).

During the past two decades some features of the original KJU model have been confirmed and others have required some modification; the status of this central feature is, as yet, unresolved. The strongest evidence in support of this notion are the findings that the digitalis glycosides inhibit transcellular Na transport and concomitantly bring about a decrease in cell K and an increase in cell Na, and that these agents are effective only when present in the inner, or serosal, solution; the presence of these glycosides only in the outer, or mucosal, solution is relatively ineffective (Ussing, 1960; Keynes, 1969; Koefoed-Johnsen, 1957; Schultz and Zalusky, 1964; Herrera, 1966, 1968; Aceves and Erlij, 1971; Lew, 1970; Mills and Ernst, 1975; Bonting, 1970). In addition, the results of histochemical and autoradiographic studies, as well as analyses of enzyme activities of fractionated epithelial cells, have localized ouabain binding sites and Na,K-ATPase activities to the inner, or basolateral, membranes; little or no activity is found in the outer, mucosal membranes (Mills and Ernst, 1975; Bonting, 1970; Farquhar and Palade, 1966; Stirling, 1972; Schmidt and Dubach, 1971; Fujita et al., 1972; Kinne et al., 1975; Murer et al., 1974). Thus, the unilateral localization of Na-K pump activity, consistent with the KJU model, has been established for a number of Na-transporting epithelia beyond reasonable doubt. These findings constitute necessary but insufficient evidence for the notion illustrated in Figure 12.1.

Several predictions emerge explicitly from the marriage between *homocellular* and *transcellular* Na transport postulated by the KJU model. First, an increase in the rate of transcellular Na transport resulting *only* from an increase in the rate of Na entry into the cell should be associated with an increase in intracellular or cytoplasmic Na concentration (or activity), an increase in the rate of K uptake across the basolateral membrane via the pump mechanism, and, all other things remaining equal, an increase in the K content of the epithelial cell. Conversely, a decrease in the rate of transcellular Na transport resulting *only* from a decrease in the rate of Na entry should be associated with a decrease in the cytoplasmic Na activity, a decrease in the rate of K uptake across the basolateral membrane, and, all other things remaining equal, a decrease in the K content of the cell. Clearly, a change in the rate of transepithelial Na transport, for whatever reason, should be paralleled by a change in the rate of K exchange across the basolateral membranes.

In this chapter, we will briefly examine some of the experimental data bearing on these direct predictions of the KJU double-membrane model, focusing on data derived from studies on rabbit ileum and summarizing related findings on other epithelia.

Transepithelial Na Transport and the Intracellular Na Transport Pool

In 1963, Schultz and Zalusky reported that the addition of a number of D-hexoses or L-amino acids to the solution bathing the mucosal surface of rabbit ileum, in vitro, resulted in an abrupt increase in the short-circuit current across the tissue and the rate of active transepithelial Na transport from mucosa to serosa (Schultz and Zalusky, 1963, 1964). The data available at that time suggested the double-membrane model illustrated in Figure 12.2, in which the enhanced rate of transcellular Na transport was attributed to an increased influx of Na into an intracellular Na pool due to coupling between Na and the mechanisms responsible for sugar and amino acid influx. Na extrusion from the cell across the basolateral membranes was attributed to an energy-dependent, ouabain-sensitive, active transport process. Coupling between Na influx and the influxes of sugars and amino acids across the mucosal membranes was subsequently confirmed directly (Curran et al., 1967; Goldner, Schultz, and Curran, 1969; Schultz and Curran, 1970), and today there is no doubt that this mechanism is physiologically operant in small intestine and renal proximal tubules from a variety of species (Schultz and Curran, 1970; Schultz, 1977). It should be stressed that there is no reason to believe that the nonelectrolytes, many of which are not metabolized, exert any *direct* effect on the basolateral pump.

However, contrary to what one might expect, the intracellular Na and K contents of mucosal strips of rabbit ileum are not significantly affected by the presence of sugars or amino acids at concentrations sufficient to *double* the rate of transepithelial Na transport. Typical data obtained on segments of rabbit ileal epithelium stripped of the underlying muscle and connective tissue layers are given in Table 12.1 (Schultz, Fuisz, and Curran, 1966). We see that intracellular Na and K contents are not significantly affected in the presence of 10 mM L-alanine. There is an increase in cell water content entirely attributable to the accumulation of the osmotically active amino acid and, if anything, a secondary *decline* in cell Na and K concentrations. Similar findings have been reported for bullfrog small intestine (Armstrong, Musselman, and Reitzug, 1970).

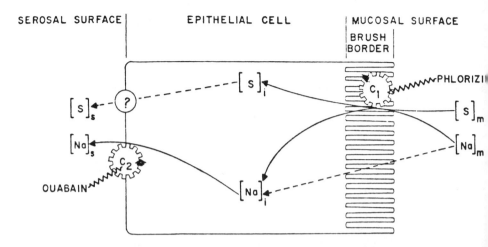

FIGURE 12.2. A double-membrane model for the coupled interaction between Na and sugar (S) transport by small intestine. (From Schultz and Zalusky, 1964.)

At first glance it might seem that these unexpected, negative findings could be easily explained if only a small number of cells in the mucosal strip were involved in Na transport stimulated by sugars or amino acids, so that an increase in the Na content of this population would be masked by Na contained within noninvolved cells. However, this explanation seems unlikely for the following reasons:

First, the intracellular alanine concentration (about 50 mM) was calculated employing the total cell water content, which was determined using inulin as an extracellular space marker; autoradiographic studies by Stirling (1972) suggest that inulin is a valid marker for the extracellular space in this tissue. If only 50 percent of the noninulin space volume of the mucosal strip were involved, the concentration of alanine in that population of cells would be approximately 100 mM; this would impose a rather severe osmotic burden on those cells. If even a smaller fraction of the total cell population is involved, the intracellular concentration of alanine would reach unreasonable levels.

Second, autoradiographic studies of [³H]-galactose uptake by hamster small intestine (Kinter and Wilson, 1965; Stirling and Kinter, 1967) clearly indicate that after only 3 minutes virtually all of the mature villus epithelial cells have accumulated this labeled sugar (Fig. 12.3). Similar results have been reported for amino acid uptake by hamster small intestine (Kinter and Wilson, 1965). Thus, it seems highly unlikely that only a small fraction of the cells in mucosal strips of rabbit and bullfrog small intestine are involved in sugar- and amino-acid-stimulated Na transport.

Third, Lee and Armstrong (1972) have examined the effect of 3-O-

TABLE 12.1. Cell water and solute contents. (From Schultz, Fuisz, and Curran, 1966.)

	Cell contents			Intracellular concentrations		
	Water (μg/mg dry weight)	Na (μmol/mg dry weight)	K	Na	K (mmol/liter H_2O)	L-alanine
Control	3.38 ± 0.09	0.209 ± 0.015	0.483 ± 0.024	62 ± 4	143 ± 6	—
L-alanine (5 mM)	3.78 ± 0.06	0.204 ± 0.012	0.465 ± 0.017	54 ± 3	123 ± 4	51 ± 1
P	<0.01	>0.5	>0.10	>0.10	<0.01	—

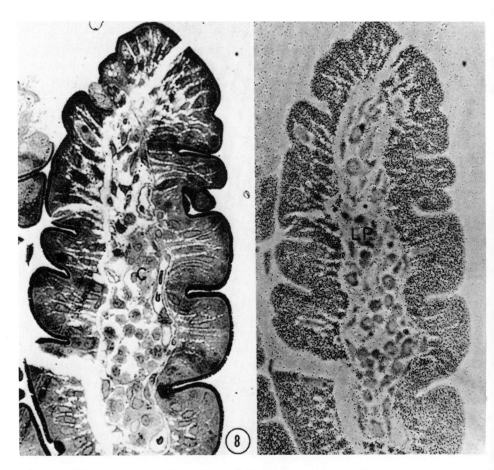

FIGURE 12.3. Autoradiograph of [³H] D-galactose accumulation by hamster small intestine (*right*). (From Kinter and Wilson, 1965.)

methylglucose on intracellular activities of Na and K in bullfrog small intestine, using cation-selective microelectrodes. In the absence of the sugar, the activity of intracellular Na was approximately 15 mM, or approximately 50 percent of the overall intracellular Na concentration (about 30 mM) determined chemically on mucosal strips using inulin to determine the extracellular space (Armstrong, Musselman, and Reitzug, 1970); this relation is similar to that found in a variety of nonepithelial cells (Lev and Armstrong, 1975). In the presence of 3-O-methylglucose, the intracellular Na activity *declined* to approximately 12 mM, in spite of the fact that transepithelial Na transport is markedly enhanced by this sugar (Quay and Armstrong, 1969). The percent decline in intracellular Na activity in the presence of 3-O-methylglucose closely parallels the percent increase in intracellular water content resulting from the accumulation of this osmotically active solute (Armstrong, Musselman, and Reitzug, 1970; Lee and Armstrong, 1972). These measurements on single cells would seem to exclude the possibility that the data reported in Table 12.1 are distorted by technical shortcomings or by the cellular heterogeneity of the mucosal strip.

Finally, the size of the intracellular Na transport pool in rabbit ileum can be roughly estimated from the following considerations. The addition of sugars or amino acids to a vigorously stirred solution bathing the mucosal surface of the rabbit ileum brings about increases in the transepithelial electrical potential difference (PD) and short-circuit current, which reach their new steady-state values within a few seconds. This increase in transepithelial PD is almost entirely attributable to an increase in the electromotive force across the basolateral membranes and is not observed in the presence of ouabain or metabolic inhibitors (Rose and Schultz, 1971). Thus, it appears that even during the rapid transient that follows the addition of the nonelectrolytes to the mucosal solution, the increase in transepithelial PD is due to rheogenic, active Na extrusion across the basolateral membranes (Schultz, 1973) and that this process achieves a new steady-state level within 10 seconds. From direct measurements of the increment in Na influx across the mucosal membranes coupled to the influxes of sugars or amino acids (Curran et al., 1967; Goldner, Schultz, and Curran, 1969; Frizzell, Nellans, and Schultz, 1973), it can be shown that the additional amount of Na that enters the cell within 10 seconds, an amount that is apparently sufficient to double the rate of active Na absorption, represents less than 1 percent of the intracellular Na and almost certainly less than 5 percent of the Na content of the villus cells that are directly involved in Na-coupled sugar and amino acid transport (Kinter and Wilson, 1965; Stirling and Kinter, 1967). These estimates are admittedly complicated by the errors involved in determining intracellular Na and water contents in epithelia having a complex geometry and by the fact that the cell layer includes several different cell types, some of which are not likely to be involved in Na-coupled sugar and amino acid transport. Nevertheless, even if these estimates are in error by several-fold, they are consistent with the repeated failures to observe an increment in cell Na content or concentration in small intestine when transepithelial Na transport is markedly enhanced.

In short, an increase in intracellular Na content or concentration associated with the marked increase in active Na transport across small intestine elicited by sugars or amino acids has not been detected to date. Although it is admittedly difficult to draw positive conclusions from negative findings, the observations cited above are difficult to reconcile with a simple double-membrane model of the intestinal cell, and the intracellular "signal" that regulates the activity of the basolateral Na pump is, at present, obscure.

The size, properties, and location of the Na transport pools in frog skin

(Cereijido et al., 1974) and toad urinary bladder (Macknight, Civan, and Leaf, 1975) have been extensively studied for a number of years, but the results, to date, do not lend themselves to a definitive conclusion. It seems certain that the Na transport pools represent but a small fraction of the Na contained within isolated epithelial cells. Macknight et al. (1975) have provided convincing evidence that only 20 percent of the Na contained within toad bladder epithelial cells is exchangeable with Na in the mucosal solution and that this pool is increased in the presence of vasopressin and decreased in the presence of amiloride. The remaining 80 percent of the Na in the epithelial cells is accessible only from the serosal solution and is not affected by vasopressin or amiloride. It is not clear whether the two pools simply reflect the properties of different cell populations or whether there are at least two discrete pools within single cells. If the former is true, the findings by Macknight et al. (1971) and Handler et al. (1972) that changes in the rates of active transepithelial Na transport brought about by vasopressin, aldosterone, or amiloride are paralleled by changes in the Na contents of isolated epithelial cells could be consistent with a double-membrane model and a close relation between homocellular and transcellular Na transport. On the other hand, Lipton and Edelman (1971) were unable to detect an increase in the Na content of isolated toad urinary bladder epithelial cells in the presence of aldosterone or vasopressin; the techniques employed by these investigators were similar to those employed by Macknight et al. (1971) and Handler et al. (1972) and these discrepant findings have not been reconciled. Further, Robinson and Macknight (1976a, b, c) have recently demonstrated that the size of the intracellular Na pool accessible from the mucosal solution is not affected when transepithelial Na transport is virtually abolished by rendering the serosal solution K-free. In addition, the size of this pool is the same under open-circuit and short-circuit conditions, in spite of the fact that the mucosa-to-serosa flux of Na under open-circuit conditions is less than half that observed when the tissue is short-circuited. It is possible that the reason why the size of the Na pool accessible from the mucosal solution does not parallel the rate of transepithelial Na transport under these conditions is that the rates of Na entry into the pool and extrusion from the pool are, somehow, coincidentally changed in the same direction and to the same extents. Clearly, the findings of Robinson and Macknight require definitive explanations; until then the precise relation between the size of the Na pool accessible from the mucosal solution and the rate of transepithelial Na transport will remain unclear.

Transepithelial Na Transport and K Exchange across the Basolateral Membranes

Recent studies aimed at directly examining the relation between transepithelial Na transport and K exchange across the basolateral membranes of rabbit ileum were prompted by three earlier observations. First, Schultz and Zalusky (1964) found that the rate of active Na transport by rabbit ileum is not affected when the tissue is bathed on both surfaces with nominally K-free media. However, the interpretation of this finding was complicated by the possibility that leakage of K from the cells into the subepithelial spaces might have resulted in a local K concentration within the villus cores sufficient to sustain a Na-K exchange across the basolateral membranes. Conclusive interpretations of the similar, "negative," findings reported by Edmonds and Marriot (1968) for rat colon and Diamond (1964) for rabbit gallbladder are precluded by the same possible complication.

Second, the intracellular K content of rabbit ileum (Schultz, Fuisz, and Curran, 1966) and bullfrog small intestine (Armstrong, Musselman, and Reitzug, 1970) is not increased when the rate of transepithelial Na transport is enhanced by the presence of sugars or amino acids (see Table 12.1). Such an increase would be expected if transepithelial Na transport were coupled to K uptake across the basolateral membranes, barring any coincidental changes in the permeability of the basolateral membrane to K or in the PD across that membrane.

Finally, Rose and Schultz (1971) demonstrated that the Na extrusion mechanism at the basolateral membranes is rheogenic and thus cannot involve a one-for-one exchange between cell Na and K in the serosal solution. However, it is now clear that the stoichiometry of the Na-K exchange pump in erythrocytes (Glynn and Karlish, 1975), nerve, and muscle (Thomas, 1972) is generally not one-for-one, so that the findings of Rose and Schultz (1971) do not exclude such an exchange.

Thus, while all of these observations taken together suggest that there may not be a direct relation between transepithelial Na transport and K exchange across the basolateral membranes of rabbit ileum, they are hardly conclusive. The results of studies (Nellans and Schultz, 1976) aimed at directly examining this question are briefly summarized in Table 12.2. These studies were performed on segments of rabbit ileum (Fig. 12.4) stripped of underlying muscle and connective tissue layers (Nellans, Frizzell, and Schultz, 1974). We see that under control conditions the unidirectional influx of K from the serosal solution into the epithelium across the basolateral membranes, J_{sc}^K, is approximately 2 μEquiv/cm^{2}hr; under these conditions the rate of transepithelial Na transport is approximately 3 μEquiv/cm^{2}hr. The addition of ouabain to the serosal solution abolishes Na transport, significantly inhibits J_{sc}^K, and brings about a marked decrease in cell K and an equivalent increase in cell Na with no change in cell water content. A similar decrease in J_{sc}^K is observed when intracellular Na is depleted by bathing the tissue by media in which Na has been completely replaced, isotonically, with choline. Finally, the results of other studies (not shown in the table) indicate that the maintenance of the high intracellular K is entirely dependent upon the presence of K in the serosal solution. Removal of K

TABLE 12.2. Unidirectional influxes of potassium across the basolateral membranes. (From Nellans and Schultz, 1976.)

Condition	$[Na]_c$ (mM)	$[K]_c$ (mM)	J_{sc}^K (μ Eq/cm^2 hr)
Ringer (control)	61 ± 1	125 ± 2	1.8 ± 0.1
Ringer + 10^{-3} M ouabain	122 ± 3	48 ± 3	0.6 ± 0.0
Na-free (choline) media	7 ± 0	126 ± 5	0.6 ± 0.1
$[Na]_m = 0$; $[Na]_s = 140$	24 ± 2	134 ± 7	1.5 ± 0.1
Ringer + 20 mM alanine	65 ± 2	125 ± 2	1.9 ± 0.2
Ringer + 10 mM glucose	62 ± 4	124 ± 6	1.7 ± 0.1

Note: $[Na]_c$ and $[K]_c$ are intracellular Na and K concentrations in mucosal strips of rabbit ileum, determined using inulin as a marker for the extracellular space. $[Na]_m$ and $[Na]_s$ designate the Na concentrations in the mucosal and serosal solutions respectively. The Na concentration in the Ringer solutions is 140 mM.

from the serosal bathing solution leads to a fall in cell K, an equivalent increase in cell Na, and no change in cell water content; removal of K from the mucosal solution alone has no effect on cell Na or K contents. These findings are consistent with the presence of a ouabain-sensitive, Na-K exchange mechanism at the basolateral membranes, which is responsible for the maintenance of the characteristic intracellular Na and K concentrations, and they are in complete accord with the results of other studies that have localized the sites of ouabain-binding and Na,K-ATPase activity of small intestine to the basolateral membranes (Stirling, 1972; Fujita, et al., 1972; Murer, et al., 1974).

However, as shown in Table 12.2, when the mucosal solution *alone* is rendered Na-free by isotonic replacement with choline, neither J_{sc}^K nor $[K]_c$ is affected in spite of the fact that transepithelial Na absorption is obviously precluded: under this condition cell water and K contents are also unaffected. Further, J_{sc}^K and $[K]_c$ are not affected by the presence of 10 mM D-glucose or 20 mM L-alanine in the mucosal solution in spite of the fact that under these conditions the rate of active transepithelial Na transport is increased approximately two-fold.

The time-course of K exchange across the basolateral membranes in the presence and absence of glucose or alanine is shown in Figure 12.5. We see that neither glucose nor alanine affects the rate of K exchange over a period of 45 minutes. Other studies have shown that only 40 percent of the total tissue K is readily exchangeable and that more than 80 percent of the readily exchangeable K equilibrates with ^{42}K in the serosal solution by 45 minutes. *Thus, a two-fold increase in the rate of transepithelial Na transport has no discernible effect on the rate at which ^{42}K in the serosal*

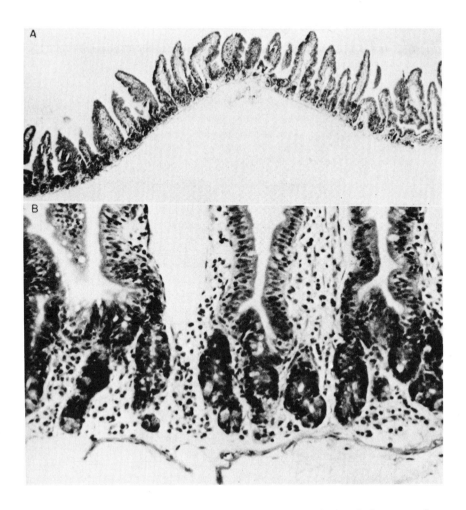

FIGURE 12.4. Photomicrographs of rabbit ileum stripped of underlying muscle and connective tissue. For the most part the mucosal strip consists of the epithelial cell layer, the lamina propria, and small discontinuous patches of muscularis mucosa. (From Nellans, Frizzell, and Schultz, 1974.)

solution exchanges with most of the readily exchangeable intracellular K. Once again, these negative findings could be reconciled with the KJU model if only a very small number of cells comprising the epithelial cell layer were involved in sugar and amino acid transport and the associated increase in transepithelial Na transport; but, as discussed above, this possibility seems remote.

Several efforts have been made to examine directly the relation between transcellular Na transport and K exchange across the inner, basolateral membranes of amphibian skin and urinary bladder; the results, to

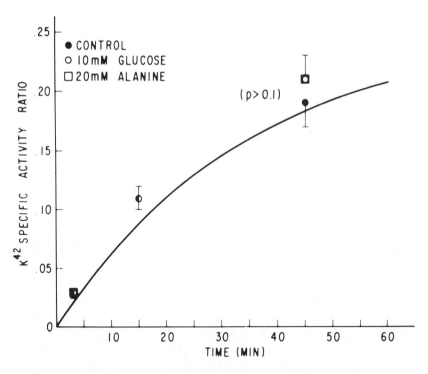

FIGURE 12.5. Kinetics of ^{42}K uptake from the serosal solution across the baso-lateral membranes of rabbit ileum under control conditions (•), in the presence of 20 mM L-alanine (□), or in the presence of 10 mM D-glucose (o). (From Nellans and Schultz, 1976.)

date, are largely negative, but at the same time are controversial and in-conclusive. For example, Curran and Cereijido (1965) and Candia and Zadunaisky (1972) could not detect a relation between transepithelial Na transport and K exchange across the inner surface of frog skin. The re-sults of Biber et al. (1972), on frog skin, are equivocal. These investiga-tors found a highly significant relation between the rate of active trans-cellular Na transport and K influx across the inner surface of the skin under control conditions and when Na absorption was stimulated by vasopressin; on the other hand, K influx was not affected when Na absorption was abolished by amiloride. However, the interpretation of all these studies is complicated by the heterogeneous nature of the epithe-lium. Isolated frog skin, even when stripped of the underlying corium, is comprised of multiple cell layers, whereas only the outermost living cell layer may be responsible for transepithelial Na transport (Ussing, Erlij, and Lassen, 1974). If so, K influx across the inner facing membranes may be markedly influenced by K pools in cells that are not involved in trans-epithelial Na transport; this complication could be particularly critical in studies involving only brief (60 second) exposure of the inner surface to ⁴²K (Biber, Aceves, and Mandel, 1972).

Perhaps the most compelling evidence against a close relation between transepithelial Na transport and the regulation of cell K composition stems from the recent studies of Macknight et al. (1971, 1975a, b, 1976a, b, c) on isolated epithelial cells from toad urinary bladder. These investigators found that only 20 percent of the intracellular K in the whole bladder is present in the epithelial cells and that only one-third of this amount is readily exchangeable with ⁴²K in the serosal solution; the remaining K in the epithelial cells exchanges so slowly that it could not possibly be linked to transepithelial Na transport. The rapidly exchangeable epithelial cell K pool is reduced by treatment of the tissue with ouabain but is not affected when transepithelial Na transport is abolished either by rendering the mucosal solution Na-free or by exposure to amiloride. The rate of K exchange with this pool is, on the average, 10 times slower than the rate of transepithelial Na transport and whereas the exchange rate was somewhat reduced in the presence of amiloride it was not affected by vasopressin at a concentration sufficient to double the rate of Na absorption. Finally, Robinson and Macknight (1976 a, b, c) have presented evidence that the inhibition of transepithelial Na transport resulting from removal of serosal K is due to a decline in intracellular K and not to the absence per se of the serosal K needed for a Na-K exchange. These observations raise the possibility that the inhibitory action of ouabain on transepithelial Na transport is an indirect effect due to the decline in cell K and is not the result of a direct interaction between this glycoside and the mechanism responsible for Na extrusion across the basolateral membrane.

The findings of Macknight et al. are in complete accord with the earlier conclusions of Essig and Leaf (1963; Essig, 1965) based on studies on whole toad bladder, but are difficult to reconcile with the findings of Finn and Nellans (1972). The latter investigators analyzed the kinetics of ⁴²K efflux from preloaded toad urinary bladders, and their analysis was consistent with a one-for-one relation between K exchange across the basolateral membranes and transepithelial Na transport. However, if the findings of Robinson and Macknight (1976a, b, c) are correct, the rapidly exchangeable K contained within the epithelial cells comprises less than 10 percent of the total K in the tissue, so that it is possible that the washout data obtained by Finn and Nellans (1972) were distorted by the behavior of K pools that were not contained within the epithelial cells responsible for transepithelial Na transport.

The Crisis of a Paradigm?

Few would question the assertion that, during the past two decades, the Koefoed-Johnsen-Ussing model established the central conceptual framework for the design and interpretation of studies on secretory and absorptive epithelia. It combines the simplicity, economy, and intellectual appeal that is often associated with "scientific truths." But, more important, it explicitly defined the direction of research, the results of which could support, challenge, prove, or disprove some of the basic tenets of the model; no model is useful unless it explicitly describes the experimental path to its "self-destruction." Kuhn has written: "In the absence of a paradigm or some candidate for a paradigm, all of the facts that could possibly pertain to the development of a given science are likely to seem equally relevant. As a result, early fact-gathering is a far more nearly random activity than the one that subsequent scientific development makes familiar" (1962). In this sense the KJU model represents the birth of a modern "epitheliology."

To date, some aspects of the original model have required modification and other aspects have been largely confirmed (Keynes, 1969; Ussing, Erlij, and Lassen, 1974). However, the central feature, namely the adequacy of a simple double-membrane representation of the epithelial cell, one which *inevitably and inextricably interweaves* homocellular and transcellular ion transport, has eluded direct, unequivocal confirmation or refutation. The major experimental problem stems from the fact that those Na-transporting epithelia that have been studied most extensively are complex, multicellular tissues. Even when techniques permit separation of the epithelial cells from underlying supportive tissues, the investigator is still faced with more than one cell type and considerable uncertainty as to which (what proportion) of these cells are engaged in transcellular active Na transport and how the observed Na and K pools are distributed within those cells.

Although small intestine, with its villous ramifications and multiple epithelial cell types, cannot be considered a "simple" epithelium, we are aided by the fact that autoradiographic studies have disclosed that *most* of the epithelial cells are capable of actively accumulating sugars and amino acids and are presumably responsible for the enhanced active Na absorption that is associated with the uptake of these nonelectrolytes. Further, results of studies with cation-selective microelectrodes suggest that the ionic composition of these cells does not differ from the "average" composition of the entire epithelium. Thus, unless these observations are deceptive, it is difficult to imagine that the Na and K contents of these cells comprise only a small fraction of the total Na and K contained in the stripped preparation, illustrated in Figure 12.4. If these judgments and line of reasoning are correct, the findings that (1) Na absorption is markedly increased in the absence of any significant increase in the Na content of the epithelium or the Na activity of single cells; and, (2) abolishing or stimulating Na transport has no effect on K influx across the basolateral membranes, or the K content of the epithelium, or the K activity of single cells, cannot be readily accommodated by a double-membrane model and a common mechanism for transcellular and homo-cellular ion transport. Instead, the observations on rabbit ileum and other epithelia suggest that, whereas a Na-K exchange pump located at the basolateral membranes is responsible for the maintenance of normal cell Na and K composition, Na that enters the cell across the mucosal membrane destined for transepithelial transport by-passes the bulk cyto-plasmic pool and may be extruded from the cell by a mechanism other than the Na-K exchange pump. If this Na transport pool is small, a change in the rate of transepithelial Na transport could be associated with changes in pool size that are undetectable using conventional tech-niques. A model incorporating these notions was proposed by Cereijido and Rotunno (1968) for isolated frog skin.

Currently, we seem to be confronted with a large body of evidence, derived from studies on a variety of epithelia, that questions the ade-quacy of a double-membrane representation of epithelial cells. Yet, such a representation continues to dominate much of our thinking, experi-mental design, and interpretations. Perhaps the reason is that the weight of confirmed and generally accepted contrary evidence is not entirely compelling and, as stated by Ussing et al. (1972): "Obviously, we do not yet possess the ideal epithelium with which to solve all the problems of Na transport and we must pool the results obtained from different sources to improve our image of the transport process." But one reason may be that given by Kuhn (1962): "Once it has achieved the status of a paradigm, a scientific theory is declared invalid only if an alternative candidate is available to take its place . . . The decision to reject one paradigm is always simultaneously the decision to accept another, and the judgment leading to that decision involves comparison of both para-digms with nature and with each other." To date, we have no attractive competitor of the "double-membrane" model that is equally amenable to direct experimental challenge. The recent findings by Voute et al. (1975), which suggest that intracellular membranous tubules, perhaps derived from the endoplasmic reticulum, may be involved in conveying Na across the "first reactive cell layer" of frog skin, might foreshadow such a competitor. These investigators found that when a hydrostatic pressure is applied to the solution bathing the inner surface of short-circuited frog skin, vacuoles (or "scalloped sacs") appear in the outermost living cell layer. The number of vacuoles is linearly related to the short-circuit cur-rent, and vacuoles are not observed if Na in the outer solution is replaced with choline. Finally, when lanthanum is added to the inner solution under these conditions, lanthanum deposits can be histologically demon-strated in the vacuoles; lanthanum added to the outer solution does not appear in the vacuoles (H. H. Ussing, personal communication). Thus, it is possible that the "scalloped sacs" observed under highly unphysiologic conditions are dilated cisternae of an intracellular tubular system that normally conveys Na through the transporting cell to the inner solution.

To be sure, the observations of Voute, Mollgard, and Ussing (1975) are preliminary, and the interpretation of these observations must be guarded. Nevertheless, they serve to focus our attention on the obvious fact that an epithelial cell is not simply a "bag of cytoplasm" contained by two limiting membranes that have different properties, and they may constitute the first clue to the existence of intracellular membranous systems involved in transepithelial transport.

REFERENCES

Aceves, J., and D. Erlij. 1971. Sodium transport across the isolated epithelium of the frog skin. *J. Physiol.* (Lond.) 212:195-210.

Armstrong, W. McD., D. L. Musselman, and H. C. Reitzug. 1970. Sodium, potassium and water content of isolated bullfrog small intestinal epithelium. *Am. J. Physiol.* 219:1023-1026.

Biber, T. U. L., J. Aceves, and L. J. Mandel. 1972. Potassium uptake across serosal surface of isolated frog skin epithelium. *Am. J. Physiol.* 222:1366-1373.

Bonting, S. L. 1970. Sodium-potassium activated adenosinetriphosphatase and cation transport. In E. E. Bittar, ed., *Membranes and ion transport.* Vol. 1, pp. 257-363. Wiley-Interscience, London.

Candia, O. A., and J. A. Zadunaisky. 1972. Potassium flux and sodium transport in the isolated frog skin. *Biochim. Biophys. Acta* 255:517-529.

Cereijido, M., C. A. Rabito, E. Rodriquez Boulan, and C. A. Rotunno. 1974. The sodium-transporting compartment of the epithelium of frog skin. *J. Physiol.* (Lond.) 237:555-571.

Cereijido, M., and C. A. Rotunno. 1968. Fluxes and distribution of sodium in frog skin: a new model. *J. Gen. Physiol.* 50:280s-289s.

Curran, P. F., and M. Cereijido. 1965. K fluxes in frog skin. *J. Gen. Physiol.* 48: 1011-1033.

Curran, P. F., S. G. Schultz, R. A. Chez, and R. E. Fuisz. 1967. Kinetic relations of the Na-amino acid interaction at the mucosal border of intestine. *J. Gen. Physiol.* 50:1261-1286.

Diamond, J. M. 1964. Transport of salt and water in rabbit and guinea pig gallbladder. *J. Gen. Physiol.* 48:1-14.

Edmonds, C. J., and J. Marriott. 1968. Factors influencing the electrical potential across the mucosa of rat colon. *J. Physiol.* (Lond.) 194:457-478.

Essig, A. 1965. Active sodium transport in toad bladder despite removal of serosal potassium. *Am. J. Physiol.* 208:401-406.

Essig, A., and A. Leaf. 1963. The role of potassium in active transport of sodium by the toad bladder. *J. Gen. Physiol.* 46:505-515.

Farquhar, M. G., and G. E. Palade. 1966. Adenosine triphosphatase localization in amphibian epidermis. *J. Cell Biol.* 30:359-379.

Finn, A. L., and H. Nellans. 1972. The kinetics and distribution of potassium in the toad bladder. *J. Membrane Biol.* 8:189-203.

Frizzell, R. A., H. N. Nellans, and S. G. Schultz. 1973. Effects of sugars and amino acids on sodium and potassium influx in rabbit ileum. *J. Clin. Invest.* 52:215-217.

Fujita, M., H. Ohta, K. Kawai, H. Matsui, and M. Nakao. 1972. Differential isolation of microvillous and baso-lateral plasma membranes from intestinal mucosa: mutually exclusive distribution of digestive enzymes and ouabain-sensitive ATPase. *Biochim. Biophys. Acta* 274:336-347.

Glynn, I. M., and S. J. D. Karlish. 1975. The sodium pump. *Annu. Rev. Physiol.* 37:13-55.

Goldner, A. M., S. G. Schultz, and P. F. Curran. 1969. Sodium and sugar fluxes across the mucosal border of rabbit ileum. *J. Gen. Physiol.* 53:362-383.

Handler, J. S., A. S. Preston, and J. Orloff. 1972. Effect of ADH, aldosterone, ouabain and amiloride on toad bladder epithelial cells. *Am. J. Physiol.* 222: 1071-1074.

Herrera, F. C. 1966. Action of ouabain on sodium transport in the toad urinary bladder. *Am. J. Physiol.* 210:980-986.

Herrera, F. C. 1968. Action of ouabain on bioelectric properties and ion content in toad urinary bladder. *Am. J. Physiol.* 215:183-189.

Keynes, R. D. 1969. From frog skin to sheep rumen: a survey of transport of salts and water across multicellular structures. *Q. Rev. Biophys.* 2:177-281.

Kinne, R., H. Murer, E. Kinne-Saffran, M. Thees, and G. Sachs. 1975. Sugar transport by renal plasma membrane vesicles. *J. Membrane Biol.* 21:375-395.

Kinter, W. B., and T. H. Wilson. 1965. Autoradiographic study of sugar and amino acid absorption by everted sacs of hamster intestine. *J. Cell Biol.* 25: 19-39.

Koefoed-Johnsen, V. 1957. The effect of g-strophanthin (ouabain) on the active transport of sodium through the isolated frog skin. *Acta Physiol. Scand.* 42 (Suppl. 145): 87-88.

Koefoed-Johnsen, V., and H. H. Ussing. 1958. The nature of the frog skin potential. *Acta Physiol. Scand.* 42:298-308.

Kuhn, T. S. 1962. *The structure of scientific revolutions.* University of Chicago Press, Chicago.

Lee, C. O., and W. McD. Armstrong. 1972. Activities of sodium and potassium ions in epithelial cells of small intestine. *Science* 175:1261-1264.

Lev, A. A., and W. McD. Armstrong. 1975. Ionic activities in cells. In F. Bronner and A. Kleinzeller, eds., *Current topics in membranes and transport.* Vol. 6, pp. 59-123. Academic Press, New York.

Lew, V. L. 1970. Short-circuit current and ionic fluxes in the isolated colonic mucosa of *Bufo arenarum. J. Physiol.* (Lond.) 206:509-528.

Lipton, P., and I. S. Edelman. 1971. Effects of aldosterone and vasopressin on electrolytes of toad bladder epithelial cells. *Am. J. Physiol.* 221:733-741.

Macknight, A. D. C., M. M. Civan, and A. Leaf. 1975a. The sodium transport pool in toad urinary bladder epithelial cells. *J. Membrane Biol.* 20:365-386.

Macknight, A. D. C., M. M. Civan, and A. Leaf. 1975b. Some effects of ouabain on cellular ions and water in epithelial cells of toad urinary bladder. *J. Membrane Biol.* 20:387-401.

Macknight, A. D. C., A. Leaf, and M. M. Civan. 1971. Effects of vasopressin on the water and ionic composition of toad bladder epithelial cells. *J. Membrane Biol.* 6:127-137.

Mills, J. W., and S. A. Ernst. 1975. Localization of sodium pump sites in frog urinary bladder. *Biochim. Biophys. Acta* 375:268-273.

Murer, H., U. Hopfer, E. Kinne-Saffran, and E. Kinne. 1974. Glucose transport in isolated brush-border and lateral basal plasma membrane vesicles from intestinal epithelial cells. *Biochim. Biophys. Acta* 345:170-179.

Nellans, H. N., R. A. Frizzell, and S. G. Schultz. 1974. Brush border processes and transepithelial Na and Cl transport by rabbit ileum. *Am. J. Physiol.* 226: 1131-1141.

Nellans, H. N., and S. G. Schultz. 1976. Relations among transepithelial sodium transport, potassium exchange and cell volume in rabbit ileum. *J. Gen. Physiol.*, in press.

Quay, J. F., and W. McD. Armstrong. 1969. Enhancement of net sodium transport in isolated bullfrog intestine by sugars and amino acids. *Proc. Soc. Exp. Biol. Med.* 131:46-51.

Robinson, B. A., and A. D. C. Macknight. 1976a. Relationships between serosal medium potassium concentration and sodium transport in toad urinary bladder. I. Effects of different medium potassium concentrations on epithelial cells composition. *J. Membrane Biol.* 26:217-238.

Robinson, B. A., and A. D. C. Macknight. 1976b. Relationships between serosal medium potassium concentration and sodium transport in toad urinary bladder. II. Effects of different medium potassium concentrations on epithelial cell composition. *J. Membrane Biol.* 26:239-268.

Robinson, B. A., and A. D. C. Macknight. 1976c. Relationships between serosal medium potassium concentration and sodium transport in toad urinary bladder. III. Exchangeability of epithelial cellular potassium. *J. Membrane Biol.* 26:269-286.

Rose, R. C., and S. G. Schultz. 1971. Studies on the electrical potential profile across rabbit ileum: effects of sugars and amino acids on transmural and transmucosal electrical potential differences. *J. Gen. Physiol.* 57:639-663.

Schmidt, U., and U. C. Dubach. 1971. Na-K stimulated adenosinetriphosphatase: intracellular localization within the proximal tubule of the rat nephron. *Pflugers Arch.* 330:265-270.

Schultz, S. G. 1973. Shunt pathway, sodium transport and the electrical potential profile across rabbit ileum. In H. H. Ussing and N. A. Thorn, eds., *Transport mechanisms in epithelia*, pp. 147-160. Munksgaard, Copenhagen.

Schultz, S. G. 1977. Ion-coupled transport across biological membranes. In T. Andreoli, D. Fanestil and J. F. Hoffman, eds., *The physiological basis for disorder of biomembranes*. Academic Press, New York, in press.

Schultz, S. G., and P. F. Curran. 1970. Coupled transport of sodium and organic solutes. *Physiol. Rev.* 50:637-718.

Schultz, S. G., P. F. Curran, R. A. Chez, and R. E. Fuisz. 1967. Alanine and sodium fluxes across the mucosal border of rabbit ileum. *J. Gen. Physiol.* 50:1241-1260.

Schultz, S. G., R. E. Fuisz, and P. F. Curran. 1966. Amino acid and sugar transport in rabbit ileum. *J. Gen. Physiol.* 49:849-866.

Schultz, S. G., and R. Zalusky. 1963. The interaction between active sodium transport and active sugar transport in the isolated rabbit ileum. *Biochim. Biophys. Acta* 71:503-505.

Schultz, S. G., and R. Zalusky. 1964. Ion transport in rabbit ileum. I. Short-circuit current and Na fluxes. *J. Gen. Physiol.* 47:567-584.

Schultz, S. G., and R. Zalusky. 1964. Ion transport in isolated rabbit ileum. II. The interaction between active sodium and active sugar transport. *J. Gen. Physiol.* 47:1043-1059.

Stirling, C. E. 1972. Radioautographic localization of sodium pump sites in rabbit intestine. *J. Cell Biol.* 53:704-714.

Stirling, C. E., and W. B. Kinter. 1967. High-resolution radioautography of galactose-^{3}H accumulation in rings of hamster intestine. *J. Cell Biol.* 35: 585-604.

Thomas, R. C. 1972. Electrogenic sodium pump in nerve and muscle cells. *Physiol. Rev.* 52:563-594.

Ussing, H. H. 1960. *The alkali metal ions in biology.* Springer-Verlag, Berlin.

Ussing, H. H., D. Erlij, and U. Lassen. 1974. Transport pathways in biological membranes. *Annu. Rev. Physiol.* 36:17-49.

Voute, C. L., K. Mollgard, and H. H. Ussing. 1975. Quantitative relationship between active sodium transport, expansion of endoplasmic reticulum and specialized vacuoles ("scalloped sacs") in the outermost living cell layer of the frog skin epithelium (Rana temporaria). *J. Membrane Biol.* 21:273-289.

13 Differentiation of Cell Faces in Epithelia

R. KINNE AND E. KINNE-SAFFRAN

Epithelial cells are capable of bringing about the net transfer of some solutes from one aqueous phase to another against, or in the absence of, external differences in electrochemical potentials. Such active transcellular or transepithelial transport processes require that the properties of the luminal and contraluminal membranes differ; that is, the properties of the two limiting membranes must exhibit functional polarities.

Suggestive evidence for such a polarity is derived from histologic studies that reveal differences in the morphologic appearances of the luminal and contraluminal membranes of some, but not all, epithelial cells. Recent studies on the sidedness and specificity of transport effectors (Brodsky, Ehrenspeck, and Kinne, 1976; Ullrich et al., 1976) and electrophysiological studies (Frömter, 1975) have suggested a general outline for approaches to the localization and characterization of steps involved in transepithelial solute transport.

However, in order to unequivocally establish the functional differences between the luminal and contraluminal membranes that bound the epithelial cell, these membranes must be isolated in pure form so that their properties related to solute transport and its humoral regulation can be examined separately.

Isolation of Luminal and Contraluminal Faces of the Plasma Membrane Envelope of Epithelial Cells

The epithelial cell is characterized by a polarity of the plasma membranes covering the apical and basal poles of the cell. This polarity is obvious from morphological, biochemical, histochemical, and immunohistochemical studies. Thus, in isolating plasma membranes from epithelia, the membranes not only have to be separated from other cellular membrane systems but, in addition, the luminal and contraluminal faces of the plasma membrane envelope have to be separated. This separation was recently achieved by applying the method of free-flow (or curtain) electrophoresis to resolve unfractionated plasma membranes of rat proximal tubule (Heidrich et al., 1972), bovine collecting duct (Schwartz et al., 1974), rat small intestine (Murer et al., 1974), and turtle urinary bladder (Brodsky, Ehrenspeck, and Kinne, 1976). An example of the technique is illustrated in Figure 13.1, where the distributions of the marker enzymes alkaline phosphatase and Na,K-ATPase and the morphological appearance of luminal and contraluminal plasma membrane fractions from rat proximal tubular epithelial cells are given. The Na,K-ATPase-containing basolateral plasma membranes migrate more rapidly towards the anode than the alkaline-phosphatase-rich brush-border microvilli, the different electrophoretic mobilities being determined by differences in surface charge density and by geometrical factors (size and shape of the membrane fragments) (Hannig et al., 1975). Similar distributions of luminal and contraluminal membranes and their respective marker enzymes have been obtained in the small intestine (alkaline phosphatase, Na,K-ATPase) (Murer et al., 1974), turtle urinary bladder (adenylate cyclase, Na,K-ATPase) (Brodsky, Ehrenspeck, and Kinne, 1976) and in the bovine collecting duct (HCO$_3$-ATPase, Ca-ATPase) (Schwartz et al., 1974). In the latter case, the luminal membranes showed the higher electrophoretic mobility, probably because the differences in size and shape were minimal and an "ideal" separation, according to the different surface charge densities, took place.

Besides the intrinsic markers of the two membranes (marker enzymes and morphology), noted above, extrinsic markers are also being used increasingly to label the membranes prior to their isolation. This approach was recently used in the intestine (Mircheff and Wright, 1976), the toad bladder (Ekblad, Strum, and Edelman, 1976), and the turtle bladder (Brodsky et al., 1976). Such extrinsic markers are especially important in instances where the morphology of the isolated plasma membrane fragments does not allow a distinction between luminal and contraluminal membranes, and where no marker enzymes for the different cell faces are known.

Asymmetrical Distribution of Transport Systems in the Epithelial Cells

ATPases

As mentioned above, Na,K-ATPase, which can be considered the biochemical equivalent of the process responsible for active sodium transport, is found only in the basolateral plasma membranes of the proximal

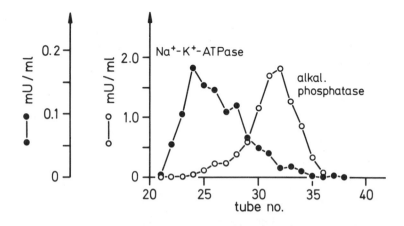

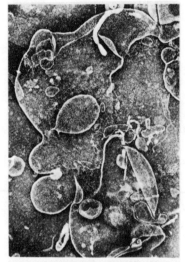

basal – lateral
plasma membranes

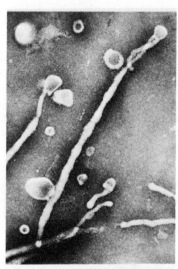

brush border
microvilli

FIGURE 13.1. Separation of luminal and contraluminal plasma membranes from rat proximal tubule by free-flow electrophoresis. Shown in the upper panel are the distributions of Na,K-ATPase (marker enzyme for basal-lateral plasma membranes) and of alkaline phosphatase (marker enzyme for brush-border microvilli) during separation of plasma membranes enriched from rat kidney cortex by differential centrifugation into luminal and contraluminal components. Shown in the lower panel are electron micrographs of the Na,K-ATPase-rich basal-lateral plasma membranes and of the alkaline-phosphatase-rich brush-border microvilli using the method of negative staining.

tubular cell; the same holds for the small intestinal epithelial cell and for the turtle bladder. This result, derived from studies with isolated membranes, agrees with the sidedness observed for the inhibition of active sodium transport in these tissues by ouabain. Ouabain, as a specific inhibitor of Na,K-ATPase, exerts its action from the contraluminal side (Györy, Brendel, and Kinne, 1972). Similarly, binding of tritiated ouabain (Stirling, 1972; Stirling and Landau, 1970) and potassium-dependent phosphatase activity (a partial reaction of Na,K-ATPase) (Ernst, 1975) is only observed at the basolateral cell face.

In analogy to the Na,K-ATPase, a Ca-ATPase and a HCO_3-ATPase have been hypothesized to be involved in active transepithelial transport of calcium and bicarbonate buffer, respectively. In the renal proximal tubular cell, the distribution of these enzyme activities would support such an assumption. As shown in Figure 13.2, Ca-ATPase is present in the basolateral plasma membranes (Kinne-Saffran and Kinne, 1974a), whereas the HCO_3-ATPase activity follows the distribution of alkaline phosphatase, suggesting a luminal localization of the HCO_3-ATPase (Kinne-Saffran and Kinne, 1974b). Thus, the ATPases would be present at those sides of the cell where an active step is required in the cellular uptake of bicarbonate (or ejection of protons) at the apical cell pole, and in the extrusion of calcium at the basal cell pole. The same is observed in the bovine collecting duct (Schwartz et al., 1974). In the small intestine the situation seems to be more complicated. Ca-ATPase activity is found both in luminal and contraluminal membrane fractions (Mircheff and Wright, 1976), whereas HCO_3-ATPase seems to be absent in plasma membranes isolated from small intestine and from flounder kidney (Kinne et al., 1976). It should be noted, in this context, that the elucidation of the cellular distribution of Ca-ATPase and HCO_3-ATPase is complicated by the fact that neither enzyme activity is very well characterized, and both are present in mitochondria as well as in plasma membranes. Furthermore, their role in transepithelial transport is difficult to evaluate because specific inhibitors of HCO_3-ATPase and Ca-ATPase are not available. There is still another difficulty, namely to reconcile the sodium-dependence of transepithelial proton transport (Ullrich, Rumrich, and Baumann, 1975) and calcium transport (Ullrich, Rumrich, and Klöss, 1976) with the insensitivity of the two enzymes to sodium. Therefore, at the moment no definite conclusions can be drawn on the role of the ATPases in proximal tubular calcium and proton transport.

Sodium Cotransport Systems

It is now generally accepted that sodium solute flux coupling is involved in the active transepithelial transport of sugars, amino acids, phosphate, and, probably, protons. The development of techniques to produce vesicles from isolated epithelial plasma membranes and to study their trans-

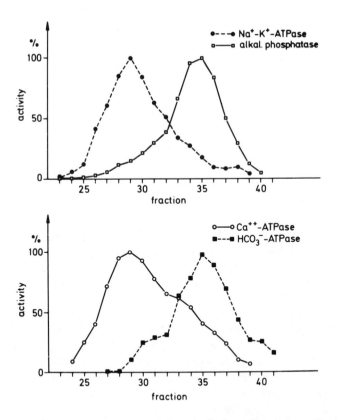

FIGURE 13.2. Distribution of Na,K-ATPase, alkaline phosphatase, Ca-ATPase, and HCO₃-ATPase after separation of luminal and contraluminal plasma membranes from renal proximal tubule. The values are given as percent of the maximal activity (mU/ml) found in the fractions. (Redrawn from Kinne-Saffran and Kinne, 1974a and b.)

port characteristics separated from the intracellular metabolism (Hopfer et al., 1973) has made it possible to investigate the cellular distribution, the driving forces, and the molecular characteristics of sodium cotransport systems.

As an example, the phosphate transport properties of renal plasma membrane vesicles (Hoffmann, Thees, and Kinne, 1976) will be considered in detail. Figure 13.3 shows the characteristics of phosphate uptake by isolated brush-border membrane vesicles and isolated basolateral plasma membrane vesicles. The uptake into brush-border membrane vesicles is stimulated by a sodium gradient across the membrane; this stimulation is only minimal in the basolateral plasma membrane vesicles. Furthermore, the uptake by brush-border membrane vesicles is inhibited by arsenate, indicating the presence of a specific, sodium-dependent phosphate transport system in the luminal but not in the contraluminal

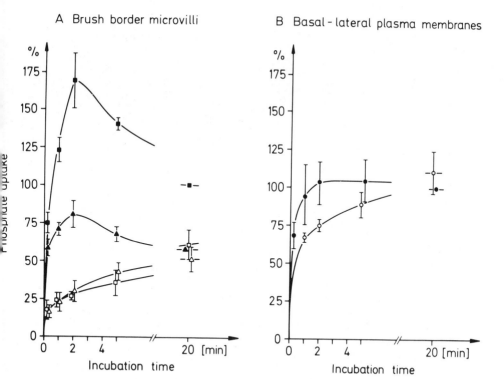

FIGURE 13.3. Uptake of inorganic phosphate by brush-border membrane vesicles (A) and by basal-lateral plasma membrane vesicles (B) isolated from rat renal cortex. The vesicles contained 100 mM mannitol and 20 mM Tris-HEPES (Tris = tris-(hydroxymethyl) aminomethane; HEPES = N-2-hydroxymethyl piperazine-N-2-ethanesulphonic acid), pH 7.4, and were incubated in a medium containing, in addition, 1 mM phosphate (^{32}P-labeled) and 100 mM NaCl (■———■, •———•), or 100 mM KCl (□———□, o———o). ▲———▲ represents the uptake in the presence of sodium and 5 mM arsenate, Δ———Δ the uptake in the presence of potassium and 5 mM arsenate. (Redrawn from Hoffmann et al., 1976.)

membranes. Such an asymmetric distribution of sodium-dependent transport systems has also been observed for D-glucose, neutral amino acids, and protons in kidney and intestine (Kinne et al., 1975a; Evers, Murer, and Kinne, 1976; Murer, Hopfer, and Kinne, 1976; Murer et al., 1974). Phosphate uptake by the brush-border membrane vesicles is characterized further by the presence of a transient intravesicular accumulation of phosphate (the amount of phosphate found in the vesicles transiently exceeds the amount present in the vesicles at equilibrium). As shown in Figure 13.4, this overshoot is not observed when the sodium gradient has been abolished by preloading the vesicles with sodium.

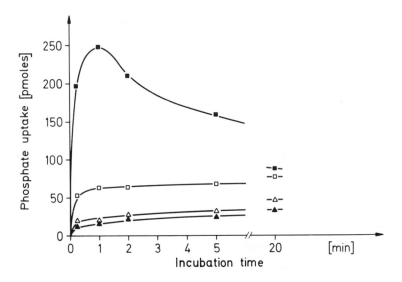

FIGURE 13.4. Effect of sodium and potassium under gradient and nongradient conditions on phosphate uptake by brush-border membrane vesicles isolated from rat kidney cortex. For studies of uptake under gradient conditions, the membranes were preincubated for 1 hour at 25°C in mannitol Tris-HEPES buffer, and the uptake was determined in media containing 100 mM mannitol, 20 mM Tris-HEPES, pH 7.4, and 100 mM NaSCN (■——■) or 100 mM KSCN (▲—— ▲), respectively. For nongradient conditions, the buffer used for the preincubation contained, in addition, 100 mM NaSCN (□——□) or 100 mM KSCN (Δ—— Δ). Intravesicular space, as determined by the amount of glucose present in the vesicles after 20 minutes, was identical under all four conditions.

Compared to the uptake in a sodium-free medium (potassium equilibrated), however, sodium still stimulates the uptake approximately four-fold. This finding can be taken as evidence that sodium interacts directly with the phosphate transport system and changes its kinetic properties. The sodium-gradient-dependent overshoot as well as the sodium-dependent stimulation of phosphate transport indicate that phosphate and sodium are transported via a cotransport system across the brush-border membrane. This assumption is further supported by the finding, shown in Figure 13.5, that a phosphate gradient transiently supports an accumulation of sodium in the brush-border vesicles.

In the isolated vesicles, the role of the membrane potential in the transepithelial transport can also be investigated. As shown in Table 13.1, the uptakes of D-glucose and phenylalanine are markedly stimulated by maneuvers that render the intravesicular space negative (Kinne et al., 1975a; Evers, Murer, and Kinne, 1976). Phosphate transfer, however, is influenced only slightly, if at all. This result suggests that the transfer of

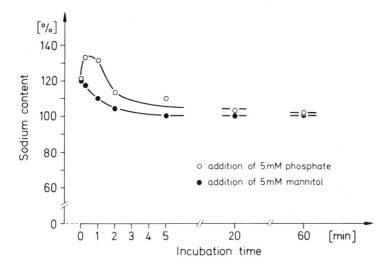

FIGURE 13.5. Phosphate-induced sodium accumulation in isolated renal brush-border membrane vesicles. The membranes were incubated for one hour at 25°C in a solution containing 100 mM mannitol, 5 mM Tris-HEPES, pH 7.4, 95 mM KSCN, and 5 mM ^{22}NaSCN. At zero time, 30 μl of a solution containing 5 mM Tris-HEPES, 95 mM KSCN, and 5 mM phosphate (o———o) or 5 mM mannitol (•———•) were added to 150 μl of the membrane suspension. The values are expressed as percent of the new equilibrium reached after 60 minutes incubation. (Redrawn from Hoffmann et al., 1976.)

sodium and phosphate via the sodium-phosphate cotransport system is electroneutral. Such studies can also provide information about the stoichiometry of the sodium-coupled solute transfer. Kinetic analysis of the sodium dependence of the phosphate transport by renal brush-border membrane vesicles has revealed, for example, that most probably two sodium ions interact with the cotransport system. Since phosphate uptake was also stimulated by increasing the pH of the incubation medium, the hypothesis was put forward that an electroneutral transfer of secondary phosphate with two sodium ions takes place in the renal brush border (Hoffmann, Thees, and Kinne, 1976). In the intestinal brush border, however, primary phosphate and one sodium ion seem to be transported together (Berner, Kinne, and Murer, 1976).

Sodium-Independent Transport Systems

Purified renal basolateral plasma membrane vesicles show sodium-independent uptake of sugars, amino acids, and phosphate (Kinne et al., 1975a; Evers, Murer, and Kinne, 1976; Hoffmann, Thees, and Kinne, 1976). In addition to the insensitivity to sodium, the sugar transport

TABLE 13.1. Influence of membrane potential on D-glucose, L-phenylalanine, and phospha
uptake by isolated renal brush-border membrane vesicles.

| Salt gradient (100 mM salt) | Membrane potential (relative to NaCl gradient) | Uptake[a] (in % of uptake in the presence of I | | |
		D-glucose	L-phenyla-lanine	Phosphate
NaSCN	Inside negative	189	181	112
NaCl	0	100	100	100
Na$_2$SO$_4$	Inside positive	56	60	87

[a]Uptake represents the amount of solute taken up during the first 15 seconds of the incubati
at 25°C.

system is characterized by a different substrate specificity and by a different sensitivity to inhibitors, as compared to the luminal sodium-dependent system (Kinne et al., 1975a). This clearly demonstrates that the sodium-independent systems are distinct molecular entities. Interestingly, highly purified brush-border membranes from kidney and intestine also show low, but significant, sodium independent uptake of glucose, with characteristics similar to the uptake observed with isolated basolateral plasma membranes (Kinne, Kleinzeller, and Murer, 1976; Hopfer, Sigrist-Nelson, and Murer, 1975). Thus, it is possible that the sodium-independent transport systems are not exclusively confined to the contraluminal membranes but are also present, though in much lower concentration, in the luminal membranes. This is in contrast to the sodium-dependent transport systems, which are present only in the luminal membranes.

In renal basolateral plasma membranes, a transport system for para-aminohippurate has also been demonstrated (Berner and Kinne, 1976); the system is inhibited competitively by probenecid and shows saturation kinetics and transstimulation. The system seems to be sodium independent and probably mediates the transport of para-aminohippurate as an anion across the membranes.

The Distribution of Transport Systems in Epithelial Cells

Figure 13.6 shows a synopsis of the cellular distribution of primary active transport systems (Na,K-ATPase, Ca-ATPase, HCO$_3$-ATPase), of sodium-solute cotransport systems (glucose, neutral amino acids, phosphate), and of sodium exchange systems (protons, calcium). With regard to the primary active transport systems, a transport function has only been demonstrated for the Na,K-ATPase; the involvement of Ca-ATPase and HCO$_3$-ATPase in transepithelial transport is still speculative, al-

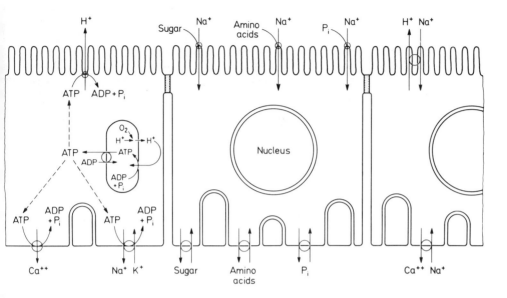

Lumen

Interstitium

FIGURE 13.6. Schematic representation of the localization of transport systems in the proximal tubule epithelial cell. Primary active transport systems are considered in the cell on the left-hand side; Na,K-ATPase is represented as the active Na-K exchange system, Ca-ATPase as the active calcium transport system, and bicarbonate ATPase as the active proton transport system. Sodium-cotransport systems and sodium-independent systems for sugars, amino acids, and phosphate are given in the middle of the figure; the cell on the right-hand side contains the sodium-proton exchange mechanism that has been found experimentally in brush-border membrane vesicles, and a hypothetical calcium-sodium exchange mechanism that has been postulated on the basis of the sodium dependence of the calcium reabsorption in the proximal tubules as revealed by micropuncture experiments and has been demonstrated recently in isolated membranes.

though their cellular localization favors involvement. It is not clear how HCO_3-ATPase and Ca-ATPase are related to sodium-proton exchange and sodium-calcium exchange. The existence of a sodium-calcium exchange system in the basolateral membranes has been postulated on the basis of the sodium dependence of transepithelial Ca-transport in the proximal tubule (Ullrich, Rumrich, and Klöss, 1976) and has been demonstrated recently.

For the sodium-dependent transport systems of glucose, amino acids, and phosphate, the luminal localization enables the cell to use the driving force of the sodium gradient across the brush border to accumulate these solutes inside the cell. The presence of sodium-independent solute transport systems at the contraluminal cell pole facilitates the exit of the solutes from the cell, driven by their own concentration gradient and, if charged, by the membrane potential. Thus, the sodium gradient can exert a driving force for secondary active transport of solutes across the epithelium without a direct involvement of the cellular metabolism. It should be noted that the sodium gradient may be used not only for transepithelial solute transport but also for the uptake of substrates into the cell; thus, the cellular uptake of glutamate and aspartate across the contraluminal membranes seems to be sodium dependent.

Asymmetrical Distribution of Hormone-Sensitive Adenylate Cyclases and cAMP-Dependent Protein Kinases in the Epithelial Cells

Adenylate Cyclase

Peptide and amine hormones are known to affect the transport properties of various epithelia. Thus, for example, parathyroid hormone inhibits phosphate reabsorption by the proximal tubule (Gekle, 1971; Agus et al., 1971; Agus et al., 1973), antidiuretic hormone increases the water permeability of the renal collecting duct (Grantham and Burg, 1966) and norepinephrine stimulates active chloride and bicarbonate transport across the turtle urinary bladder (Brodsky, Schilb, and Parker, 1976). It has been postulated that all three hormones exert their action on their target cell by stimulation of adenylate cyclase, which initiates the chain of events that lead to the final effector process.

The cellular distribution of the PTH-sensitive adenylate cyclase in the proximal tubule and of the ADH-sensitive adenylate cyclase in the bovine collecting duct is shown in Figure 13.7. Adenylate cyclase activity is found, almost exclusively, in the membranes derived from the basal cell pole (Shlatz et al., 1975; Schwartz et al., 1974) (as indicated by the parallel distribution with Na,K-ATPase and Ca-ATPase, respectively), and it is absent from the luminal membranes of the two epithelia. In contrast, in the turtle bladder the membrane fractions with a high Na,K-ATPase activity—presumably representing basolateral membranes of the epithelial cells—show no norepinephrine-sensitive adenylate cyclase activity (Brodsky et al., 1976) (Fig. 13.8).

cAMP-Dependent, Membrane-Bound Protein Kinase Activity

Since, in the proximal tubule (Agus et al., 1971; Baumann et al., 1976) and in the collecting duct (Grantham and Burg, 1966), the action of the

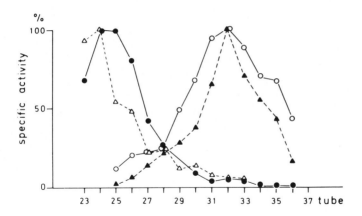

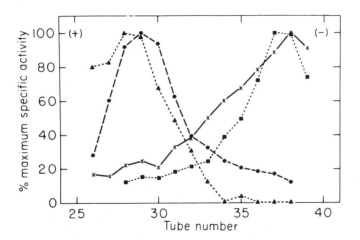

FIGURE 13.7. Distribution of adenylate cyclase and protein kinase activity during separation of luminal and contraluminal plasma membranes by free flow electrophoresis. The upper panel shows results obtained with membranes from the proximal tubule. The distribution of Na,K-ATPase (• ——— •) (marker enzyme for basolateral plasma membranes), PTH-stimulated adenylate cyclase (Δ ——— Δ), alkaline phosphatase (o ——— o) (marker enzyme for brush-border microvilli), and protein kinase activity (▲ ——— ▲) is given. In the lower panel, results obtained with membranes from bovine papillary collecting duct are demonstrated. The distribution of Ca-ATPase (■ ——— ■) (marker enzyme for contraluminal membranes), ADH-stimulated adenylate cyclase (x ——— x), HCO_3-ATPase (▲ · · ▲) (marker enzyme for luminal membranes), and protein kinase activity (• ——— •) of the membranes is given. (Redrawn from Kinne et al., 1975b, Shlatz et al., 1975, and Schwartz et al., 1974.)

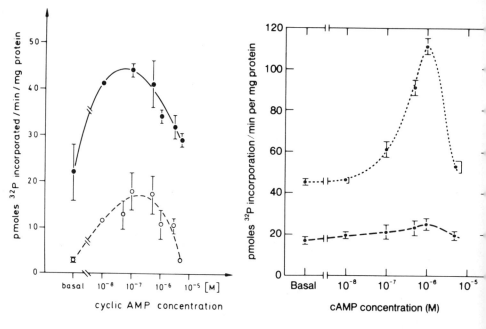

FIGURE 13.8. Effect of cAMP on the protein kinase activity of luminal (•———•) and contraluminal (o ——— o) plasma membranes isolated from rat proximal tubule (*left*) and of luminal (• · · · •) and contraluminal (• ——— •) plasma membranes isolated from collecting duct (*right*). (Redrawn from Kinne et al., 1975b, and Schwartz et al., 1974.)

peptide hormone can be mimicked by cAMP, it seems reasonable to assume that cAMP-dependent systems in the plasma membrane envelope may be responsible for the alteration of the transport properties of the membranes in response to the peptide hormones. In isolated membranes of epithelial cells, a cAMP-dependent self-phosphorylation system has been recently demonstrated that catalyzes the transfer of ^{32}P from γ-labeled ATP to membrane components. Interestingly, these systems do not seem to be distributed uniformly within the plasma membrane but are concentrated in those membranes where, as suggested by physiological observations, the terminal event of the hormone action takes place. Thus, as shown in Figure 13.9, the luminal membranes of the renal proximal tubule (Kinne et al., 1975b) and the bovine collecting duct (Schwartz et al., 1974) show the highest degree of phosphorylation in the presence of optimal concentrations of cAMP. Similarly, in the turtle bladder the cAMP-induced increment of phosphorylation is confined to the Na,K-ATPase-poor, probably luminal, membranes of the epithelial cells (Davidson et al., 1976) (Fig. 13.8).

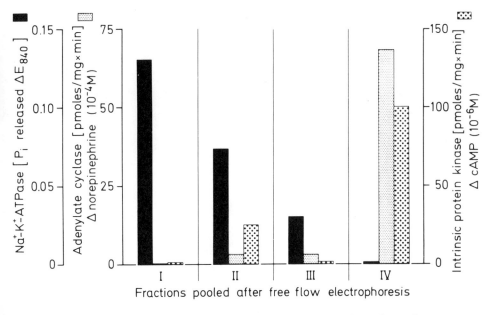

FIGURE 13.9. Distribution of Na,K-ATPase, norepinephrine-dependent adenylate cyclase, and cAMP-dependent intrinsic protein kinase activity in four fractions obtained after free-flow electrophoresis of turtle urinary bladder microsomes (Davidson et al., 1976, Brodsky et al., 1976).

The Distribution of Adenylate Cyclase and Membrane-Bound cAMP-Dependent Protein Kinase in Epithelial Cells

Figure 13.10 shows that two types of epithelia may be distinguished by their cellular distributions of adenylate cyclase and membrane-bound kinase. One type, represented by the proximal tubule and the collecting duct, is characterized by a spatial separation of the site of cAMP generation and the site where cAMP seems to exert its final action. In the other type, represented by the turtle bladder, adenylate cyclase and membrane self-phosphorylation are present in the same membrane. One might speculate that in the former type of epithelium, cAMP not only exerts its action on the membranes but also influences intracellularly located systems during its passage through the cytoplasm.

Figure 13.10 also depicts the interaction of the luminal membrane component that is phosphorylated by the self-phosphorylation system with a phosphoproteinphosphatase. In the proximal tubule the luminal membrane contains both the protein kinase and the protein phosphatase (Kinne et al., 1975b); thus, membrane-bound kinase, membrane-bound phosphoproteinphosphatase, and their substrates seem to exist in a delicate arrangement that enables the cell to alter, selectively and reversibly, the

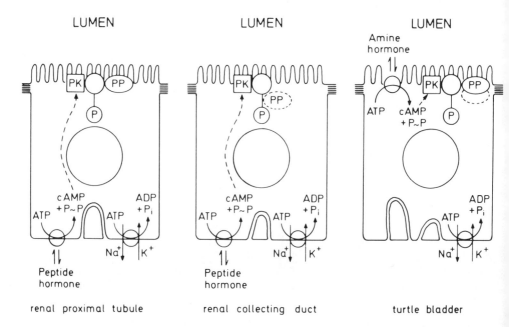

FIGURE 13.10. Schematic representation of the cellular distribution of peptide (or amine) hormone-sensitive adenylate cyclase and membrane-bound cAMP-dependent protein kinase (*PK*) in renal proximal tubule, renal collecting duct, and turtle urinary bladder. In addition, the interaction of membrane-bound or cytosolic phosphoprotein phosphatase (*PP*) with the phosphorylated membrane component (circle attached to *P*) is shown.

transport properties of the membrane. In the luminal membranes of the collecting duct no phosphoprotein phosphatase activity could be detected (Schwartz et al., 1974). Therefore, in order for a phosphorylation-dephosphorylation cycle to be completed, an interaction of a cytosolic (soluble) phosphatase with the membranes must be postulated. A soluble phosphatase that acts on phosphorylated plasma membranes has been described in bovine renal collecting-duct epithelial cells (Doŭsa, Sands, and Hechter, 1972). In the turtle bladder it is not clear whether the membranes contain phosphoproteinphosphatase activity, therefore both possibilities are indicated in Figure 13.10.

Asymmetrical Distribution of Antigens in Epithelial Cells As Revealed by Immunohistochemistry

Recently, immunohistochemistry at the electronmicroscopic level of resolution has been used to localize Na,K-ATPase in the cells of the proximal and distal tubule (Kyte, 1976a, b). This method seems to be very promising for elucidating the localization of transport systems, especially in those cells where the isolation of luminal and contraluminal membranes is not possible, because of technical difficulties or the lack of appropriate markers. However, one of the prerequisites of the method is that highly purified antigens should be available in order to produce highly specific antibodies. Another difficulty of the method is that all membrane surfaces have to be exposed to the antibody solution; this can be achieved by using ultrathin cryostat sections of the tissue or tangential sections of tissue blocks.

In our laboratory, the localization of aminopeptidase M—an enzyme that, from studies on isolated membranes (Pockrandt-Hemstedt et al., 1972), is thought to be located exclusively at the surface of the brush border microvilli—has been studied using immunohistochemical techniques (Pockrandt-Hemstedt et al., 1976). Further studies are in progress on the localization of the Na,K-ATPase and the phlorhizin-sensitive glucose-binding protein that can be enriched from rat kidney (Thomas, 1973).

Figure 13.11 gives an example of the interaction of peroxidase-conjugated antibodies, raised against highly purified rat kidney aminopeptidase M, with the cell membranes of the proximal tubule. As indicated by the electron-dense product of the peroxidase reaction, antigen-antibody complexes are observed exclusively at the outside of the luminal membranes; the basolateral membranes show no reaction with the antibody. Assuming that all aminopeptidase present in the membranes was accessible to the antibodies—the intracellular reaction with some membrane systems (Golgi vesicles?) suggests that indeed all membrane surfaces have been exposed during the experiment—one might conclude that aminopeptidase M is exclusively localized in the brush-border membranes.

As depicted in Figure 13.11c, the tight junctions seem to prevent an intermingling of the different components of the luminal and contraluminal membranes; a strong antigen-antibody reaction is observed at the luminal membranes (above the tight junctions) but none at the basal-lateral plasma membranes (below the tight junctions).

The Differentiation of the Luminal and Contraluminal Cell Face in the Renal Proximal Tubule

The cellular distribution of the membrane-bound enzymes and transport systems discussed in this contribution are depicted schematically in Figure 13.12. Exclusive luminal localization is assumed for aminopeptidase M, alkaline phosphatase, HCO_3,-ATPase, and the sodium-dependent transport systems for sugars, neutral amino acids, phosphate, and protons on the basis of the immunohistochemical results and the results obtained with isolated luminal membranes. Similarly Na,K-ATPase, Ca-ATPase, and adenylate cyclase are assumed to be localized exclusively in

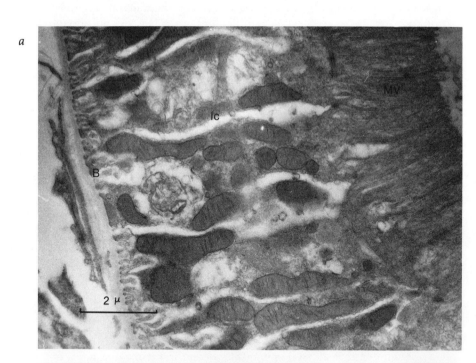

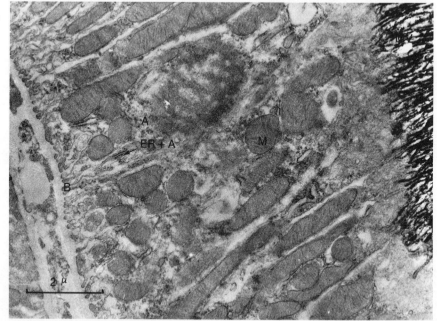

c

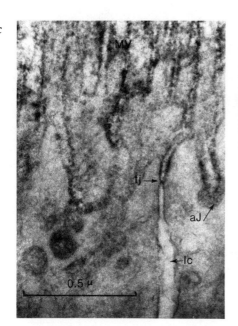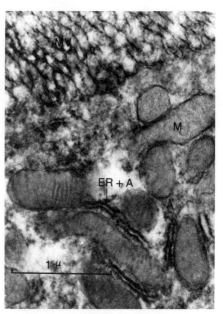

FIGURE 13.11. Distribution of aminopeptidase M in rat kidney proximal tubule epithelial cells, as revealed by antigen-antibody reaction of the tissue with peroxidase-conjugated y-globulins raised in rabbits against aminopeptidase M isolated from rat kidney. (*a*) Peroxidase reaction in tissue incubated with peroxidase-conjugated y-globulins isolated from nonimmunized rabbits (control). (*b*) Peroxidase reaction in tissue incubated with peroxidase-conjugated y-globulins isolated from rabbits immunized with isolated aminopeptidase. (*c*) Higher magnification of the tight junction region and the mitochondrial region of the cell shown in *b* (Pockrandt-Hemstedt et al., 1976). Abbreviations: *A*, antibodies; *B*, basal infoldings; *ER*, endoplasmic reticulum; *M*, mitochondria; *MV*, microvilli; *lc*, lateral cell border; *tj*, tight junction.

the basolateral plasma membranes. Predominant contraluminal localization is proposed for the sodium-independent transport systems for sugars (and probably also for neutral amino acids) and predominant luminal localization for Mg-ATPase and the cAMP-dependent self-phosphorylation system. However, for the latter two systems, it is not clear whether the enzyme activities, found in the luminal as well as in the contraluminal membrane fractions, represent only one or several enzyme entities. Thus it is possible that several systems are present that cannot as yet be distinguished and whose intracellular distribution, therefore, cannot be determined with certainty.

In Figure 13.12, we also attempt to distinguish between enzyme activities and transport systems found in nonpolarized cells (erythrocytes, fat cells, tumor cells) as compared to the polarized, epithelial cell. It is inter-

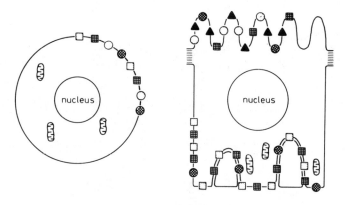

systems present in polarized and non polarized cells luminal localization		systems present exclusively in luminal membranes of polarized cells	systems present in polarized and non polarized cells contraluminal localization	
○ exclusive	◉ predominant	▲	◻ exclusive	▦ predominant
alkaline phosphatase aminopeptidase M			Na^+-K^+-ATPase	
HCO_3^--ATPase	Mg^{++}-ATPase		Ca^{++}-ATPase	
cAMP-dependent protein kinase			adenylate cyclase	
sodium-dependent transport systems for neutral amino acids, phosphate and protons		sodium-dependent transport system for sugars		cation-independent transport systems for sugars

FIGURE 13.12. Schematic representation of the distribution of enzymes and transport systems in a nonpolarized cell and in a polarized epithelial cell. A distinction is attempted between exclusive and predominant luminal or contraluminal localization and between systems that are present both in the nonpolarized and the polarized cell and those that are present exclusively in epithelial cells.

esting to note that almost all systems that are distributed unevenly in the plasma membrane envelope of the epithelial cell are also found in the plasma membranes of the nonpolarized cells. One exception is the sodium-dependent transport system for monosaccharides, which seems to be specific for the luminal membranes of the polarized cell. In view of this fact, one might consider the luminal membrane as the most differentiated part of the plasma membrane envelope of the epithelial cell.

REFERENCES

Agus, Z. S., L. B. Gardner, L. H. Beck, and M. Goldberg. 1973. Effects of parathyroid hormone on renal tubular reabsorption of calcium, sodium and phosphate. *Amer. J. Physiol.* 224: 1143-1148.

Agus, Z. S., J. B. Puschett, D. Senesky, and M. Goldberg. 1971. Mode of action of parathyroid hormone and cyclic adenosine 3', 5'-monophosphate on renal tubular phosphate reabsorption in the dog. *J. Clin. Invest.* 50: 617-626.

Baumann, K., Y. L. Chan, F. Bode, and F. Papavassiliou. 1976. Effect of PTH and cyclic AMP on isotonic fluid reabsorption: polarity of proximal cells. *Kidney Internat.*, submitted for publication.

Berner, W., and R. Kinne. 1976. Transport of p-aminohippuric acid by plasma membrane vesicles isolated from rat kidney cortex. *Pflügers Arch.* 361: 269-277.

Berner, W., R. Kinne, and H. Murer. 1976. Phosphate transport into brush border membrane vesicles isolated from rat small intestine. *Biochem. J.*, in press.

Brodsky, W. A., J. Cabantchik, G. Ehrenspeck, E. Kinne-Saffran, and R. Kinne. 1976. In preparation.

Brodsky, W. A., G. Ehrenspeck, and R. Kinne. 1976. Anion- and cation-selective paths in turtle bladder membranes. *Fed. Proc.* 35: 702.

Brodsky, W. A., Th. P. Schilb, and J. L. Parker. 1976. Moderators of anion transport in the isolated turtle bladder. In D. Kasbekar, G. Sachs, and W. Rehm, eds., *Gastric hydrogen ion secretion*. Marcel Deckker, Inc., New York, in press.

Davidson, N., E. Ehrenspeck, R. Kinne, and W. A. Brodsky. 1976. In preparation.

Doŭsa, T. P., H. Sands, and O. Hechter. 1972. Cyclic AMP-dependent reversible phosphorylation of renal medullary plasma membrane protein. *Endocrinology* 91: 757-763.

Ekblad, E. B., J. M. Strum, and I. S. Edelman. 1976. Differential covalent labeling of apical and basal-lateral membranes of the epithelium of toad bladder. *J. Membrane Biol.* 26: 301-317.

Ernst, S. A. 1975. Transport ATPase cytochemistry: ultrastructural localization of potassium-dependent and potassium-independent phosphatase activities in rat kidney cortex. *J. Cell Biol.* 66: 586-608.

Evers, J., H. Murer, and R. Kinne. 1976. Phenylalanine uptake in isolated renal brush border vesicles. *Biochim. Biophys. Acta* 426: 598-615.

Frömter, E. 1975. Ion transport across renal proximal tubule: analysis of luminal, contraluminal and paracellular transport steps. In A. Wessing, ed., *Excretion*, pp. 248-260. Gustav Fischer Verlag, Stuttgart.

Gekle, D. 1971. Der Einfluss von Parathormon auf die Nierenfunktion. 1. Tierexperimentelle Untersuchungen. *Pflügers Arch.* 323: 96-120.

Grantham, J. J., and M. B. Burg. 1966. Effect of vasopressin and cyclic AMP on permeability of isolated collecting tubules. *Amer. J. Physiol.* 211: 255-259.

Györy, A. Z., U. Brendel, and R. Kinne. 1972. Effect of cardiac glycosides and sodium ethacrynate on transepithelial sodium transport in *in vitro* micropuncture experiments and on isolated plasma membrane Na-K-ATPase *in vitro* of the rat. *Pflügers Arch.* 335: 287-296.

Hannig, K., H. Wirth, B.-H. Meyer, and K. Zeiller. 1975. Free-flow electrophoresis I. Theoretical and experimental investigations of the influence of mechanical and electrokinetic variables on the efficiency of the method. *Hoppe Seyler's Z. Physiol. Chem.* 356: 1209-1223.

Heidrich, H. G., R. Kinne, E. Kinne-Saffran, and K. Hannig. 1972. The polarity of the proximal tubule cell in rat kidney: different surface charges for the brush-border microvilli and plasma membranes from basal infoldings. *J. Cell Biol.* 54: 232-245.

Hoffmann, N., M. Thees, and R. Kinne. 1976. Phosphate transport by isolated renal brush border vesicles. *Pflügers Arch.* 362: 147-156.

Hopfer, U., K. Nelson, J. Perrotto, and K. J. Isselbacher. 1973. Glucose transport in isolated brush border membane from rat small intestine. *J. Biol. Chem.* 248: 25-32.

Hopfer, U., K. Sigrist-Nelson, and H. Murer. 1975. Intestinal sugar transport: studies with isolated plasma membranes. *Ann. N.Y. Acad. Sci.* 264: 414-427.

Kinne, R. 1976. Membrane-molecular aspects of tubular transport. In K. Thurau, ed., *International review of physiology*, vol. 2, pp. 145-210. Butterworths, London; University Park Press, Baltimore.

Kinne, R., J. Eveloff, K. Karnaky, and W. B. Kinter. 1976. In preparation.

Kinne, R., A. Kleinzeller, and H. Murer. 1976. Unpublished observations.

Kinne, R., H. Murer, E. Kinne-Saffran, M. Thees, and G. Sachs. 1975a. Sugar transport by renal plasma membrane vesicles: characterization of the systems in the brush-border microvilli and basal-lateral plasma membranes. *J. Membrane Biol.* 21:375-395.

Kinne, R., L. J. Shlatz, E. Kinne-Saffran, and I. L. Schwartz. 1975b. Distribution of membrane-bound cyclic AMP-dependent protein kinase in plasma membranes of cells of the kidney cortex. *J. Membrane Biol.* 24: 145-159.

Kinne-Saffran, E., and R. Kinne. 1974a. Localization of a calcium stimulated ATPase in the basal-lateral plasma membranes of the proximal tubule of rat kidney. *J. Membrane Biol.* 17: 263-274.

Kinne-Saffran, E., and R. Kinne. 1974b. Presence of bicarbonate stimulated ATPase in the brush border microvillus membranes of the proximal tubule. *Proc. Soc. Exp. Biol. Med.* 146: 751-753.

Kyte, J. 1976a. Immunoferritin determination of the distribution of $(Na^+ + K^+)$ ATPase over the plasma membranes of renal convoluted tubules. I. Distal segment. *J. Cell Biol.* 68:287-303.

Kyte, J. 1976b. Immunoferritin determination of the distribution of (Na^+-K^+) ATPase over the plasma membranes of renal convoluted tubules. II. Proximal segment. *J. Cell Biol.* 68:304-318.

Mircheff, A. K., and E. M. Wright. 1976. Analytical isolation of plasma membranes of intestinal epithelial cells: identification of Na,K-ATPase rich membranes and the distribution of enzyme activities. *J. Membrane Biol.* 28:309-333

Murer, H., U. Hopfer, and R. Kinne. 1976. Sodium/proton antiport in brush-border-membrane vesicles isolated from rat small intestine and kidney. *Biochem. J.* 154: 597-604.

Murer, H., U. Hopfer, E. Kinne-Saffran, and R. Kinne. 1974. Glucose transport in isolated brush-border and lateral-basal plasma-membrane vesicles from intestinal epithelial cells. *Biochim. Biophys. Acta* 345: 170-179.

Pockrandt-Hemstedt, H., E. Kinne-Saffran, H. Koepsell, W. Haase, and R. Kinne. 1976. In preparation.

Pockrandt-Hemstedt, H., J. E. Schmitz, E. Kinne-Saffran, and R. Kinne. 1972. Morphologische und biochemische Untersuchungen über die Oberflächenstruktur der Bürstensaummembran der Rattenniere. *Pflügers Arch.* 333: 297-313.

Schwartz, I. L., L. J. Shlatz, E. Kinne-Saffran, and R. Kinne. 1974. Target cell polarity and membrane phosphorylation in relation to the mechanism of action of antidiuretic hormone. *Proc. Natl. Acad. Sci. USA* 71: 2595-2599.

Shlatz, L. J., I. L. Schwartz, E. Kinne-Saffran, and R. Kinne. 1975. Distribution of parathyroidhormone-stimulated adenylate cyclase in plasma membranes of cells of the kidney cortex. *J. Membrane Biol.* 24: 131-144.

Stirling, C. E. 1972. Radioautographic localization of sodium pump sites in rabbit intestine. *J. Cell Biol.* 53: 704-714.

Stirling, C. E., and B. R. Landau. 1970. High resolution radioautography of ouabain-^{3}H and phlorizin-^{3}H in in vitro kidney and intestine. *Fed. Proc.* 29:595 Abs.

Thomas, L. 1973. Isolation of N-ethylmaleimide-labelled phlorizin-sensitive D-glucose binding protein of brush border membrane from rat kidney cortex. *Biochim. Biophys. Acta* 291: 454-464.

Ullrich, K. J., G. Capasso, G. Rumrich, and K. Sato. 1976. Effect of p-chloromercuribenzoate (pCMB), ouabain and 4-acetamido-4'-isothiocyanatostilbene-2, 2'-disulfonic acid (SITS) on proximal tubular transport processes. Ninth Conference on Environmental Toxicity, Rochester. Plenum Press, New York, in press.

Ullrich, K. J., G. Rumrich, and K. Baumann. 1975. Renal proximal tubular buffer-(glycodiazine) transport: inhomogeneity of local transport rate, dependence on sodium, effect of inhibitors and chronic adaptation. *Pflügers Arch.* 357: 149-163.

Ullrich, K. J., G. Rumrich, and S. Klöss. 1976. Active Ca^{2+} reabsorption in the proximal tubule of the rat kidney. Dependence on sodium and buffer transport. *Pflügers Arch.*, 364: 223-228.

14 Solute-Coupled Water Transport in the Kidney

EMILE L. BOULPAEP

Some water movement occurs across all tubular segments of the kidney despite wide differences in water permeability among different renal epithelia. The composition of the fluid absorbed from or secreted into the tubular lumen varies greatly with the segment of the nephron. Active transport of water has never been found, and fluid movement appears to be the consequence of either primary solute transport or of transepithelial effective osmotic and hydrostatic pressure differences. The proximal tubule of the kidney transports large amounts of salt and water in the absence of any transepithelial osmotic gradient, salt activity, or hydrostatic pressure differences. Moreover, the absorbate closely mimics the identical solutions that bathe both sides of the epithelium. Solute and water transport across the proximal tubule is similar to that of the gall bladder, small intestine, choroid plexus, and other epithelia, and raises essentially two questions: What drives solvent across a barrier separating two compartments of identical composition? How is the reabsorbate exactly or nearly exactly isotonic?

The first question has been studied with reference to the small intestine (Curran and Solomon, 1957), renal proximal tubule (Windhager et al., 1959), and gall bladder (Diamond, 1962), and these studies have led to the establishment of the primacy of solute transport, water movement being secondary. The studies indicated that solvent movement was virtually abolished in the absence of solute, that is, sodium chloride, movement. This is illustrated in Figure 14.1 for *Necturus* proximal tubule. Since normally no large external osmotic or hydrostatic gradients exist across these epithelia, osmosis and ultrafiltration are ruled out as driving forces of water movement. Based on the existence of streaming potentials in

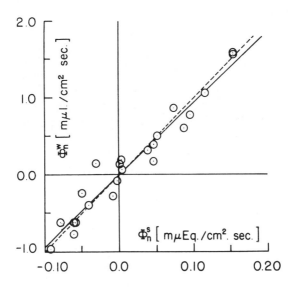

FIGURE 14.1. Relationship between solute flux Φ^s and solvent flux Φ^w in Necturus proximal tubule. (From Windhager et al., 1959.)

these leaky epithelia (Smyth and Wright, 1966; Diamond and Harrison, 1966; Boulpaep and Seely, 1971) electroosmosis may be involved, although electrokinetic phenomena might also be generated by boundary layer potentials (Wedner and Diamond, 1969). When, however, the epithelium is treated as a single barrier, it is manifest that the sign and magnitude of the open circuit transepithelial potential and the apparent electroosmotic coefficient of these epithelia are not of adequate magnitude to provide a sufficiently large driving force for normal isotonic fluid absorption. Thus, active transport of solute is the primary driving force for water movement. In the absence of an obvious external osmotic driving force, it remains to establish the site of solute solvent coupling within the epithelium itself.

The description of a double-membrane model for water transport by Curran (1960), Curran and McIntosh (1962) and Ogilvie, McIntosh, and Curran (1963) provided a basis for the direct link of fluid transfer to active solute transport. Figure 14.2 illustrates the principle of this model, together with a diagrammatic representation of what we believe today are the anatomical counterparts of the various compartments. The three-compartmental model has two membranes, a and b, in series, with different properties; in the artificial model of Curran and McIntosh (1962) these membranes are represented by a cellophane-membrane and sintered-glass disk, respectively. Solute is actively transported into the volume between the two membranes, thereby raising the solute concentration in the middle space. If the reflection coefficients (σ) are $\sigma_a > \sigma_b$, the

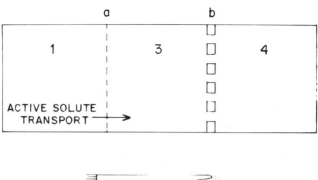

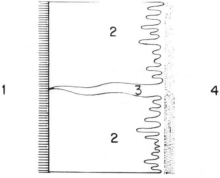

FIGURE 14.2. Diagram of the double-membrane model with three compartments. *Bottom,* pair of proximal tubular cells illustrating different compartments.

effective osmotic pressure across *a* will exceed that across *b*. Water will be drawn into the noncompliant middle compartment *3*, an event that will raise the hydrostatic pressure in the space. If the filtration coefficients (L_p) are such that $L_p^a < L_p^b$, the effective pressure driving force across *b* is greater than across *a* and net water flow will proceed from *1* to *4* in the absence of any external driving force or even against an osmotic gradient. The numbering of the compartments in Figure 14.2 clearly indicates our belief that the middle compartment is located in the lateral intercellular spaces, whereas the cell compartment *2* is not indicated in the three-compartmental model. The constituents of barrier *a* are the apical cell membrane, the cytoplasm, and the basolateral membrane, in series, all of which are in a parallel arrangement with the zonula occludens. As an effective outflow resistance to the intercellular channel barrier, *b* comprises the basement membrane, the peritubular interstitium, and the capillary endothelium.

Evidence that the paracellular pathway, that is, the intercellular space between the lateral membranes of cells, is a plausible site for solute-solvent interaction is mainly indirect. With respect to the kidney, morphological studies have indicated the presence of large interspaces in transporting epithelia (Schmidt-Nielsen and Davis, 1968). Electrophysiological investigations (Windhager, Boulpaep, and Giebisch, 1966; Boulpaep, 1970; Boulpaep, 1972; Boulpaep, 1976a) have emphasized the existence and physiological role of a low-resistance paracellular shunt in proximal renal tubules. Collection of hypertonic fluid from intercellular regions of insect rectal pads represents the only direct evidence for the paracellular compartment as the site of solute accumulation (Wall, Oshman, and Schmidt-Nielson, 1970). Thus, both theoretical and experimental data are available in order to answer the first question raised.

The second problem is to explain what causes the active transport of one solute molecule to effect the movement of several hundred molecules of water in a constant isosmotic ratio across a composite system with multiple barriers. The double-membrane model (Curran and McIntosh, 1962) does not require that the ratio between solute and solvent movement be set at a particular osmolarity, since this would depend on the specific membrane coefficients of both barriers. Patlak, Goldstein, and Hoffman (1963) have derived equations for the calculation of the ratio of net solute to solvent flux. In the case of the proximal tubule, one may introduce the following simplifying assumptions: luminal solute concentration (C_1) and peritubular concentration (C_4) are identical, $C_1 = C_4$, and the solute reflection coefficient of barrier b is probably zero, since basement membranes are even permeable to protein molecules (Welling and Grantham, 1972), that is, $\sigma_b = 0$. The emergent concentration C_{emerg} or the ratio of solute flux J_s) over solvent flux (J_v) becomes

$$\frac{J_s}{J_v} = \frac{C_3 + C_4}{2} + \frac{2\omega_b RT (C_3 - C_4)}{J_v} \qquad (1)$$

where ω_b is the solute permeability coefficient of barrier b, R the gas constant, and T the absolute temperature. If $C_3 > C_4$, then the first term of equation (1), $(C_3 + C_4)/2$, is $> C_4$ and the second term is > 0. Hence $J_s/J_v > C_4 = C_1$. Theoretically, a double-membrane model could yield almost any ratio, but under conditions of symmetrical solutions and virtually no solute reflection at the second barrier, the absorbate is predicted to be hyperosmotic to either outside compartment.

In order to solve the requirement of isosmotic absorption, Diamond proposed a local osmosis theory (Diamond, 1964) that was essentially similar to the three-compartment, two-barrier model of Curran and McIntosh (1962), with the difference that an osmotically discriminating barrier a ($\sigma_a = 1$) was followed by a restricted homogeneous middle region 3 continuous with the bathing solution 4 by means of a negligible purely hydraulic resistance ($\sigma_b = 0$). Clearly, application of the Curran model to the paracellular pathway, as indicated above for the proximal tubule, implied that barrier b is probably a negligible hydraulic resistance, or $L_p^b >> L_p^a$, and mainly not reflective for solute $\sigma_b = 0$. Diamond claimed that the reabsorbate in such a model would follow more closely the bathing osmolarity when the latter is varied than in the double-membrane model, since actively transported solute could completely equilibrate with water. Although it was experimentally shown in the gall bladder that the transported fluid was always close to a sodium chloride solution nearly isosmotic to whatever solution was used in compartment 1 (Diamond, 1964), this observation constitutes no evidence against the double-membrane model. Indeed, when σ_b is zero, a very realistic assumption for the proximal tubule, the double-membrane model and that of local osmosis cannot be distinguished. With respect to reabsorbate osmolarity, local osmosis assumes that $C_{emerg} = C_3$. The value of C_3 can be readily obtained if $\sigma_a = 1$, $\sigma_b = 0$, $L_p^b >> L_p^a$. Thus,

$$\frac{J_s}{J_v} = C_3 = \frac{C_1}{2} + \sqrt{\frac{-C_1^{\,2}}{4} + \frac{J_s}{RTL_p^a}} \tag{2}$$

Hence again $J_s/J_v > C_4 = C_1$ as long as the term $J_s/RTL_p^a > 0$

Except for the degenerate case where no transport occurs, $J_s = 0$, or an infinite filtration coefficient of the barrier a, there will always be an hyperosmotic absorbate in local osmosis.

Diamond and Bossert (1967) refined this model in what they called the standing-gradient model, which treats the middle compartment not as a volume with uniform concentration but as an unstirred continuum. Intuitively, in order to meet the isosmotic reabsorption requirement, any means that retards the movement of solute relative to the movement of solvent either in barrier b or in compartment 3 might balance the J_s/J_v ratio to the desired (that is, isosmotic) value. Curran's model could accomplish this by reflecting solute at barrier b ($\sigma_b > 0$), but we have already pointed out the lack of a structural basis for that assumption. Diamond and Bossert (1967) attempted to somehow delay solute transport, compared with solvent movement, by making use of the length of compartment 3 as an appreciable distance for solute diffusion. Long and narrow intercellular channels formed the basis for this assumption (Tormey and Diamond, 1967). The striking features of the Diamond and Bossert (1967) hypothesis are illustrated in Figure 14.3. Active solute transport across the lateral cell membrane, taking place close to the apical end of the intercellular channel, creates a hyperosmotic zone. This hypertonicity becomes dissipated along the diffusion distance of the channel, with water inflow into this conduit occurring osmotically along the entire lateral length of the intercellular space. Osmotic equilibration presumably is completed by the end, that is, the mouth, of the channel.

The predicted solute concentration profiles along channel length in the

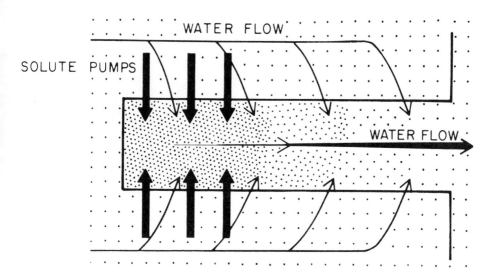

FIGURE 14.3. Principle of the standing-gradient system, consisting of a narrow channel closed at one end. Active solute pumping is indicated by the thick arrows, water flow by the thin arrows. The density of the dots indicates the solute concentration. (From Diamond and Bossert, 1968.)

standing-gradient model had the form described in Figure 14.4. Dissipation of the concentration in the model varies, depending on the assumed solute transport rates and the assumed L_p or osmotic permeability of the lateral cell membrane.

With respect to the osmolarity of the emergent reabsorbate, or J_s/J_v, it is important to realize that in the absence of any resistance to flow of solvent or solute at the mouth of the channel, two conditions have to be fulfilled for $J_s/J_v = C_4$. First, the concentration reached at the end of the channel should be isosmotic, such that the convection term in the emergent fluid carries solute at the exact concentration C_4. Second, the derivative of concentration with distance at the end of the channel should be zero, in order to avoid any additional diffusive term that would enhance J_s over J_v. In fact, all four curves in Figure 14.4, taken from Diamond and Bossert, exhibit at 100-percent length a negative slope $\frac{dC}{dx}\big|_L < 0$ and hence the solute leaving the mouth of the waterway by diffusion will exceed that carried by bulk water movement, with the result that the emergent concentration will exceed the isomotic value (by 1 percent for the lower curve and up to 200-300 percent for the upper curve). A numerical reanalysis of the Diamond and Bossert model (Sackin and Boulpaep, 1975) predicted exact isotonicity in only the degenerate cases of either zero active transport or infinite channel-wall water permeability, as was already apparent from the analysis of the local osmosis or

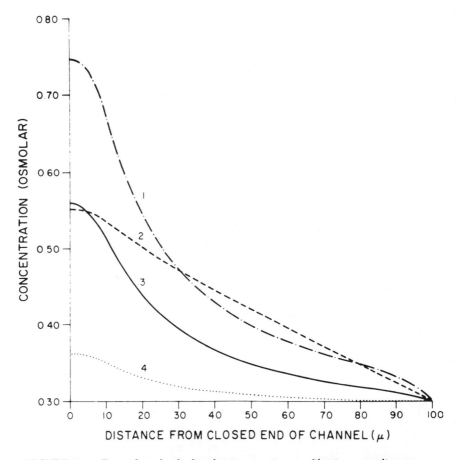

FIGURE 14.4. Examples of calculated concentration profiles in a standing-gradient flow system. Solute concentration is plotted against distance along the channel. Active solute transport was assumed to occur over the first 10 μ only. Curves 1, 3, and 4 differ by the use of a decreasing solute transport rate. Curves 2 and 4 have the same assumed solute transport rate but curve 2 assumes a lower water permeability in the lateral membranes than curve 4. The reabsorbate in all four curves was hypertonic, curve 1 = 342 m osM, curve 2 = 803 m osM, curve 3 = 318 m osM, and curve 4 = 304 m osM, against an isotonic value of 300 m osM. (From Diamond and Bossert, 1967.)

double-membrane model. Exact isotonicity is never reached entirely in the Diamond-Bossert model with any physically meaningful set of parameters. Although Diamond and Bossert's views have received wide acceptance, several challenges have been raised recently that invalidate some of the principal concepts of the model. Five general criticisms will be discussed.

Criticisms and New Developments

Need of an Unstirred Layer at the End of the Channel

Diamond and Bossert (1967) postulated an infinite sink outside the mouth of the channel, that is, a well-mixed compartment at isosmotic concentration C_O. However, it is possible to demonstrate formally that such a system may not work. Instead, an unstirred layer at the end of the interspace is required. An analysis of the solute and solvent fluxes at the basal end of the interspace tests the correctness of this requirement (Sackin and Boulpaep, 1975). Figure 14.5 illustrates two concentration profiles along the intercellular space, which differ mainly by the conditions assumed to exist at the apical end. If we concentrate on what happens at the basal end of the interspace ($X=L$), all profiles appear to have a negative slope (as in Figure 14.4), that is, $\frac{dC}{dx} < 0$ at L. Whichever profile exists at $x=L$, the profile must be continuous about point L; thus, C_{L^-}, the concentration at the left-sided limit of L, and C_L+, the concentration at the right-sided limit of L, should be identical and equal to C_O.

The equation for conservation of volume at L is

$$A_{L^-} \cdot V_{L^-} = A_L+ \cdot V_L+ \tag{3}$$

where A is cross-sectional area and V linear velocity in the channel. The equation for solute conservation at L is:

$$A_{L^-}[(V_{L^-} \cdot C_{L^-})-(D_{L^-} \cdot \frac{dC}{dx}\Big|_{L^-})] = A_L+ [(V_L+ \cdot C_L+)-(D_L+ \cdot \frac{dC}{dx}\Big|_{L^+})] \tag{4}$$

where D is the solute diffusion coefficient. One may conclude that in the event the derivative on the left-sided limit is negative ($\frac{dC}{dx}\big|_{L^-} < 0$), this should also be the case for the right-sided limit. Indeed, also $\frac{dC}{dx}\big|_{L}+ < 0$, unless the cross-sectional area on the right-sided limit A_L+ or the diffusion coefficient on the right-sided limit D_L+ should suddenly break off to an infinite value.

In conclusion, the concentration profile in the space immediately outside the mouth of the channel cannot be flat or uniformly C_O, since $\frac{dC}{dx}\big|_L+ < 0$. Instead, an unstirred layer is required. If the region outside L were well mixed, as postulated by Diamond and Bossert, the system would have to violate mass balance (Sackin and Boulpaep, 1975). As shown in Figure 14.5, the concentration would continue to decrease beyond the intercellular space, and the concentration in the peritubular interstitium (between L and M) would fall below the isosmotic line. Just outside the capillary wall at $x=M^-$, the concentration would be less than $C_L+ = C_O$. In view of the effects of mass balance on volume flow, the convective term for solute flow should be greater at L^+ than M^+. How-

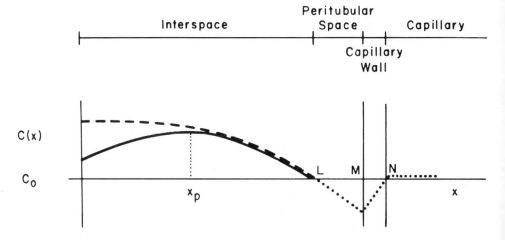

FIGURE 14.5. Concentration profiles expected for uniform distributions of solute transport, using the boundary condition similar to Figure 14.4, that is, with C_L at the mouth of the channel equal to C_o. The assumption that there is an unstirred layer in the peritubular space leads to a concentration profile in the interval L-M as indicated and a violation of mass balance at M. In the interval 0-L, the curves differ depending on whether a closed tight junction (interrupted line) or a permeable tight junction (continuous line) is assumed.

ever, the capillary solute concentration $C_n = C_o$ and thus, the concentration derivative in the capillary wall should be positive. Hence, the diffusional term for solute flow at M^+ should subtract from the convective term, and total solute flow at M^+ would end up to be far less than at L^+, which would violate overall mass balance of solute. Weinbaum and Goldgraben (1972) also indicated the theoretical need for an unstirred layer. Actually, the peritubular interstitium in vivo may not be considered anatomically as a perfectly mixed compartment.

Uniform Distribution of Solute Input Sites

The second criticism of the standing-gradient model concerns the assumption that active transport should be confined to a small region at the apical end of the interspace, whereas addition of fluid occurs over the entire length of the channel. The importance of this assumption for effecting isosmotic transport more easily is intuitively clear. Indeed, it is only another variety in a class of proposals that would somehow allow diffusion of solute to lag behind movement of solvent. However, recent experimental evidence has not confirmed the assertion of very localized transport sites in epithelial interspaces.

Farquhar and Palade (1964) used a lead precipitation technique to mark ATPase in frog skin and showed uniform staining along the inter-cellular spaces throughout the epidermis. By means of a more refined procedure, localizing the K-dependent phosphatase activity of ATPase and substituting strontium for lead as the capture ion for the released phosphate, Ernst (1972) found K-nitrophenyl-phosphatase activity in the avian salt gland on all basal and lateral membranes of secretory cells. Stirling (1972), utilizing radio autographs of ouabain binding sites in rab-bit intestine, also demonstrated a uniform localization along the lateral cell borders. Finally, immunoferritin binding of Na,K-ATPase (Kyte, 1976) indicated again high concentrations of the enzyme undiversified along the intercellular spaces of dog proximal tubule. Thus, a selective localization of the solute pumps at the luminal end of the interspace is very unlikely.

If solute pumps coexist uniformly with sites of water permeability along the interspace, models of the kind used in the standing-gradient hypothesis always give a hyperosmotic absorbate. First, taking the specific numerical example of curve 1 in Figure 14.4, where solute flux was limited to the first 10 percent of total channel length, the predicted ratio J_s/J_v or C_{emerg} is 307 mosM where $C_O = 300$ mosM. If the same total active transport were distributed over the entire length, the emer-gent concentration would be 385 mosM. Second, the importance of the distribution of solute pumps was also recognized by Segel (1970), who provided an analytical approximation of the emergent osmolarity, in-cluding a term λ that is the ratio of the channel length to the solute input length, and later by Hill (1975), who used an analytical approximation similar to Segel's. Finally, we have recently given a general analytical proof that the end point derivative $\frac{dC}{dx}\big|_L$ will always be negative if trans-port is uniform, and thus result in hyperosmotic transport (Sackin and Boulpaep, 1975).

Figure 14.5 illustrates the proof by means of two curves that differ only in that either an impermeable or permeable tight junction at $x = 0$ is con-sidered. We assumed that the concentration is continuous in the closed interval $[0,L]$ and the concentration derivative dC/dx continuous in the open interval $0,L$. If water is to be attracted from cell or lumen to the interspace there must be a region x' between 0 and L where the concentra-tion exceeds C_O. It is possible to show that the curve at the start of the channel, $x = 0$, must be either flat or have a positive slope, that is, $(dC/dx)_{x = 0} > 0$. If the system is further constrained such that the end point concentration is isomotic with lumen and plasma, these three con-ditions, $(dC/dx)_{x=0} > 0$, $C_x > C_O$, and $C_{(x=L)} = C_O$, require that the possible concentration profiles are concentration curves as in Figure 14.5: beyond a certain point x_p these curves are monotonically decreasing and concave downward in the interval $x_p \to L^-$ (Sackin and Boulpaep, 1975). As is evident from Figure 14.5, all these concentration curves have a negative concentration derivative at the mouth of the channel, L. Since

$C_{(x=L)} = C_O$ it may be concluded that the emergent concentration is always hyperosmotic, or

$$\frac{J_s}{J_v} = C_O - \frac{D_{L^-}}{v_{L^-}} \frac{dC}{dx}\bigg|_{L^-} > C_O \qquad (5)$$

Leaky Zonula Occludens

Diamond and Bossert's model (1967) postulated that the intercellular spaces are closed at the apical or luminal end. For the proximal tubule it has been firmly established that a low-resistance paracellular shunt dominates the transepithelial permeability properties. This paracellular pathway, made up by the tight junction or zonula occludens and the intercellular space, provides a pathway of leakage for ions, small non-electrolytes, and water between lumen and blood.

The evidence with respect to the high paracellular conductance is best illustrated by means of an equivalent circuit, shown in Figure 14.6. A pair of cells is represented with peritubular and luminal cell borders and, in addition, the paracellular path between the two cells. The peritubular cell membrane has a membrane potential difference V_1 of opposite sign and larger magnitude than the luminal cell membrane V_2. The overall transepithelial potential difference V_3 is the sum of V_1 and V_2. Characteristic for a leaky epithelium such as the proximal tubule are the following properties. (1) Transepithelial potential difference V_3 is small. (2) Each diffusion barrier is represented with its particular set of electromotive forces, shown by E_1, E_2, and E_3, where a single comprehensive emf of undefined sign illustrates the net equivalent of all ionic electromotive forces due to the diffusional pathways of that barrier. In addition, the ionic resistances are also pooled as R_1, R_2, R_3, representing all the ionic diffusion pathways within the ionic batteries, their series resistances, and all undefined leak resistances within that same barrier in parallel (Boulpaep, 1967; Boulpaep, 1976b). Total transepithelial resistance resulting from the combination of these three resistors is small (Boulpaep, 1972). (3) The two cellular resistances, R_1 and R_2, in series are two orders of magnitude larger than the transepithelial resistance (Windhager, Boulpaep, and Giebisch, 1966). Hence, the low transepithelial resistance reflects essentially the resistance of the shunt R_3. (4) Electrical interaction between the two membrane potentials V_1 and V_2 occurs when only a single emf, E_1 or E_2, is altered, thus indicating the existence of a low resistance path, R_3 (Boulpaep, 1967). (5) Several additional indirect lines of evidence have confirmed the view that ions may easily bypass the transcellular route (Boulpaep, 1970; Whittembury, Rawlins, and Boulpaep, 1973). Leakiness of the zonula occludens to small nonelectrolytes such as raffinose (Boulpaep, 1972) and sucrose (Berry and Boulpaep, 1975) has been demonstrated for *Necturus* proximal tubule.

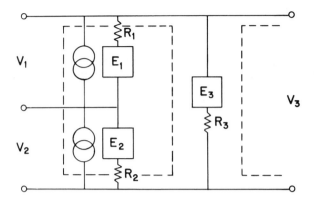

FIGURE 14.6. Electrical equivalent circuit of the proximal tubular cells of the kidney: V_1, peritubular cell membrane potential difference; V_2, luminal cell membrane potential difference; V_3, transepithelial potential differences. E_1, E_2, E_3, R_1, R_2, R_3 represent the diffusional emf's and passive resistances of, respectively, peritubular cell membrane, luminal cell membrane, and paracellular pathway. In addition constant current sources are shown, illustrating the possibility of rheogenic pumps. (From Boulpaep, 1976b.)

The phenomenon of solute entrainment of sucrose via the tight junction has also been shown for *Necturus* proximal tubule, indicating that at least a fraction of water can enter the lateral intercellular space directly across the tight junction (Berry and Boulpaep, 1975). Little information is available to assess quantitatively the moieties of fluid that use a transcellular, transjunctional, and interspace route in the translocation between lumen and blood. A comparison of apparent overall filtration coefficients (L_p) across the epithelium and cellular osmotic permeability coefficients also allows the prediction that water permeates across the seal between cells (Whittembury, Sugino, and Solomon, 1960; Whittembury, 1967). In conclusion, a moderate degree of junctional solute and solvent permeability should be included, contrary to the assumptions made by Diamond and Bossert (1967) and the analysis provided by Segel (1970) and Hill (1975). Several authors have recently introduced properties of leaky tight junctions in a variety of models for transport across convoluted and straight proximal tubule (Sackin and Boulpaep, 1975; Schafer, Patlak, and Andreoli, 1975; Huss and Marsh, 1975).

Permeability of the Lateral Cell Membrane to Water

The isotonic convection approximation used by Segel (1970) for the analytical expression of the emergent osmolarity shows clearly that the ratio J_s/J_v or the emergent concentration depends critically on the following four terms.

1. λ, the ratio of total channel length L to solute input length.
2. D, the diffusion coefficient for the solute. Usually, free solution diffusion coefficients are assumed within the channel.
3. L_p or P_{osm}, the osmotic water permeability of the lateral wall of the channel.
4. The dimensional parameter L^2/r, or the ratio of channel length squared to channel radius.

Criticism of the Diamond-Bossert model has therefore been focused on the numerical values for these parameters, particularly L_p and L^2/r. For the osmotic water permeability, Diamond and Bossert (1967) used a range of values from 10^{-6} to 10^{-4} cm s^{-1} osM^{-1}. Hill (1975) in his analysis of the standing gradient theory claims that, using a particular set of values for the other parameters, values of 10^{-1} cm s^{-1} osM^{-1} would be required for the lateral cell membrane's permeability if the system were to transport isotonically.

These discrepancies have to be interpreted in the light of probable artefacts induced by the determination of water permeability. In *Necturus* proximal tubule, L_p across the entire epithelium was determined using three different driving forces: (1) a change in colloid osmotic pressure in the capillary, (2) a change in luminal hydrostatic pressure, or (3) a change in peritubular capillary hydrostatic pressure or a change in peritubular capillary hydrostatic pressure (Grandchamp and Boulpaep, 1974). The apparent transepithelial L_p using a colloid osmotic pressure difference was 3-6 x 10^{-4} cm s^{-1} osM^{-1}; with luminal hydrostatic pressure displacements 10^{-3} cm s^{-1} osM^{-1}; with peritubular hydrostatic pressure changes from 2 x 10^{-3} cm s^{-1} osM^{-1} to 3 x 10^{-2} cm s^{-1} osM^{-1}. For the measurement of accurate and realistic values of osmotic or hydraulic conductivities, minute and instantaneous displacements of the driving force should be employed, with measurements of the corresponding volume flow rates taken over very short time intervals. This has not yet been realized in renal tubules. One may conclude that presently available L_p measurements are probably grossly inaccurate and that structural modifications of the epithelium are induced by large displacements of hydrostatic pressure (Maunsbach and Boulpaep, 1975).

Whichever of the reported values are chosen for transepithelial L_p, all exceed by one to three orders of magnitude the cellular water permeability found by Whittembury, Sugino, and Solomon (1960). Finally, since changes in L_p correlate well with changes in paracellular morphology (Maunsbach and Boulpaep, 1975), and since transepithelial L_p exceeds expected cell membrane L_p, it seems most likely that at present a comparison between theoretically required and observed L_p values is not possible. Both Diamond and Bossert (1967) and Hill (1975) have ignored the possibility of a high junctional water permeability and have simply deduced lateral cell membrane permeability as if the epithelium were made up of two membranes in series. In contrast, we have recently chosen to represent, first, the cellular pathway by a single effective barrier with an effective area that is equal for apical and basolateral membrane, and, second, the tight junction as a parallel barrier with different L_p and different effective area. Using Kedem and Katchalsky's (1962b) analysis of parallel membranes we calculated for cell membrane L_p a range of 7×10^{-6} cm s^{-1} osM^{-1} to 7×10^{-5} cm s^{-1} osM^{-1}, and for the tight junction L_p 10^{-1} cm s^{-1} osM^{-1} to 6 cm s^{-1} osM^{-1} (Sackin and Boulpaep, 1975).

Lateral Interspace Dimensions

The crucial term influencing the emergent concentration in the case of a cylindrical channel is L^2/r and for a parallel slit L^2/W, where L is the length of the channel, r the radius, and W the width. Hill (1975), in his analysis of the application of Diamond and Bossert to a variety of epithelia, lists physiological values for L^2/r as fractions of a centimeter, whereas several centimeters would be required for an isosmotic emergent solution. For *Necturus* proximal tubule, we described large changes in interspace volume as compared to cellular volume calculated from morphometric analysis of thin section electron micrographs (Maunsbach and Boulpaep, 1975). A value of about 2.5 cm results for L^2/W in *Necturus* proximal tubule.

Two Models for Solute and Water Coupling

Among the various weaknesses of Diamond and Bossert's model the first three mentioned appear quite fundamental: neglecting the need for an unstirred layer at the end of the channel, assuming a nonuniform distribution of solute input sites, and assuming a closed zonula occludens. The last two weaknesses—assuming unrealistic lateral cell membrane water permeabilities and lateral interspace dimensions—are mainly a matter of controversial numerical parameters. It is in view of the three fundamental difficulties that we developed two alternative models for coupling of salt and water, specifically using the appropriate parameters of the *Necturus* proximal tubule. The models can easily be extended to epithelial transport in the gall bladder and small intestine.

Continuous Model

This model, illustrated in Figure 14.7b, is somewhat similar to that of Diamond and Bossert, with the following differences:
    (1) A different boundary condition was used. Instead of assuming that

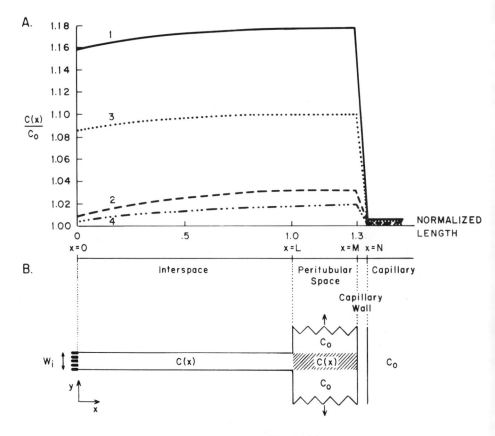

A.

FIGURE 14.7. (A) Solute concentration profiles from a one-dimensional analysis of the continuous model. Uniform distribution of solute pumps, a permeable tight junction, and no solute dispersion in the peritubular space were assumed for Necturus proximal tubule: curve 1, control conditions and low L_p of the tight junction; curve 2, control conditions and high L_p of the tight junction; curve 3, volume expanded animal and low L_p; curve 4, volume expanded animal and high L_p.
(B) The schematic geometry used for the calculation of the peritubular concentration profiles in a one-dimensional analysis. There is assumed to be no interaction or mixing between the shaded band of fluid and the surrounding bulk of peritubular space, which is taken as isosmotic to capillary plasma. (From Sackin and Boulpaep, 1975.)

the concentration at the end of the interspace equaled C_o at $x=L$, the boundary condition was assumed to be such that the capillary concentration at $x=N$ was equal to C_o.

(2) Active transport of solute was distributed uniformly along the lateral cell membrane.

(3) The zonula occludens was considered permeable to solute and

water with the hydraulic permeability L_p, the reflection coefficient σ, and solute permeability ω estimated from overall transepithelial values. These determine the boundary conditions C and dC/dx at the apical end of the channel.

(4) Lateral membrane osmotic water permeability values were taken from data of Whittembury, Sugino, and Solomon (1960). In addition, channel walls were considered to have a moderate solute permeability not considered previously.

(5) A rectangular geometry was used for the intercellular spaces with a L^2/W ratio of 2.5 cm.

(6) In addition, net volume flux leaving the interspace was made to conform to experimental values. This contrasts with Diamond and Bossert's approach, where no attempt was made to match net volume flux with that characteristic for a particular epithelium.

(7) The analysis was one-dimensional, as in Diamond and Bossert's treatment, but in the region of the interstitium an abrupt widening in the y axis occurs at the transition between the narrow interspace and the peritubular space. A simple one-dimensional approximation was chosen. The heterogeneous peritubular space was divided into two homogeneous noninteracting regions, as shown in Figure 14.7b. In the shaded area the solute concentration was assumed to be independent of y. Its x dependence was calculated as an extension of the one-dimensional analysis of the interspace proper. However, the surrounding area outside the shaded area was assumed to be at C_O concentration and not to mix with the shaded area.

Figure 14.7a gives four different profiles of concentration against interspace length. The main features of these profiles are:

(a) A remarkably flat shape of the curve compared to those of Figure 14.4.

(b) A slight dip at the apical end of the interspace, caused by a permeable tight junction.

(c) A flat peritubular space concentration profile with a steep drop across the capillary wall. Note that this concentration drop applies to only a small fraction of the capillary endothelium, and is only apparent since it results from the one-dimensional approximation chosen.

(d) The solute concentration within the interspace was from 116 to 118 mM for low L_p of the tight junction (curve 1 of Fig. 14.7a) and from about 101 to 103 mM for a high L_p of the tight junction (curve 2). C_O was set at 100 mM. Fluxes across tight junction and cell were predicted for solute and solvent. It is interesting to note that the low tight junctional L_p predicted 42 percent of fluid movement through the zonula occludens, whereas a high L_p gave a value of 92 percent for transjunctional fluid flow. The emergent concentration of curve (1) was 115.2mM, of curve (2) 102.6 mM.

A one-dimensional analysis of the kind given in Figure 14.7 is not appropriate to describe the mixing of solute that occurs as fluid from the

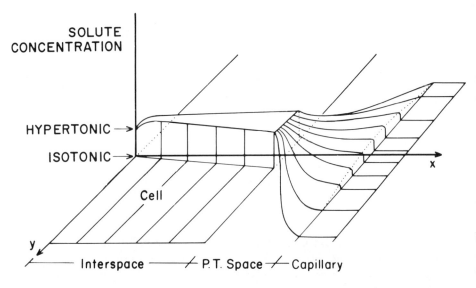

SOLUTE
CONCENTRATION

HYPERTONIC →

ISOTONIC →

x

Cell

y

Interspace ——— P.T. Space — Capillary

FIGURE 14.8. Solute concentration profiles from a two dimensional analysis. Height above a particular point in the x-y plane denotes the solute concentration at that point. Solute dispersion in the peritubular space leads to a lower average concentration drop across more of the capillary surface than in a one-dimensional analysis.

interspace enters the peritubular space. An approximate solution was calculated for the dispersion of solute in the peritubular space, as shown in Figure 14.8 (Sackin and Boulpaep, 1975). Here, the drop in concentration across the capillary is reduced by the lateral spreading of solutes into the peritubular space. It is of particular importance for the discussion of the ratio of solute to solvent flux, J_s/J_v, that this value be identical in the one- and two-dimensional analyses.

Compartment Model

In view of the small solute-concentration differences between apical and basal ends of the interspace, an approximation may be used that assumes the interspace of the middle compartment to be well stirred and homogeneous. Consequently we developed a series-parallel model with five compartments: (1) lumen, (2) cell, (3) interspace, (4) peritubular space, and (5) capillary; five barriers: (a) tight junction, (β) baso-lateral cell membrane, (γ) luminal cell membrane, (δ) basement membrane, and (ϵ) capillary endothelium (see Fig. 14.9). In addition to the basic driving forces of the continuous model, electrical driving forces were included. The latter permitted the calculation of ion fluxes instead of neutral salt fluxes. Equations similar to the formulations of Kedem and Katchalsky

A.

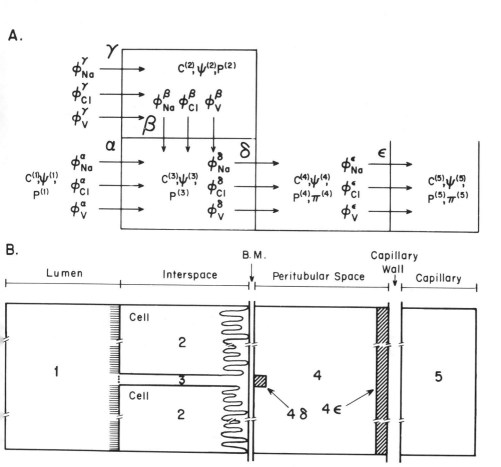

B.

FIGURE 14.9. (A) The five-compartment model for NaCl and water. a denotes the tight junction region, β the lateral cell membrane, γ the luminal membrane, δ the open end of the interspace and basement membrane, ε the capillary endothelium. C_j denotes concentration of the ion j. P denotes hydrostatic pressure, Ψ electrical potential, π colloid osmotic pressure. The direction of the arrows indicates the convention chosen for positive fluxes, ϕj. (B) Illustration of the location of the subcompartments used.

(1962a) for ion and volume flux were written and solved using basic principles of volume balance, solute balance, and electroneutrality.

The results of the compartment model are quite compatible with those of the distributed model. The average solute concentration in the interspace for a low tight-junction water permeability was 118mM, and for a high tight-junction permeability, the average concentration was 101mM. Volume flow was 48 percent and 98 percent across the junction for low and high water permeability, L_p^a, respectively. The emergent concentra-

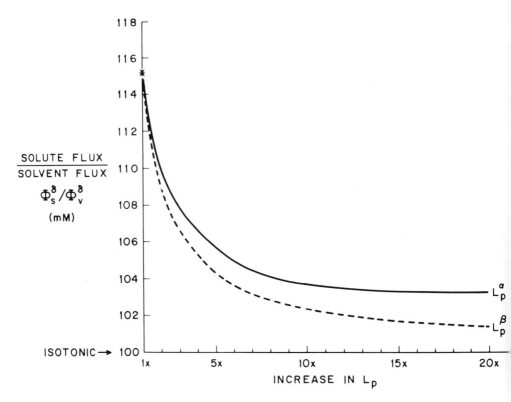

FIGURE 14.10. Comparison of the relative effects of tight junction water permeability L_p^a and lateral membrane water permeability L_p^β on the solute-to-solvent flux ratio (emergent concentration) for the continuous model.

tion was 118 mM and 101 mM, also for low and high L_p values, respectively.

Both for the continuous and the compartment models, the transport parameters that had the largest effect on the flux ratio J_s/J_v were the hydraulic conductivity of the tight junction, L_p^a, the hydraulic conductivity of the lateral cell membrane L_p^β, the interspace length and the reflection coefficient of the tubular basement membrane (Sackin and Boulpaep, 1975).

Figure 14.10 compares the relative effectiveness of an increase in tight junctional L_p^a and lateral membrane L_p^β hydraulic conductivity on the emergent concentration. It is apparent that relative lateral cell membrane permeability increments are only slightly more powerful in effecting near isosmotic J_s/J_v ratios.

Two conclusions from these epithelial models are particularly important. First, if solute pumps are homogeneously distributed along the lateral cell membrane and the zonula occludens is permeable to salt and water, standing gradients along the axis of the intercellular channel do not exist. Concentration differences between different points within an interspace have never been described experimentally. Second, none of the models, even with very high water permeabilities, predict exact isotonicity of transport. A high tight-junction hydraulic conductivity may greatly contribute to osmotic equilibration but never accomplishes perfect isosmotic absorption. It remains to be seen whether such small deviations from isotonicity as predicted may be detected experimentally.

REFERENCES

Berry, C., and E. L. Boulpaep. 1975. Nonelectrolyte permeability of the paracellular pathway in *Necturus* proximal tubule. *Am. J. Physiol.* 228: 581-595.

Boulpaep, E. L. 1967. Ion permeability of the peritubular and luminal membrane of the tubular cell. In F. Krück, ed., *Transport und Funktion intracellulärer Elektrolyte*, pp. 98-107.

Boulpaep, E. L., 1970. Electrophysiological properties of the proximal tubule: importance of cellular and intercellular transport pathways. In *Electrophysiology of epithelial cells*, Symp. Med. Hoechst., pp. 91-118. F. K. Schattauer Verlag, Stuttgart.

Boulpaep, E. L. 1972. Permeability changes of the proximal tubule of *Necturus* during saline loading. *Am. J. Physiol.* 222: 517-531.

Boulpaep, E. L. 1976a. Recent advances in electrophysiology of the nephron. *Annu. Rev. Physiol.* 38: 20-36.

Boulpaep, E. L. 1976b. Electrical phenomena in the nephron. *Kidney Int.* 9: 88-102.

Boulpaep, E. L., and J. F. Seely. 1971. Electrophysiology of proximal and distal tubules in the autoperfused dog kidney. *Am. J. Physiol.* 221: 1084-1096.

Curran, P. F. 1960. Na, Cl and water transport by rat ileum in vitro. *J. Gen. Physiol.* 43: 1137-1148.

Curran, P. F., and J. R. McIntosh. 1962. A model system for biological water transport. *Nature* 193: 347-348.

Curran, P. F., and A. K. Solomon. 1957. Ion and water fluxes in the ileum of rats. *J. Gen. Physiol.* 41: 143-168.

Diamond, J. M. 1962. The mechanism of water transport by the gall bladder. *J. Physiol.* 161: 503-527.

Diamond, J. M. 1964. The mechanism of isotonic water transport. *J. Gen. Physiol.* 48: 15-42.

Diamond, J. M., and W. H. Bossert. 1967. Standing-gradient osmotic flow: a mechanism for coupling of water and solute transport in epithelia. *J. Gen. Physiol.* 50: 2061-2083.

Diamond, J. M., and W. H. Bossert. 1968. Functional consequences of ultrastructure geometry in backwards fluid-transporting epithelia. *J. Cell. Biol.* 37: 694-702.

Diamond, J. M., and S. C. Harrison. 1966. The effect of membrane fixed charges on diffusion potentials and streaming potentials. *J. Physiol.* 183: 37-57.

Ernst, S. A. 1972. Transport adenosine triphosphatase cytochemistry. II. Cytochemical localization of ouabain-sensitive, potassium-dependent phosphatase activity in the secretory epithelium of the avian salt gland. *J. Histochem. Cytochem.* 20: 23-38.

Farquhar, M. G., and G. E. Palade. 1964. Functional organization of amphibian skin. *Proc. Natl. Acad. Sci.* 51: 569-577.

Grandchamp, A., and E. L. Boulpaep. 1974. Pressure control of sodium reabsorption and intercellular backflux across proximal kidney tubule. *J. Clin. Invest.* 54: 69-82.

Hill, A. E., 1975. Solute-solvent coupling in epithelia: a critical examination of the standing-gradient osmotic flow theory. *Proc. R. Soc. Lond.* B 190: 99-114.

Huss, R. E., and D. J. Marsh. 1975. A model of NaCl and water flow through paracellular pathways of renal proximal tubules. *J. Membrane Biol.* 23: 305-347.

Kedem, O., and J. A. Katchalsky. 1962a. Permeability of composite membranes. Part I. Electric current, volume flow and flow of solute through membranes. *Trans. Farad. Soc.* 59: 1918-1930.

Kedem, O., and A. Katchalsky. 1962b. Permeability of composite membranes. Part 2. Parallel elements. *Trans. Farad. Soc.* 59: 1931-1940.

Kyte, J. 1976. Immunoferritin determination of the distribution of $(Na^+ + K^+)$ ATPase over the plasma membranes of renal convoluted tubules. II. Proximal segment. *J. Cell. Biol.* 68: 304-318.

Maunsbach, A. B., and E. L. Boulpaep. 1975. Correlation of hydrostatic pressure changes and ultrastructure of the paracellular shunt in proximal tubule. In *Abstracts*, 6th Intern. Congr. Nephrol., no. 147.

Ogilvie, J. T., J. R. McIntosh, and P. F. Curran. 1963. Volume flow in a series-membrane system. *Biochim. Biophys. Acta* 66: 441-444.

Patlak, C. S., D. A. Goldstein, and J. F. Hoffman. 1963. The flow of solute and solvent across a two-membrane system. *J. Theoret. Biol.* 5:426-442.

Sackin, H., and E. L. Boulpaep. 1975. Models for coupling of salt and water transport: proximal tubular reabsorption in *Necturus* kidney. *J. Gen. Physiol.* 66: 671-733.

Schafer, J. A., C. S. Patlak, and T. E Andreoli. 1975. A component of fluid absorption linked to passive ion flows in the superficial pars recta. *J. Gen. Physiol.* 66: 445-471.

Schmidt-Nielsen, B., and L. E. Davis. 1968. Fluid transport and tubular intercellular spaces in reptilian kidneys. *Science* 159: 1105-1108.

Segel, L. A. 1970. Standing gradient flows driven by active solute transport. *J. Theor. Biol.* 29: 223-250.

Smyth, D. H., and E. M. Wright. 1966. Streaming potentials in the rat small intestine. *J. Physiol.* 182: 591-602.

Stirling, E. S. 1972. Radioautographic localization of sodium pump sites in rabbit intestine. *J. Cell. Biol.* 53: 704-714.

Tormey, J. D., and J. M. Diamond. 1967. The ultrastructural route of fluid transport in rabbit gall bladder. *J. Gen. Physiol.* 50: 2031-2060.

Wall, B. J., J. L. Oschmann, and B. Schmidt-Nielsen. 1970. Fluid transport: concentration of the intercellular compartment. *Science* 167: 1497-1498.

Wedner, H. J., and J. M. Diamond. 1969. Contributions of unstirred layer effects to apparent electrokinetic phenomena in the gall bladder. *J. Membrane. Biol.* 1: 92-108.

Weinbaum, S., and J. R. Goldgraben. 1972. On the movement of water and solute in extracellular channels with filtration, osmosis and active transport. *J. Fluid Mech.* 53: 481-512.

Welling, L. W., and J. J. Grantham. 1972. Physical properties of isolated perfused renal tubules and tubular basement membranes. *J. Clin. Invest.* 51: 1063-1075.

Whittembury, G. 1967. Sobre los mecanismos de absorción en el tubo proximal del riñón. *Act. Cient. Venezolana.* Suppl. 3: 71-83.

Whittembury, G., F. A. Rawlins, and E. L. Boulpaep. 1973. Paracellular pathway in kidney tubules: electrophysiological and morphological evidence. In H. H. Ussing and N. A. Thorn, eds., *Transport mechanisms in epithelia*, pp. 577-595. Munksgaard, Copenhagen.

Whittembury, G., N. Sugino, and A. K. Solomon. 1960. Effects of antidiuretic hormone and calcium on the equivalent pore radius of slices from *Necturus*. *Nature* 187: 699-701.

Windhager, E. E., E. L. Boulpaep, and G. Giebisch. 1966. Electrophysiological studies on single nephrons. In *Proceedings of the Third International Congress of Nephrology, Washington.* Vol. 1, pp. 35-47.

Windhager, E. E., G. Whittembury, D. E. Oken, H. J. Schatzmann, and A. K. Solomon. 1959. Single proximal tubules of the Necturus kidney. III. Dependence of H_2O movement on NaCl concentration. *Am. J. Physiol.* 197: 313-318.

15 Modes of Sodium Transport in the Kidney

GUILLERMO WHITTEMBURY,
MADALINA CONDRESCU-GUIDI,
MARGARITA PEREZ-GONZALEZ DE LA MANNA,
AND F. PROVERBIO

Na is reabsorbed in the kidney in what is thought to be a two-step process. In the first step, Na that is to be reabsorbed (after being supplied by the glomerular filtrate to the tubular lumen) enters the cells from the tubular lumen by moving down an electrochemical potential gradient. The second step involves the activity of the Na pump situated at the peritubular boundary of tubule cells. This pump extrudes Na out of the cell (actively, against the electrochemical potential gradient) into the peritubular spaces. From there, Na enters the peritubular capillaries and is removed by the blood stream.

The renal Na pump is thought to be a Na-for-K exchange pump in the sense that Na ions are expelled out of the cell into the peritubular space at the same time that K ions are taken back from the peritubular spaces into the cell. The Na pump is sensitive to the action of ouabain. The energy for its activity comes from the hydrolysis of ATP. Therefore, activity (or inhibition, or both) of a Na,K-ATPase preparation of cell membranes should correlate with the activity (or inhibition) of Na and K movements across the peritubular cell membrane.

To test the possibility that the postulated Na-for-K exchange pump is responsible for Na reabsorption, direct measurements of unidirectional Na and K movements across the peritubular membrane have to be related to Na reabsorption. Measurements of ion fluxes across the peritubular membrane are at present technically difficult. Hence, one has to resort to an indirect approach. One approach is to carry out experiments trying to examine whether net Na reabsorption correlates with peritubular K concentration in proximal tubules perfused via the lumen and the peritubular capillaries. Indeed, Na reabsorption has been found to be proportional to the peritubular K concentration in the mammalian kidney (Windhager, 1974). A similar relationship has been found to hold in the perfused amphibian kidney. In the latter, furthermore, net Na reabsorption has been shown to be directly proportional to peritubular K influx (Giebisch, Sullivan, and Whittembury, 1973).

Other types of experiments have been performed with kidney slices and isolated tubules. In these preparations high concentrations of inhibitors can be used, and drastic changes in cellular and bathing fluid ionic concentrations can be induced to alter the normal function. The effects of these perturbations can be correlated with the observed ionic distribution and movements. With these experimental techniques, several lines of evidence have been obtained showing that the Na pump exchanges Na for K. It has been found that, in the first place, Na and K move in opposite directions across tubular cell membranes. Thus, in slices of renal tissue, cell Na decreases when cell K increases; that is, a clear relationship can be established between net movements of Na from cell to extracellular space and net entry of K into the cells, and vice versa (Mudge, 1951; Whittembury, 1965; Willis, 1968; Kleinzeller, 1972; Whittembury and Grantham, 1976). Also, net Na extrusion depends on the K concentration in the extracellular medium (Whittam and Willis, 1963; Whittembury, 1965, 1968), and net K entry depends on intracellular Na concentrations (Whittam and Willis, 1963). In the second place, ouabain inhibits extrusion of Na and uptake of K such that the ratio of net Na extrusion inhibited to net K entry inhibited is one-to-one (Whittam and Willis, 1963; Whittembury, 1968; Whittembury and Proverbio, 1970). In the third place, cells of kidney slices can also extrude Na in the presence of Rb and of Cs instead of K in the extracellular fluid. These ionic movements are in part inhibited by ouabain. Again, the magnitude of the net Rb or net Cs entry inhibited by ouabain is in a one-to-one ratio to the magnitude of the net efflux of Na inhibited by ouabain (Proverbio and Whittembury, unpublished). Finally, as illustrated in Figure 15.1, the dose-response curves of the inhibitory action of ouabain on the Na-K exchange, just described, is closely paralleled by a similar inhibitory action of ouabain upon the Na, K-ATPase activity of a cell membrane preparation from the same tissue (Whittam and Wheeler, 1961; Willis, 1968; Proverbio, Robinson, and Whittembury, 1970).

There are further reasons to link ATP and Na,K-ATPase with Na-K transport in the kidney. Since Na-K exchange is inhibited by DNP and by anoxia, one can conclude that the Na-K pump requires renewed synthesis of ATP. Furthermore, Na,K-ATPase is found in very high concentrations in the kidney, and it seems strategically located at the peritubular part of the tubular cells where one expects the Na-K pump to be located (Schmidt et al., 1974). Also, the ATPase activity increases when animals have been adapted to conditions in which the activity of the Na pump is expected to be enhanced because the kidney reabsorbs more Na (that is, after administration of adrenal steroids, unilateral nephrectomy, high-

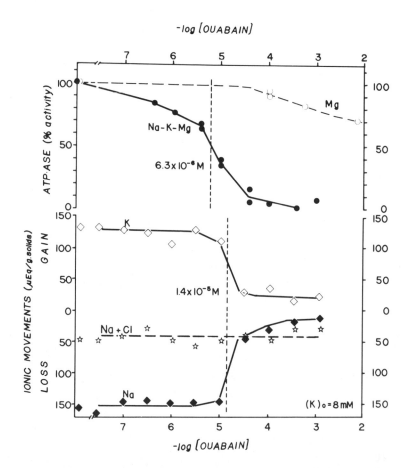

FIGURE 15.1. Dose-response curves of Na,K-ATPase and Mg-ATPase activity (*top*) of a kidney cortex microsomal preparation, and ionic movements (exchange of Na for K and extrusion of Na with Cl) observed in guinea pig kidney cortex slices bathed in a medium containing 8 mm K (*bottom*), as a function of the ouabain concentration. (From Proverbio, Robinson, and Whittembury, 1970, with permission of the authors and *Biochim. Biophys. Acta.*)

protein diet) or excretes more K (chronic K loading) (Epstein and Silva, 1974). Finally, the ATPase activity decreases when the activity of the Na-K pump is reduced, that is, after adrenalectomy (Epstein and Silva, 1974) and during extracellular volume expansion (Kramer and Gonick, 1974)—which diminishes reabsorption in the kidney—and in post-obstructive diuresis in the rat (Wilson et al., 1974).

These observations support the view that energy for the function of the Na pump is derived from the hydrolysis of ATP at the level of the cell membrane, stimulated by Na from inside and by K from the outside, and sensitive to the cardiac glycoside ouabain. However, existing observations do not completely link Na-extruding mechanisms and ATP hydrolysis. For example, it has not been possible in the kidney—as it has been in the red cell and in the axon—to test whether the transport system shows sidedness (that is, whether it indeed requires Na within the cell and K outside the cell) or vectoriality, or whether ATP is synthetized within the cell when the Na-K pump is made to run backwards.

Furthermore, a number of observations do not fit easily the classical view outlined above. Some of these observations emerge from studies of Na reabsorption in the whole kidney. Others result from studies of volume regulation in kidney cells under isosmotic conditions or from experiments on hypoosmotic volume regulation in isolated tubules. These observations indicate that the classical scheme of the Na-K pump should be modified or complemented.

Na Reabsorption in the Whole Kidney

Studies in the mammalian kidney indicate that ouabain does not completely inhibit Na reabsorption. If there was only one system to effect Na reabsorption, complete inhibition of the energy supply to the system (that is, inhibition of Na,K-ATPase activity by ouabain) should block Na reabsorption of the kidney completely. However, about one-half to one-third of the filtered Na (and Cl) continues to be reabsorbed in the mammalian kidney in vivo and in the perfused rat kidney in vitro under conditions in which the Na,K-ATPase activity is 100 percent inhibited. This remaining moiety of Na reabsorption can be blocked with cyanide and iodoacetate. Consequently, this ouabain-resistant Na reabsorption is dependent on metabolic energy. One may conclude that reabsorptive mechanisms other than the fully inhibited Na-K pump must be operative to account for that fraction of sodium that continues to be reabsorbed (Epstein and Silva, 1974; Ross et al., 1974).

Other evidence suggesting transport mechanisms in addition to the classical pump stems from experiments performed in the perfused amphibian kidney. It has already been mentioned that Na reabsorption in the amphibian kidney under control conditions correlates well with peritubular K uptake. However, this relationship can be markedly changed by different manipulations: near normal Na reabsorption may coexist with very low K uptake and, under other conditions, negligible Na and Cl reabsorption can be observed with near-normal K uptake (Giebisch et al., 1973).

Isosmotic Volume Regulation in Kidney Slices

Cells of mammalian kidney slices swell and become loaded with Na and Cl and lose K if they are leached in isosmotic media under appropriate conditions, for example, in the cold (Mudge, 1951; Kleinzeller, 1972; Whittembury and Grantham, 1976). Cell volume can be restored if the tissue is subsequently rewarmed; in this phase of the experiment, if the external medium contains K, cells gain K and extrude Na. Clearly, if regulation of cell volume and maintenance of normal intracellular ionic concentration were due entirely to the activity of a single pump or mechanism, one would expect volume regulation to be inhibited in the presence of ouabain concentrations that fully inhibit Na for K exchange. However, it has been observed that, even in the presence of ouabain concentrations that completely inhibit the exchange of Na for K, a sizable amount of Na can be extruded with Cl with the restoration of cell volume. Thus, it appears that the ouabain-sensitive exchange of Na for K is independent of extrusion of Na with Cl and water, that is, unrelated to volume regulation (Kleinzeller, 1972; Whittembury and Grantham, 1976).

From these observations, the description of two modes of Na extrusion has emerged. (A) *Na can be expelled in exchange for K.* This mode of Na extrusion is very sensitive to ouabain; it seems little affected by ethacrynic acid (Whittembury, 1968; Whittembury and Proverbio, 1970; Proverbio et al., 1970), furosemide, and triflocin (Perez-Gonzalez, Proverbio, and Whittembury, unpublished). (B) *Na can be expelled with Cl and water.* This mode of Na extrusion (which is essential to volume regulation) proceeds in the absence of K in the bath and in the presence of ouabain at concentrations that clearly inhibit the exchange of Na for K. Mode B is curtailed by ethacrynic acid (Whittembury and Proverbio, 1970; de Jairala et al., 1972), furosemide (de Jairala et al., personal communication) and by triflocin (Perez-Gonzalez et al., unpublished). It is stimulated (even in the presence of ouabain) by angiotensin (Munday, Parsons, and Poat, 1971). Both modes of Na extrusion are inhibited if ouabain is added to either ethacrynic acid (Whittembury and Proverbio, 1970), furosemide, or triflocin (Perez-Gonzalez et al., unpublished). Since both modes of Na extrusion are inhibited by dinitrophenol and anoxia, it may be concluded that hydrolysis of ATP is their source of energy (Whittembury, 1968; Whittembury and Proverbio, 1970).

Three explanations, each proposing separate mechanisms working in parallel, have been offered to account for the observations just described (Kleinzeller, 1972; Whittembury and Grantham, 1976). They are illustrated in Figure 15.2. The cryptic pump hypothesis (Fig. 15.2b) maintains that ouabain is not inhibiting all Na pumps (Whittembury, 1967; Willis, 1968). According to this view, the Na pumps located on the surface of the kidney in the peritubular basement membrane are easily accessible from the external bathing solution. Therefore, these pumps are inhibitable by ouabain. In contrast, those pumps situated deep in the crypts in the peritubular membrane of kidney cells are thought to be inaccessible from the external solution. Hence, ouabain does not reach

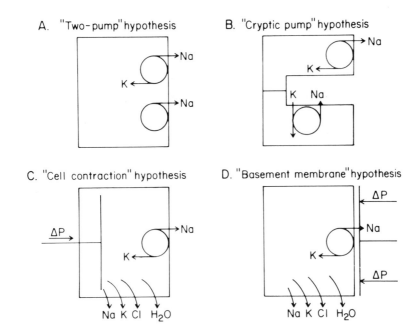

FIGURE 15.2. Working hypotheses to complement the classical Na pump.

them. Na extrusion in a medium without K can also be accounted for, because the cellular K that is lost into the crypts does not diffuse into the bulk bathing solution but is readily taken up by the Na pump. As a consequence the K concentration is not really zero within the crypts. This allows Na to be expelled from the cells even in the presence of nominally zero K concentration in the external medium.

Thus, Na extrusion could continue in the presence of ouabain, since ouabain would not reach the Na pump sites in the depth of the crypts. This view has received recent support from work in isolated tubules. It was found that while both modes of Na extrusion could normally be described, and extrusion of Na with Cl could be observed in the presence of ouabain, no extrusion of Na could be observed in the presence of ouabain if the tubules had been dialyzed so as to lower their K concentration (Podevin and Boumendil-Podevin, 1972). Although we had considered this possibility previously (Whittembury, 1967), we found it difficult to understand how ouabain could be sufficiently excluded from the crypts to leave even a fraction of the Na pumps uninhibited. Exchange of Na for K can be maximally inhibited by ouabain at much lower levels than those required to inhibit extrusion of Na with Cl (Whittembury, 1968; Whittembury and Proverbio, 1970). One of the limitations of the experiments of Podevin and Boumendil-Podevin is that extrusion of Na with Cl and water failed to occur under conditions in which the intracellular ionic composition was so severely distorted that the supply of metabolic energy for the ouabain-resistant mode of extrusion of Na might have been severely interfered with.

Kleinzeller (1972) has provided evidence for the existence of a mechanochemical (contractile) mechanism, with physical properties determined by Ca and ATP metabolism, that may counterbalance a major hydrostatic pressure gradient in kidney cells (Fig. 15.2c). This mechanism can qualitatively explain most of the observations quoted above, namely extrusion of Na or Li with Cl and water, the relationship between Ca, ATP, and volume regulation, the finding of a Ca-ATPase, and presence of filamentous contractile structures at the basal aspects of the cells.

Finally, the existence of a second Na pump (Whittembury, 1968; Whittembury and Proverbio, 1970; Robinson, 1972) or a volume regulating mechanism that would extrude Na has also been considered (Fig. 15.2a) (Macknight, 1968; Macknight, Pilgrim, and Robinson, 1974). The main observations to support this idea are: (1) Mode B of Na extrusion proceeds even in the absence of K in the bathing medium and in the presence of ouabain. (2) The inhibitory action of ouabain differs from the inhibitory actions of ethacrynic acid, triflocin, and furosemide. Ouabain inhibits mainly mode A of Na extrusion and the latter group of inhibitors mainly mode B of Na extrusion. (3) Angiotensin stimulates mode B when mode A is inhibited by ouabain (Munday et al., 1971). Although the cryptic pump hypothesis seems the most reasonable choice when only the action of ouabain is considered, the observation of two distinct types of inhibitors is more likely to be consistent with the existence of two types of pumps.

At present, it seems that further experiments are required to resolve these problems. One of the limitations is that highly artificial conditions (of Na-loaded cells) are used to demonstrate different modes of Na extrusion and it is difficult to assess the relative role of both modes of Na extrusion under physiological conditions.

A Ouabain-Insensitive Na-Stimulated Mg-Dependent ATPase

In view of the possible existence of a second Na-extruding mechanism— which would also derive its energy from hydrolysis of ATP—Proverbio, Guidi, and Whittembury (1975) decided to look for a ouabain-insensitive, Na-stimulated ATPase in microsomal preparations of kidney tissue. Fresh kidney preparations showed the well known Mg-dependent ATPase and Na,K-ATPase activities with ouabain sensitivity (Proverbio et al, 1970). The preparation was aged, since aging diminishes the Mg-dependent activity and enhances the Na-K stimulation (Robinson, 1970). When Na was added to test for the presence of a Na-stimulated Mg-dependent ATPase, Na showed an inhibitory action in the fresh preparation, a result in agreement with work by Gutman, Wald, and Czaczkes

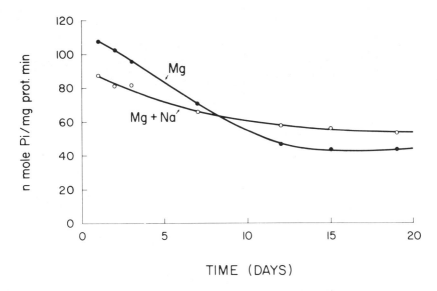

EFFECT OF STORAGE ON ATPase ACTIVITY

FIGURE 15.3. Activity of Mg-dependent ATPase as a function of days of storage, studied in the absence (•————•) and in the presence (o————o) of 100 mM Na. (From Proverbio, Guidi, and Whittembury, 1975, with permission of the authors and *Biochim. Biophys. Acta.*)

(1973) and by Cole and Dirks (1971). However, the degree of Na-induced inhibition decreased with days of storage, so that after about 7-10 days no inhibition by Na was present. With further storage, a small but consistent stimulation by Na was noticed. This is shown in Figure 15.3. Encouragingly, this stimulation was insensitive to the addition of up to 10 mM ouabain. Experiments have ruled out the possibility that the ouabain-insensitive Na-stimulation is related to the activity of a Ca-ATPase or to the presence of mitochondrial Mg-ATPases, or to other forms of Na stimulation of the ATPase activity that are sensitive to the action of ouabain (Proverbio et al., 1975).

In short, this new form of ATPase activity has the following characteristics: (1) It is stimulated by Na and by Li, and not by other cations (Table 15.1). (2) The Na stimulation is independent of the anion accompanying Na. (3) It is insensitive to the presence of ouabain. (4) It does not require K for its activity. (5) It is more sensitive to inhibition by ethacrynic acid than the activity of Na,K-ATPase or Mg-ATPase (Table 15.2). (6) It is inhibited by 1 mM triflocin, which does not inhibit Mg-ATPase or Na,K-ATPase activity (Table 15.2). (7) It is inhibited by 10^{-5}M furosemide, a concentration that leaves the other two ATPase activities

TABLE 15.1. Action of different monovalent cations on the Mg-ATPase activity of a microsomal preparation. (From Proverbio, Guidi, and Whittembury, 1975.)

Incubation medium	ATPase activity (10^- mol P_i per mg protein per min)
Mg alone	52
Mg + Na	65[a]
Mg + Li	60[a]
Mg + K	55
Mg + Rb	53
Mg + Cs	56
Mg + NH_4	52
Mg + Choline	55

Note: Preparation was aged 12 days; 5 mM Mg, 100 mM of Cl salt of the cations given, and 10^{-3} ouabain were used.

[a]Only the results with Na and with Li were statistically different ($p < 0.001$) from those obtained with Mg alone.

uninhibited (Table 15.2). Ethacrynic acid, triflocin, and furosemide inhibit preferentially the extrusion of Na with Cl and water, and they inhibit only slightly the exchange of Na for K in the kidney (Perez-Gonzalez et al., unpublished). (8) This Na-stimulated ATPase activity appears clearly after 14 days of storage of the microsomal preparation from untreated kidney tissue. The stimulation by Na appears earlier (about 7 days storage) if the tissue has been treated with angiotensin II, an agent that has been shown to increase ouabain-resistant Na extrusion in the kidney (Munday et al, 1971). The pH optimum for the activity of the Na-stimulated ATPase is 6.8-7.0. It seems that spontaneous acidification of the buffer (originally at pH 7.2) also accelerates the appearance of the stimulation by Na (Proverbio, Guidi, and Whittembury, unpublished).

Some of the characteristics listed above are consistent with the idea that this ATPase is the energy source of the system responsible for the proposed mode B of Na extrusion. Much work remains to be performed to solve unanswered questions. Can the yield of this new ATPase activity be increased? What is the possible physiological role of these observations? We also need to know whether this Na-stimulated ATPase is present in vivo. We have to rule out the possibility that the Na-stimulated ATPase activity is due to other endoplasmic reticulum contaminants and is not present in the cell membranes. Other questions are, does intracellular Na indeed activate this ATPase in the intact cell? Can changes produced in the living organism in mode B of Na extrusion be correlated with changes in the activity of the Na-stimulated ATPase?

TABLE 15.2. Inhibitors and ATPases. (From Perez-Gonzalez et al., unpublished.)

	Mg^{++}	Na^+	$Na^+ + K^+$
Control	69.3 ± 2.3	23.7 ± 2.4	80.7 ± 4.7
Furosemide 10^{-5} M	68.5 ± 2.5	3.4 ± 3.9	80.1 ± 4.9
Control	64.5 ± 3.0	17.1 ± 3.2	51.6 ± 3.4
Triflocin 10^{-3} M	64.1 ± 0.9	0.1 ± 1.4	51.4 ± 1.4
Control	39.0 ± 0.1	12.5 ± 1.8	47.0 ± 1.3
Ethacrynic A. 10^{-3} M	35.0 ± 0.6	3.5 ± 1.0	31.0 ± 1.6
Ethacrynic A. 2×10^{-3} M	27.5 ± 1.0	0.0 ± 0.6	28.0 ± 1.0

Note: Table shows action of inhibitors on the Mg-ATPase activity, on the Na-stimulated ATPase activity (Mg + Na minus Mg activity), and on the Na + K-stimulated ATPase activity (Mg + Na + K minus Mg activity) of three preparations. Mg concentration was 5 mM, Na concentration (as NaCl) was 100 mM, K concentration (as KCl) was 10 mM. Ouabain (1 mM) was present in the assay of the Mg and Mg + Na activities. No ouabain was added when the Mg + Na + K activity was assayed. N = 3-5. Similar results with each agent were obtained when the experiment was repeated on three different lots of animals (preparations). (ATP hydrolyzed is given as nmole P_i per mg protein per min.)

Finally, if there are two Na pumps, what is their relative role in kidney function?

Regulation of Cell Volume in Anisosmotic Media

Recent studies suggest that kidney cells are "good osmometers." Thus, Dellasega and Grantham (1973) have observed that isolated rabbit renal tubules change volume rapidly in response to sudden alterations in the osmolality of the external bathing medium (Fig. 15.4). Importantly, swollen cells rapidly adjust their volume towards the initial resting volume. For their studies these authors dissected straight portions of proximal tubule out of the rabbit kidney. The tubules were not perfused and their lumina were collapsed. With this technique, as with kidney slices, changes in cell volume should be due mainly to movements of fluid across the peritubular membrane, and the mechanism of volume regulation should presumably be located in the basolateral plasma membrane (Whittembury and Grantham, 1976). As shown in Figure 15.4 (dashed line) in experiments carried out at 37°C, tubule cell volume increased rapidly when the bath was changed from isosmotic to hyposmotic medium. Although not shown in the figure, it was found that the peak swell-

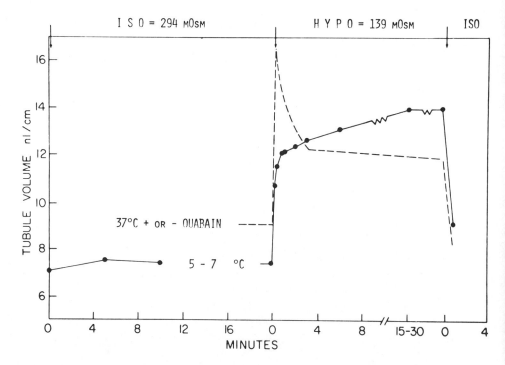

FIGURE 15.4. Effect of hyposmotic medium on proximal straight tubular cell volume. Note the sudden increase in tubular cell volume (dashed line) followed by a period in which the cell volume returned towards the initial value despite continuing in hyposmotic medium. Tubules immersed in isosmotic medium containing 10^{-4}M ouabain for 20 minutes before being immersed in hyposmotic medium with the same ouabain concentration showed similar behavior. They are also represented by the dashed line. The continuous line represents the effect of cooling on volume regulation of proximal straight tubules in hyposmotic medium. (Modified from Dellasega and Grantham, 1973, with permission of the authors and the *American Journal of Physiology.*)

ing response was inversely related to bath osmolality, indicating osmometric behavior of this nephron segment. Importantly, the tubule cells adjusted their volume towards normal within 3-5 minutes when kept in the hyposmotic bath. On the other hand, reduction of bath temperature to 5-7°C prior to and during the initial osmometric phase inhibited markedly the volume regulatory phase (Fig. 15.4, continuous line). Grantham and Dellasega (unpublished) have recently studied the pattern of intracellular Na and K content during this impaired regulatory response at low temperature. As shown in Table 15.3, net loss of K was detected as in tubules incubated in hyposmotic medium at 37°C. However, in contrast to the normothermic state, in hypothermia the intracellular Na content increased rather than decreased. Thus, the failure to volume-regulate at

TABLE 15.3. Effect of hypothermia on electrolyte content of proximal straight tubules in hyposmotic media. (From Grantham et al., unpublished.)

	Content (10^{-12} Eq/mm)		
Electrolyte	Control Isosmotic 37°C	Hyposmotic 37°C	Hyposmotic 10°C
Na	28.9	−3.9	+20.2
K	100.0	−12.4	−12.5

low temperature in the hyposmotic medium is probably due to the failure to extrude intracellular Na. As a consequence, the hypothermia experiments may be interpreted to indicate that Na transport out of the cell (presumably by an active mechanism) plays an important role in volume regulation under these circumstances (Grantham et al, 1976).

If active Na transport is importantly related to volume regulation in hyposmotic medium, then ouabain, the classical inhibitor of the ATPase-dependent component of Na transport, should block volume regulation. However, the dashed line in Figure 15.4 illustrates that placement of ouabain-treated tubules (10^{-4}M ouabain) in hyposmotic media caused swelling, followed by a volume regulatory phase quite similar to that of the control tubules. Thus, ouabain failed to abolish volume regulation of renal tubules in a hyposmotic medium.

Recent experiments of Grantham et al. (1977), measuring Na and K content of tubules in the presence of ouabain, indicate that K loss is not of primary importance in volume regulation. As illustrated in Figure 15.5, in ouabain-treated tubules the volume regulation was accompanied by the unexpected loss of intracellular Na rather than by the expected loss of K (which is the case in noninhibited tubules). Thus, it is clear that cation loss (either of K or of Na or of both) is involved in the contraction of cell volume in hyposmotic medium (Grantham et al, 1977).

Clearly, in the ouabain-treated Na-loaded cell, hyposmotic volume adjustment is principally due to the loss of Na and anion. This suggests that a ouabain-insensitive Na efflux mechanism may be a factor in hyposmotic volume regulation, as it is in the isosmotic cell volume regulation referred to above. However, in the opinion of Grantham and co-workers, there are at least two alternative explanations (besides that of a ouabain-insensitive Na pump) of hyposmotic volume regulation in sodium-rich cells. One is the possibility that residual active Na transport by the classical Na pump is incompletely inhibited by the glycoside (Fig. 15.2b). The second derives from their observation that, in the presence of oua-

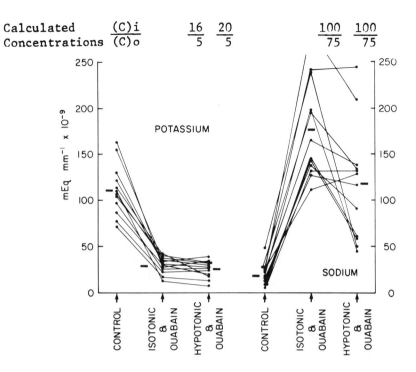

Calculated $\frac{(C)i}{(C)o}$ ... $\frac{16}{5}$ $\frac{20}{5}$ ... $\frac{100}{75}$ $\frac{100}{75}$
Concentrations

FIGURE 15.5. Effect of hyposomotic medium on ionic Na and K content of proximal straight tubules. Results of control experiments performed without ouabain are compared with experiments performed after immersion of the tubules for 30 minutes in isosmotic solution containing 10^{-4}M ouabain. The hyposmotic solution also contained the same ouabain concentration. At the top of the figure are given intra- to extracellular concentration ratios. At the peak of swelling, $[K_i]$ / $[K_o]$ would be 29/1.8/5=16/5 and $[Na_i]$ / $[Na_o]$ would be 177/1.8/75= 100/75. Once volume regulation occurred, the ionic ratios would be 25/5 and 100/75, for K and Na, respectively. (From Grantham et al., 1977, with permission of the authors and the *American Journal of Physiology*.)

bain, fluid and salts may be ultrafiltered across the peritubular membrane in response to oncotic gradients equivalent to 20-40 mm Hg. It is conceivable that hydrostatic pressure generated by swelling of cells within the constraint of the cylindrical basement membrane would promote the ultrafiltration of Na and Cl from cytoplasm to bath (Grantham, personal communication). As quoted above, Kleinzeller (1972) has proposed a similar mechanism, but in Kleinzeller's view the plasma membrane—instead of the basement membrane—participates in the generation of transmembrane hydrostatic pressure gradients. This is illustrated in Figure 15.2 *c-d*. What is hard to understand at present is the observation that this postulated mechanism residing within the basement membrane is inhibited by cold.

The examination of the role of active Na extrusion in volume regulation under the experimental conditions just described requires also a careful assessment of the electrochemical potential differences across the cell membrane. Direct measurements of cell concentrations and electrical potentials are not available at present. However, as a first approximation, one can make an estimate of these values using the data of Grantham and his co-workers. From Figure 15.4 one can estimate tubular volumes to be 1.8 nl/mm and 1.2 nl/mm, respectively, in the hyposmotic medium at the peak of swelling and after balance when a "regulated volume" has been attained. From these values and the figures for Na and K content given in Figure 15.5, the cellular Na and K concentrations can be calculated. They are given at the top of Figure 15.5. It can be seen that at the peak of swelling the intracellular-to-extracellular Na concentration ratio would be 100:75. This concentration ratio would favor net Na efflux only if the membrane potential (cell inside negative) is less negative than -8 mv (-61.5 log 100/75). The possible actual value of the cell membrane may be calculated from the K concentration ratio to range between -30 mv (-61.5 log 16/5) and -40 mv (-61.5 log 25/5). These values for the membrane potential would outweigh the opposite Na concentration ratio and should result in retention of Na within the cell. Hence, the actually observed net Na efflux would have to be viewed as an active process, provided that other driving forces are absent. Obviously, these conclusions will have to be reassessed when direct information becomes available.

In summary, several lines of evidence suggest that a single Na pump conceived along the classical lines is insufficient as the sole explanation of a large number of experimental results. These are: incomplete blockage of Na reabsorption by ouabain in the perfused mammalian kidney, dissociation of Na transport from K uptake in the amphibian kidney, and incomplete blockage of isosmotic volume regulation in kidney slices and of hyposmotic regulation in isolated tubules by inhibitors of the "classical" Na-K exchange pump. There is as yet no compelling evidence that the "classical" pump can be completely inhibited by ouabain, cold, or deletion of external potassium. In order to explain sodium extrusion in presumed absence of the classical Na-K exchange pump, other explanations have been put forward, for example, the existence of a separate mechanism (or Na pump) that expels Na from the cell and that regulates cell volume, or the possibility that the basement membrane or the plasma membrane intervenes in expelling fluid and salts out of the cell. Further studies are required to choose among these alternatives; we favor the idea of a second pump mechanism. In addition, the physiological role of these different sodium extrusion processes in normal renal function remains to be explored.

ACKNOWLEDGMENTS

It is a pleasure to thank Dr. J. J. Grantham for his kind permission to quote unpublished experiments from his laboratory. Thanks are due to Mrs. M. A. Silva and to Mr. Henry Linares for excellent assistance. Part of this research was supported with grants for Projects 0296 and 31.26.S2-0173 from CONICIT and with a grant from the Regional Program for Scientific and Technological Development of OEA.

REFERENCES

Burg, M. B. 1976. Tubular chloride transport and the mode of action of some diuretics. *Kidney Int.* 9: 189-197.

Cole, C. H., and J. H. Dirks. 1971. A comparison of Na$^+$ activation of ATPase in the red cell, renal cortex and renal medulla. *Can. J. Physiol. Pharmacol.* 49: 63-69.

de Jairala, S. W., O. Saravalli, J. Palazzi, A. P. Gardia, and E. Vergara. 1975. Influence of furosemide, ethacrynic acid and ouabain on the oxygen consumption and electrolyte content of dog kidney cortex slices. In preparation.

de Jairala, S. W., A. Vieyra, and M. MacLaughlin. 1972. Influence of ethacrynic acid and ouabain on the oxygen consumption and potassium and sodium content of the kidney external medulla of the dog. *Biochim. Biophys. Acta.* 279: 320-330.

Dellasega, M., and J. J. Grantham. 1973. Regulation of renal tubule cell volume in hypotonic media. *Am. J. Physiol.* 224: 1288-1294.

Epstein, F. H., and P. Silva. 1974. Role of Na, K-ATPase in renal function. *Ann. N. Y. Acad. Sci.* 242: 519-526.

Giebisch, G., L. P. Sullivan, and G. Whittembury. 1973. Relationship between tubular net sodium reabsorption and peritubular potassium uptake in the perfused Necturus kidney. *J. Physiol.* 230: 51-74.

Grantham, J. J., C. Lowe, M. Dellasega, and B. Cole. 1977. Effect of hypotonic medium on K and Na content of proximal renal tubules. *Am. J. Physiol.* 232: F42-F49

Gutman, Y., H. Wald, and W. Czaczkes. 1973. Urea and sodium: effect on microsomal ATPase in different parts of the kidney. *Pflügers Arch.* 345: 81-92.

Kleinzeller, A., 1972. Cellular transport of water. In *Metabolic pathways: metabolic transport.* Vol. 6, pp. 91-131. Academic Press, New York.

Kramer, H. J., and H. C. Gonick. 1974. Effect of extracellular volume expansion on renal Na-K-ATPase and cell metabolism. *Nephron* 12: 281-296.

Macknight, A. D. C. 1968. Water and electrolyte contents of rat renal cortical slices incubated in potassium-free media and in media containing ouabain. *Biochim. Biophys. Acta* 150: 263-270.

Macknight, A. D. C., J. P. Pilgrim, and B. A. Robinson. 1974. The regulation of cellular volume in liver slices. *J. Physiol.* 238: 279-294.

Mudge, G. H. 1951. Studies on potassium accumulation by rabbit kidney slices: effect of metabolic activity. *Am. J. Physiol.* 165: 113-127.

Munday, K. A., B. J. Parsons, and J. A. Poat. 1971. The effect of angiotensin on cation transport by rat kidney cortex slices. *J. Physiol.* 215: 269-282.

Podevin, R. A., and E. F. Boumendil-Podevin. 1972. Effects of temperature, medium K$^+$, ouabain and ethacrynic acid on transport of electrolytes and water by separated renal tubules. *Biochim. Biophys. Acta* 282: 234-249.

Proverbio, F., M. Condrescu-Guidi, and G. Whittembury. 1975. Ouabain-insensitive Na$^+$-stimulation of an Mg$^+$-dependent ATPase in kidney tissue. *Biochim. Biophys. Acta* 394: 281-292.

Proverbio, F., J. W. L. Robinson, and G. Whittembury. 1970. Sensitivities of Na$^+$-K$^+$-ATPase and Na$^+$ extrusion mechanisms to ouabain and ethacrynic acid in the cortex of the guinea-pig kidney. *Biochim. Biophys. Acta* 211: 327-336.

Robinson, J. W. L. 1970. The difference in sensitivity to cardiac steroids of (Na$^+$ + K$^+$)-stimulated ATPase and amino acid transport in the intestinal mucosa of the rat and other species. *J. Physiol.* 206: 41-60.

Robinson, J. W. L. 1972. The inhibition by glycine and beta-methyl glucoside transport in dog kidney cortex slices by ouabain and ethacrynic acid: contribution to the understanding of the sodium pumping mechanism. *Comp. Gen. Pharmacol.* 3: 145-157.

Ross, B., A. Leaf, P. Silva, and F. H. Epstein. 1974. Na-K-ATPase in sodium transport by the perfused rat kidney. *Amer. J. Physiol.* 226: 624-629.

Schmidt, U., H. Schmidt, B. Funk, and U. C. Dubach. 1974. The function of Na, K-ATPase in single portions of the rat nephron. *Ann. N. Y. Acad. Sci.* 242: 489-500.

Whittam, R., and K. P. Wheeler. 1961. The sensitivity of a kidney ATPase to ouabain and to sodium and potassium. *Biochim. Biophys. Acta* 51: 622-624.

Whittam, R., and J. S. Willis. 1963. Ion movements and oxygen consumption in kidney cortex slices. *J. Physiol.* 168: 158-177.

Whittembury, G. 1965. Sodium extrusion and potassium uptake in guinea-pig kidney cortex slices. *J. Gen. Physiol.* 48: 699-717.

Whittembury, G. 1967. Sobre los mecanismos de absorción en el tubo proximal del riñón. *Acta Cient. Venezolana* 3 (Suppl.): 71-83.

Whittembury, G. 1968. Sodium and water transport in kidney proximal tubular cells. *J. Gen. Physiol.* 51: 303s-313s.

Whittembury, G., and J. J. Grantham. 1976. Cellular aspects of renal sodium transport and cell volume regulation. *Kidney Int.* 9: 103-120.

Whittembury, G., and F. Proverbio. 1970. Two modes of Na extrusion in cells from guinea-pig kidney cortex slices. *Pflügers Arch.* 316: 1-25.

Willis, J. S. 1968. The interaction of K$^+$, ouabain and Na$^+$ on the cation transport and respiration of renal cortical cells of hamster and ground squirrel. *Biochim. Biophys. Acta* 163: 516-530.

Wilson, D. R., W. Knox, E. Hall, and K. Sen. 1974. Renal sodium and potassium activated adenosine triphosphatase deficiency during postobstructive diuresis in the rat. *Can. J. Physiol. Pharmacol.* 52: 105-113.

Windhager, E. E., 1974. Some aspects of proximal tubular salt reabsorption. *Fed. Proc.* 33: 21-24.

Index

Acetylcholine, binding to membrane receptors, 99-123

Acetylcholinesterase activity, and functions of red cell band 3 protein, 131, 137, 139

Adenylate cyclase, distribution in epithelial cell membranes, 282, 285-286

Alanine, and intracellular Na and K content, 255-256, 262

Aldolase, association with red cell band 3, 137

Aldosterone, and transepithelial sodium transport, 260

Alga *Chlamydomonas reinhardtii*, thylakoid membranes of. *See* Thylakoid membranes

Amiloride, and transepithelial sodium transport, 260, 264, 265

Amino acids: sodium-dependent transport of, 275, 277; sodium-independent transport of, 279; and sodium transport in ileum, 255-259, 261; transport by membrane vesicles, 162, 163, 165, 171-172

p-Aminohippurate transport, in renal plasma membranes, 280

Aminopeptidase M, in renal proximal tubule epithelial cells, 287

Ammonium compounds, binding to membrane cholinergic receptors, 105

AMP, cyclic, and protein kinase activity of plasma membranes, 284, 285

Anesthetics: and affinity of membrane-bound receptors for cholinergic ligands, 110, 122; binding to receptor-rich membranes, 116-121

Angiotensin, and sodium transport in kidney, 324

1-Anilino-8-naphthalene sulfonate fluorescence, in inverted membrane vesicles, 165

Anion transport, in erythrocyte membranes, 229-248; and action of DAS, 239-240; and band 3 protein binding sites, 231-233; and binding sites for DNFB, 231-233, 238, 240, 242-243, 245, 246, 247; and binding sites for SITS, 231-233; and functions of band 3 protein, 131, 137, 139, 142-153; and functions of membrane constituents, 238-247; and inhibition and binding by DIDS, 234-236, 238; membrane proteins in, 230-238; number of sites for, 233; studies with APMB, 240-247, 248

Antidiuretic hormone. *See* Vasopressin

Antigens in epithelial cells, distribution of, 286-287

APMB, binding sites on red cell membrane, 240-247, 248

Arthrobacter pyridinolis, hexose transport in membrane vesicles of, 166

Ascorbate, in studies of transbilayer migration of lipids, 84, 92

Asymmetry
—in epithelial cells: adenylate cyclase activity, 285-286; protein kinase activity, 282-286; transport systems in, 273-282
—and mechanism of red cell Na-K pump, 191-207
—of membrane components, 61; and arrangement of band 3 protein in red cells, 134, 151; and arrangement of cytochrome oxidase subunits in mitochondria, 64-65; cholesterol, 79, 83; in glycoprotein distribution, 23-24; glycosphingolipids, 79, 80; lipids, 32, 78-95; phospholipids, 79, 80-83; in protein composition, 32, 95; and sidedness of vesicles, 163-167; and transbilayer migration of lipids, 84-95

ATP: and mechanism of red cell Na-K pump, 192, 194-195, 213-218; proton-driven synthesis of, 224-225; and sodium pump activity, 316, 318-319, 323

ATPase: distribution in renal tubular membranes, 303; in luminal and contraluminal membranes of epithelial cells, 273-275; magnesium-dependent, and ouabain-insensitive sodium stimulation, 322-325; mitochondrial, 62; in sarcoplasmic reticulum, lipid-protein association in, 42-44; and sodium and potassium transport in red cells, 191, 192-193, 212-213; and sodium transport in kidney, 317-318

Bacillus megaterium membrane, phospholipid distribution in, 81

Bacillus subtilis membrane, transport in vesicles of, 165

Bacterial membranes: lipid in, 22, 45; transport studies in vesicles of, 160-163

Bacteriochlorophyll *a* protein complex, 46-48, 56

Band 3 protein of red cells, 128-153; and acetylcholinesterase activity, 131, 137, 139; and anion transport, 131, 137, 139, 142-153, 231-236 (and concept of conformational change, 151-153); arrangement in membrane, 133-136, 137, 151; association with intramembrane particles, 135, 136; association with other membrane proteins, 136-137;

binding site for DAS, 240; carbohydrate content of, 129; chemical and physical properties of, 129-132; dimer form of, 135, 151; functions of, 137-142; heterogeneity of, 131, 141; percentage of total membrane protein, 129-131; and sodium and potassium transport, 137; spanning of membrane, 133-136, 151, 233; and sugar transport, 131, 137, 139, 141, 152-153; as transmembrane aqueous channel, 136, 137, 151; and water transfer, 131, 137, 139, 152-153

Benzoylcholine, binding to membrane cholinergic receptors, 105

Bicarbonate, transepithelial transport of, 275; norepinephrine affecting, 282

Biogenesis of membranes, 3-25; continuity principles in, 4, 25; in cytoplasm, and transport of components, 5, 12, 24; de novo assembly in, 6, 17; and dispersive insertion of molecules, 6, 16, 22; and expansion of membrane, 5-6, 16, 25; modes of assembly in, 5, 14, 15, 24; in polysomes attached to membranes, 5, 12, 17-22, 23; and regional insertion of molecules, 6; and sites of synthesis and assembly, 4-5; in thylakoids, 7-22; and topography of assembly, 5-6; and transport from site of synthesis to assembly site, 22-23, 24

Bladder, urinary: epithelial cells of (exchangeable potassium pool in, 265; sodium transport pools in, 260); luminal and contraluminal membranes in, 273

Boundary lipid, 34-45, 54; compared to annular lipid, 43, 44; exchange with fluid bilayer, 39-40; immobilization by protein, 36; surrounded by bilayer halo, 53-54, 56

Calcium: and mechanism of red cell Na-K pump, 205-206; transport of (in membrane vesicles, 165; transepithelial, 275)

Carbamylcholine: and binding of dimethisoquin to receptor-rich membranes, 119-120; interaction with cholinergic receptors, 109; and membrane permeability, 100; receptor affinity for, 107

Cardiolipin: and cytochrome *c* activity, 49-51; and cytochrome oxidase activity, 41, 47

Chemiosmotic theory, of transport in

membranes, 167-172, 222

Chlamydomonas reinhardtii, thylakoid membranes of. *See* Thylakoid membranes

Chloride: and NAP-taurine binding to membrane proteins, 148, 149; transport of (in bladder, norepinephrine affecting, 282; in red cell membranes, 142-143, 229-230, 234)

Chlorobium limicola, bacteriochlorophyll *a* protein complex in, 46-48, 56

Chlorophyll-protein associations, 46-48

Chloroplasts, thylakoid membrane biosynthesis in, 7-22. *See also* Thylakoid membranes

Cholesterol: asymmetric localization in membranes, 79, 83; transbilayer migration in phospholipid vesicles, 90-91

Cholinergic receptors, membrane-bound, 99-123; anesthetics binding to, 116-121; binding functions in cholate solutions, 110, 122; binding of ligands at equilibrium, 104-112, 121; conformation and function of, 121-122; desensitization of, 101, 103, 115 (kinetics of, 121-122); dissociation constants for binding to, 104-109, 121; electron microscopy of, 103; factors affecting affinities of, 109-112, 122; polypeptide composition of, 103; ratio of agonist to antagonist binding sites in, 105; and time dependence of acetylcholine binding, 112-115

Chymotrypsin, red cell exposure to, 235

Colon, rat, sodium transport in, 261

Continuity principles, membrane, 4, 25

Cytochrome b_5, and lipids, 40, 52-53

Cytochrome *c*: association with lipids, 49-51, 56; binding to cytochrome oxidase, 65-67; oxidation affected by phospholipids, 41

Cytochrome oxidase: arrangement of subunits in mitochondrial inner membrane, 64-65; from beef heart mitochondria, 34-40, 62, 64; biosynthesis of, 67-71; lack of, in yeast mutants, 67-70; mitochondrially made, function of, 74-75; polypeptide composition of, 62-64; subunits binding to cytochrome *c*, 65-67; and two-dimensional analysis of mitochondrial inner membrane, 71-74; from yeast, 62-71

Cytochromes, reduction by electron donors, and transport in membrane vesicles, 162, 163, 164

Cytoplasmic ribosomes: attached to endoplasmic reticulum membrane of hepatocytes, 20; attached to mitochondrial membrane of yeast cells, 22; and cytochrome oxidase synthesis, 67

DAS, binding sites on red cell membrane, 239-240

DASA, interaction with red cell band 3 protein, 133, 142

Decamethonium, interaction with cholinergic receptors, 109

Desensitization, of membrane cholinergic receptors, 101, 103, 115, 121-122

DIDS: binding sites on red cell membrane, 234-238; interaction with red cell band 3, 133, 135, 136, 137, 139, 142-143, 145-146, 147, 148, 152

Dimethisoquin, binding to receptor-rich membranes, 117-121

Disulfonic acid derivatives, in anion transport studies, 231-248

DNFB, binding sites on red cell membrane, 231-233, 238, 240, 242-243, 245, 246, 247

DNS-chol, interaction with membrane cholinergic receptors, 105-107

Electric tissue, cholinergic mechanisms in, 99-123

Electrical potential difference, transepithelial, affecting transport, 259, 278-279, 304

Electrophorus tissue, cholinergic receptors in, 100, 102, 103, 107

Endoplasmic reticulum membrane of hepatocytes: multistep assembly in production of, 15; protein synthesis in attached cytoplasmic ribosomes of, 20-21, 23; rapid growth of, 7; synthesis of components in, 5; topography of assembly of, 16-17

Epithelial cells: asymmetrical distribution in, 273-282, 286-287; luminal and contraluminal cell face in differentiation in renal proximal tubule, 287-290 (isolation of, 272-273); transepithelial ion transport, 253-268 (*see also* Transepithelial transport)

Erythrocytes
—anion transport system in membranes, 229-248. *See also* Anion transport
—band 3 protein of, 128-153. *See also* Band 3 protein

—cholesterol localization in, 83
—phospholipid asymmetry in, 80, 81
—sodium-potassium pump in, 191-207; calcium affecting, 205-206; conformational changes of, 193-194, 201, 219; and effects of Na and K 195-198; magnesium affecting, 191, 201-205, 213; molecular weight of, 192; ouabain affecting, 193-207, 213; phosphorylation states of, 201; protein components of, 193; and relationship between ouabain binding rate and transport rate, 197; and simultaneous or consecutive occupancy of Na and K ions, 206, 212-220; thiol groups in, 194; two forms of glycoside binding site in, 194, 201, 206
—transbilayer migration of lipids in membranes of, 84-93, 94
Escherichia coli membrane vesicles, transport in, 162-166, 168, 171, 173, 177
Ethacrynic acid, as inhibitor of ATPase activity, 323, 324, 325

FDNB, interaction with red cell band 3, 142
Flow dialysis, for membrane transport studies, 161, 168, 180, 182
Flow hypothesis, membrane, 24
FMMP, interaction with red cell band 3, 133
Furosemide, as inhibitor of ATPase activity, 323, 324, 325

Galactopyranosides, substituted, in studies of transport in membrane vesicles, 176-183
Galactose uptake, in small intestine, 256
β-Galactosides: fluorescent, 176; photoreactive, 177
Gallamine, interaction with cholinergic receptors, 109
Gallbladder, transport in, 261, 294, 298
Glycophorin, protein regions in, 45
Glycoproteins, membrane, synthesis of, 23
Glycosides: and mechanism of red cell Na-K pump, 193-207, 213; and transcellular sodium transport, 254. *See also* Ouabain
Glycosphingolipids, asymmetric localization in membranes, 79, 80

Gramicidin, and channel formation in membranes, 101
Halides, transport in red cell membranes, 230
Halo, of perturbed lipid bilayer, 53-54, 56
Halobacterium halobium: purple membrane protein of, 32; transport in membrane vesicles of, 173
Hepatocytes. *See* Endoplasmic reticulum membrane of hepatocytes
Hexamethonium, interaction with cholinergic receptors, 109
Hormones, and transport properties of epithelia, 282

IBSA, interaction with red cell band 3, 133, 142
Ileum, transepithelial sodium transport in, 255-263, 266-267
Immunohistochemistry, of antigens in epithelial cells, 286-287
Influenza virus: cholesterol distribution in, 83; phospholipid asymmetry in, 80; transbilayer migration of lipids in membrane of, 93
Intestines: sodium transport in rat colon, 261; transport in brush border membranes, 279, 280. *See also* Small intestine
Isethionyl acetimidate, interaction with red cell band 3, 133

Kidney: aminopeptidase M in proximal tubule epithelial cells, 287; luminal and contraluminal membranes of, 273, 287-290; phosphate transport in plasma membrane vesicles, 276-279; sodium transport in, 316-329; transport in brush border membranes, 279, 280; water transport in, 294-312

Lac carrier protein, and transport in membrane vesicles, 173-174, 176-183
D-Lactate, and active transport in membrane vesicles, 162, 163-167, 168, 173, 178, 180
D-Lactate dehydrogenase, and transport in membrane vesicles, 162, 164, 165-166; in reconstituted vesicles, 166, 173-175
Lactose transport. *See* Sugar transport
Ligand-binding properties, of membrane

cholinergic receptors, 99-123
Light exposure: and activity in thylakoid membranes, 7-8, 12, 18; and glutamate transport in membrane vesicles, 173; and NAP-taurine binding to red cell band 3, 147
Lipid(s): asymmetric localization in membranes, 78-95 (and maintenance of stability, 84-95); immobilization by protein, 36 (see also Boundary lipid); motional characteristics of, 35 (in perturbed bilayers, 54, 56); specificity of, 41; of thylakoid membranes, synthesis in morning, 8
Lipid bilayers in membranes, 3; band 3 protein segments in, 134, 136, 151; exchange with boundary lipid, 39, 40; kinetics of transbilayer lipid migration, 84-95; model systems of, 79; perturbed, halo of, 53-54, 56; potassium transport across, 234; protein associations with, 31-56; protein-free, 35; turnover of components in, 25
Lipid-protein associations, 31-56; boundary lipid in, 34-45, 54; captive lipid in, 46-49, 55; in cytochrome b_5, 52-53; in cytochrome c, 49-51, 56; in cytochrome oxidase, 34-40, 54; and halo of perturbed bilayer, 53-54, 56; hydrophobic interface in, 34-45; integral proteins in, 34-49; in liver microsomal membranes, 44-45; peripheral proteins in, 49-51, 56; and properties of interface, 40-41; and properties of membranes, 78-79; in sarcoplasmic reticulum ATPase, 42-44, 54
Lithium, and transport in membrane vesicles, 172-173
Lysine transport, by membrane vesicles, 167

Magnesium, and mechanism of red cell Na-K pump, 192, 201-205, 213
3-0-Methylglucose, and intracellular sodium activity, 258
Methyl 1-thio-β-D-galactopyranoside, transport in membrane vesicles, 172-173
Mitochondria: ATPase complex of, 62; cytochrome oxidase of (in beef heart, 34-40, 64; synthesis of subunits, 74-75; in yeast, 62-71); polysomes attached to membrane of, protein synthesis in, 20, 22, 23; proton translocation in sub-

mitochondrial vesicles, 222-224; transbilayer migration of lipids in membranes, 94; two-dimensional analysis of inner membrane, 71-74
Molecular architecture, of membrane proteins, 61-75

NADH, and transport in membrane vesicles, 162, 164-165
NAP-taurine, interaction with red cell band 3, 133, 139, 142, 146-149
Necturus proximal tubule, water transport in, 294, 304, 305, 306, 307
Neurospora: cytochrome oxidase from, 62; protein synthesis in mitochondrial membrane of, 20
Neurotoxins, binding to cholinergic receptors, 100, 104, 107, 112
Nicotinic cholinergic receptors, membrane-bound, 99-123
Nigericin, and transport in membrane vesicles, 169-171
Norepinephrine, and anion transport in bladder, 282

Osmolality, and cell volume in renal tubules, 325-329
Osmosis, local, and water transport, 297-298
Ouabain: binding sites in epithelial cells, 254, 262; and cell volume regulation in hyposmotic medium, 327; and mechanisms of red cell Na-K pump, 193-206, 213; and potassium exchange across basolateral membranes of ileum, 261; and sodium pump activity, 316, 317, 319, 320-321; and transepithelial sodium transport, 265, 275

Parathyroid hormone, and renal tubular phosphate reabsorption, 282
Permeability control, and function of membrane cholinergic receptors, 99-123
pH: and activity of sodium-stimulated ATPase, 324; and ATP synthesis in mitochondria, 224-225; and transport in membranes, 167-173, 222, 230
Phenazine methosulfate (PMS), and active transport in membrane vesicles, 162, 166, 168, 175, 180
Phlorizin, red cell exposure to, 239, 240
Phosphate transport, in renal plasma membrane vesicles, 276-279

Phosphatidylcholine: asymmetric distribution in membranes, 80-81; transbilayer exchange with phosphatidylethanolamine, 85-88

Phosphatidylethanolamine: asymmetric distribution in membranes, 80-81; transbilayer exchange with phosphatidylcholine, 85-88

Phosphatidylserine, asymmetric distribution in membranes, 80

Phospholipids: asymmetric localization in membranes, 79, 80-83; and cytochrome oxidase activity, 40; and membrane protein function, 32

Phosphorylation: in membranes, cAMP-induced, 284, 285, 289; oxidative, in mitochondria, 223-225; of sodium-potassium pump, and conformational changes, 201, 206

Polypeptides, of membrane cholinergic receptors, 103

Polysomes, attached to membranes, protein synthesis in, 5, 12, 17-22, 23

Potassium transport: and exchange across basolateral membranes related to transepithelial Na transport, 260-265; and functions of red cell band 3 protein, 137; in lipid bilayer, 234; in membrane vesicles, 163, 167; in red cells, 191-207. *See also* Sodium-potassium pump

Pox viruses, membrane formation in, 17

Prilocaine, interaction with cholinergic receptors, 110, 122

Proline transport, in membrane vesicles, 165, 168, 171-172

Pronase, red cell exposure to, 235-236

Protein: asymmetric localization in membranes, 95; bacteriochlorophyll *a* complex, 46-48, 56; band 3 protein of red cells, 128-153; chlorophyll-protein associations, 46-48; and immobilization of lipids, 36; in lipid bilayers of membranes, 3; lipid-protein associations, 31-56; synthesis by polysomes attached to membranes, 5, 12-14, 17-22, 23; in thylakoid membranes, 8-12

Protein kinase, distribution in epithelial cell membranes, 282-286

Protons: stoichiometries with transport substrates affected by pH, 171-172; transepithelial transport of, sodium-dependence of, 275, 277; translocation

by reconstituted segments of respiratory chain, 225-227; translocation in submitochondrial vesicles, 222-224

Purple membrane protein, of *Halobacterium halobium*, 32

Pyocyanine, and active transport in membrane vesicles, 162

Pyridoxal phosphate (PDP), interaction with red cell band 3, 133, 139, 142, 143-146, 152

Receptors, cholinergic, membrane-bound, 99-123

Red cells. *See* Erythrocytes

Retinal rod membrane, phospholipid distribution in, 80-81

Ribosomes
—chloroplastic, protein synthesis in, 12, 17-22
—cytoplasmic: attached to endoplasmic reticulum membrane of hepatocytes, 20; attached to mitochondrial membrane of yeast cells, 22; and cytochrome oxidase synthesis, 67
—microsomal, and cytochrome oxidase synthesis, 67

Salmonella typhimurium membrane vesicles, transport in, 162, 164, 172, 173

Sarcoplasmic reticulum ATPase, lipid-protein association in, 42-44

Sialoglycoproteins, association with band 3 protein of red cells, 135, 136

SITS, as inhibitor of anion transport, 231-233

Skin of frog, transepithelial sodium transport in, 259, 263-264, 267

Small intestine: luminal and contraluminal membranes in, 273; transepithelial sodium transport in ileum, 255-263, 266-267; water transport in, 294

Sodium-independent transport systems, 279-280, 289

Sodium-potassium pump: in epithelial cells, 253, 254; in erythrocytes, 191-207 (*see also* Erythrocytes); renal, 316-319

Sodium transport
—cotransport systems in, 172, 275-279
—intracellular tubular system in, 267-268
—in kidney, 316-329; angiotensin affecting, 324; cell contraction hypothesis of, 322; and cell volume

regulation in anisosmotic media, 325-329; cryptic pump hypothesis of, 320-322; and isosmotic volume regulation in kidney slices, 319-322; membrane potential affecting, 329; and ouabain-insensitive magnesium-dependent ATPase activity, 322-325; ouabain sensitivity affecting, 316, 317, 319, 320-321; and peritubular potassium concentrations, 316-317; pH affecting, 324; reabsorption in whole kidney, 319; two-pump hypothesis of, 322
—in red cells, 191-207; and functions of band 3 protein, 137
—transepithelial, 253-268; in bladder, 260, 265; in colon, 261; double-membrane model for, 253-268; in gallbladder, 261; in ileum, 255-263, 266-267; and intracellular sodium transport pool, 255-260; and potassium exchange across basolateral membranes, 260-265; in skin, 259, 263-264, 267

Sphingomyelin, asymmetric distribution in membranes, 80-81
Staphylococcus aureus, active transport in membrane vesicles, 163
Stilbene disulfonic acid derivatives, in anion transport studies, 231-248
Succinate, and transport in mutant membrane vesicles, 165
Sugar transport: interaction with sodium transport in ileum, 255-259, 261; by membrane vesicles, 162, 168, 171-172 (in anaerobic conditions, 163; ubiquinone affecting, 165); in red cells, and functions of band 3 protein, 131, 137, 139, 141, 152-153; sodium dependence of, 275, 277; sodium-independent system in, 279-280, 289
Sulfate transport, in red cell membranes, 229-246

Temperature: and anion transport in red cell membranes, 230; and cell volume regulation in hyposmotic medium, 326-327
Thylakoid membranes, 7-22; activity in light and dark exposure, 7-8, 12, 18; chloroplastic polysomes attached to, 12, 17-22; mode of assembly of, 14-16; proteins in, 8-12 (association with chlorophyll *a*, 48); specific gravity

during greening, 16
Time dependence: and assembly of thylakoid membranes, 14; and binding of AcCh to receptor, 112-115; and cholinergic permeability response, 101
TMG, transport in membrane vesicles, 172-173
TNBS, interaction with red cell band 3, 133, 142
Torpedo electric tissue, membrane-bound cholinergic receptors in, 99-123
Toxins in venoms, binding to cholinergic receptors, 100
Transbilayer migration of lipids, 84-95
Transepithelial transport: ATPase system in, 273-275; double-membrane model for, 253-268, 295, 297; hormones affecting, 282; membrane potential affecting, 259, 278-279, 304; and polarities of limiting membranes, 272; of sodium, *see* Sodium; sodium-independent systems in, 279-280, 289; of water in kidney, 294-312
Transport: of anions in membranes, 229-248; and band 3 as transmembrane aqueous channel, 136, 137, 151; chemiosmotic hypothesis of, 167-172, 222; of components for membrane biogenesis, 5, 12, 22-23, 24; energetics of, in membranes, 167-173; of lipids, *see* Transbilayer migration of lipids; of potassium, *see* Potassium transport; of sodium, *see* Sodium transport; transepithelial, *see* Transepithelial transport; vesicle system in, 160-167 (*See also* Vesicle system); of water, in kidney, 294-312
Triflocin, as inhibitor of ATPase activity, 323, 324, 325
Trypsin, red cell exposure to, 231-233
Tubocurarine: binding sites in membrane, 100, 105; interaction with cholinergic receptors, 109

Ubiquinone: and proton translocation in respiratory chain, 227; and stimulation of transport in membrane vesicles, 162, 165

Valinomycin: and active transport in membrane vesicles, 163, 168, 169-171; and potassium transport across lipid bilayers, 234
Vasopressin: and transepithelial sodium

transport, 260, 264, 265; and water permeability of renal collecting duct, 282

Vesicle system in membrane transport activity, 160-167; anisotropy studies of, 179-181, 183; and carrier function at molecular levels, 175-183; cotransport mechanisms in, 172; and energy-dependent binding of compounds, 177-181; energy source for, 162; in eukaryotic cells, 172; flow dialysis studies of, 161, 168, 180, 182; in inverted vesicles, 164-167; *lac* carrier protein in, 173-174, 176-183; D-lactate role in, 162, 163-167; and proton: substrate stoichiometries affected by pH, 171-172; quenching studies with N-methylpicolinium perchlorate, 182-183; in reconstituted vesicles, 166, 173-175; specificity of electron donors in, 163-167; studies with azidophenylgalactosides, 177, 180-181; studies with dansylgalactosides, 176-183; symport mechanisms in, 168-172; types of vesicles in, 162

Vinylglycolate, transport by membrane vesicles, 166, 174

Viruses: cholesterol distribution in membranes, 83; enveloped, glycoproteins of, 23; phospholipid asymmetry in membranes, 80, 82; pox virus membrane formation, 17; transbilayer migration of lipids in membranes, 84-93, 94

Water transport
—and functions of red cell band 3 protein, 131, 137, 139, 152-153
—renal, 294-313; antidiuretic hormone affecting, 282; compartment model for solute and water coupling, 310-312; continuous model for solute and water coupling, 307-310; double-membrane model for, 295-296, 297; local osmosis in, 297-298; paracellular pathway for, 296-297, 304; standing-gradient model for, 298-299, 301-307

Yeast: cytochrome oxidase from, 62-71; cytoplasmic ribosomes attached to membrane of cells, 22; mutants lacking cytochrome oxidase, 67-70